A Key to AMPHIBIANS & REPTILES of the Continental United States and Canada

This publication sponsored by

The Center for North American Amphibians and Reptiles

and

Western Resources

A Key to AMPHIBIANS & REPTILES of the Continental United States and Canada

Robert Powell

Joseph T. Collins

Errol D. Hooper, Jr.

UNIVERSITY PRESS OF KANSAS

Published by the University Press of Kansas (Lawrence, Kansas 66049), which was organized by the Kansas Board of Regents and is operated and funded by Emporia State University, Fort Hays State University, Kansas State University, Pittsburg State University, the University of Kansas, and Wichita State University

Library of Congress Cataloging-in-Publication Data

Powell, Robert, 1948–
A key to amphibians and reptiles of the continental United States and Canada / Robert Powell, Joseph T. Collins, Errol D. Hooper, Jr.
p. cm.
Includes bibliographical references (p.).
ISBN 0-7006-0929-6 (pbk. : alk. paper)
1. Reptiles—United States—Identification. 2. Reptiles—Canada—Identification. I. Collins, Joseph T. II. Hooper, Errol D.
III. Title.
QL652.P69 1998
597.9'0973—dc21 98-28187

British Library Cataloguing in Publication Data is available

Printed in the United States of America

10 9 8 7 6 5 4 3

The paper used in this publication meets the minimum requirements of the American National Standard for Permanence of Paper for Printed Library Materials Z39.48-1984.

To Dean E. Metter, my mentor in all "things herpetological," to my students, in whose lives I hope I have made a similarly significant contribution, to Joseph P. Ward, who guided me into herpetology (albeit inadvertently), and to my wife, Beverly, whose tolerance of and patience with my unusual interests have exceeded my greatest hopes—she so seldom complains and hardly ever makes me feel guilty.

R.P.

To my maternal grandparents, Helen (Taylor) Aichele and George Aichele, and my paternal grandparents, Ida May (Coss) Collins and Joseph Rueben Collins, Ohio Buckeyes all.

J.T.C.

To Hague Lindsey, professor emeritus, The University of Tulsa, for a most excellent and meaningful field zoology course that stirred the craving for nature in my spirit and directed me toward a professional life with nature; to Linda Trueb, for opening up the world of scientific illustration to me, for her patience and tutelage, and for making her office and illustrating equipment available, enabling the completion of this project; and to Hobart M. Smith, renowned herpetologist, for the most genuine act of kindness ever experienced by this admirer and fan.

E.D.H.

CONTENTS

INTRODUCTION

This key is designed primarily for use in college-level herpetology or vertebrate biology courses. A dichotomous key is essential in teaching the principles of taxonomy as well as in introducing students to the systematics of any one taxon. Yet in attempting to do just that in herpetology classes over the years, we have been frustrated with the inadequacies of existing keys to amphibians and reptiles, although they were often excellent for their time and scope. Blair (1968) and Cagle (1968) are out of date and their use necessitates numerous corrections, confusing to students who often are struggling already to become acquainted with what must seem an endless number of strange and foreign names. Smith (1978) and Smith and Brodie (1982) are also dated, use only common names (often quite different than the standards presented most recently by Collins, 1997), rely heavily on characteristics inappropriate for use with preserved specimens, and omit dichotomies for many groups and species. Most recently, Ballinger and Lynch (1983) have prepared a key more useful than its predecessors, but it also has become dated (most noticibly in its exclusion of many recently described and introduced forms) and fails to provide dichotomies to all levels of taxa, the latter so necessary in teaching systematics. Other available keys are either dated or are regional and lacking in the scope necessary for use in collegiate level classes. This key attempts to address these deficiencies, while still recognizing the changing nature of systematics. Throughout this key, we point out many instances of uncertainty or disagreement among authorities.

Though this key will serve to identify recently captured specimens in the field, its major role is to classify preserved materials whose geographic origins are known. Still, to address all possible applications, references to characteristics of live specimens (such as color) are included when necessary or useful in working with preserved animals or if otherwise of particular interest. Both endemic and alien taxa are included. For a complete list of these taxa (excluding those subsequently described from or introduced to the area covered by this key), see Collins (1997), which has also served as a source of most currently accepted scientific names. Alien taxa are indicated by an **(A)** after the names listed in the key.

To avoid prolonged descriptions of species and their ranges, this text will, when possible, refer the user to the excellent descriptions, maps, and illustrations in the Peterson Field Guide Series: *A Field Guide to Western Reptiles and Amphibians* by Stebbins (1985) and *A Field Guide to Reptiles and Amphibians of Eastern and Central North America* by Conant and Collins (1998, third edition, expanded). These references are indicated by **S** (for Stebbins) or **CC** (for Conant and Collins), followed first by the range map number (Stebbins) or page (Conant and Collins), and then the plate (pl) or page (p) number(s) on which the reader may find an illustration of that taxon. For students who seek to learn more about a particular taxon, references are also made to accounts (when available) published in the *Catalogue of American Amphibians and Reptiles* by the Society for the Study of Amphibians and Reptiles. References to the *Catalogue* are indicated by **CAAR** and include the account number and date of publication. The keys and Guides mentioned above, various regional guides, and pertinent articles in various scientific journals may in many instances be of considerable utility and interest.

A WORD TO THE USER OF THIS KEY

We have made every possible effort to create a user-friendly key (some biologists would consider this an oxymoron). Still, the complexity of amphibian and reptilian taxonomy, the increased reliance in systematics on molecular techniques that elucidate characters not always expressed morphologically, the avoidance of characteristics normally seen only in living specimens, and the dynamic nature of our understanding of the diversity of amphibians and reptiles have resulted in a number of very complex couplets. Therefore, the user must be careful to read completely and comprehend entirely each couplet, noting especially the use of "and" versus "or" in the text; carelessness in this area will lead inevitably to errors, which in turn will result in confusion and frustration, because only rarely will the proper identification be possible once a mistake has been made.

Users of this key, especially students engaging in such an endeavor for the first time, must recognize several facts:

1. **No key can identify every specimen.** Individuals of any species vary considerably; geographic origin, sex, age, and even pathologies may play a role in this variation. Ages and developmental stages are particularly important in identifying amphibians. Most North American species have larval stages, which frequently are very different than adults in body form and lifestyle. As a result, the correct identification of every individual specimen will not always be possible. We have not provided a key to larval frogs and toads (tadpoles) or larval salamanders, and even some immature reptiles may not be identified properly using the keys provided here.

Because some taxa (genus or species) vary considerably, the proper identification of such forms may be achieved by following more than one sequence of couplets. When this occurs, the name of the taxon is followed by "(part)" to indicate that alternative pathways will lead to the same taxon. In most cases, the forms so indicated demonstrate considerable variation, usually geographic in nature.

Finally, the user should remember that the use of the Peterson Field Guides can help in ways other than by providing range maps and illustrations. Descriptions of species in these guides often note common variations. By using these guides, or by referring to the CAAR accounts (when available) and references cited therein, the difficulties in dealing with variants that confound the key will be lessened.

2. **Geography is important.** Sophisticated genetic and biochemical techniques are being used increasingly to identify species which are not always morphologically distinct. In such instances, because DNA analyses and/or electrophoretic examinations are rarely practical, the only sure way to determine the identity of some animals is to know the exact locality where they were collected. Many species cannot be identified any other way. Particularly difficult are distinctions among species in complexes of closely related forms.

3. **Scientific names are not now, and never will be, constant.** As we come to better understand the complexities of relationships that exist among amphibian and reptilian populations, classifications must change to accommodate new knowledge. Consequently, some scientific names used in older books and articles will not be found here, and some recently published names appear here in the context of a key for the first time.

4. **Science is dynamic and ever-changing.** Because classification by category reflects our best efforts to portray true relationships that exist in nature, not all herpetologists agree at any one time as to which names or combinations of names most accurately reflect that reality. In various places throughout this key, we have noted some of these differences of opinion. However, one should always keep in mind that amphibians and reptiles do not read our books and articles, and do not feel obligated to conform to any single paradigm. As we learn more about amphibian and reptilian groups whose taxonomic status is in dispute, we must adjust again and again to the evidence that most closely approaches scientific truth.

5. **America north of México does not exist in isolation.** Although the continental United States and Canada is the area addressed by this key, amphibians and reptiles do not recognize political, and sometimes even natural, boundaries. Especially in the southern portions of the United States, new species are being introduced, and some will become entrenched (i.e., establish breeding, self-perpetuating populations). We have made every effort to include in this key all alien species now established in the continental United States and Canada; however, uncertainty regarding the breeding status of some species and the fact that more may, at this moment, be on their way to becoming a part of our fauna, preclude anything more than a diligent effort on our part to cover the entire herpetofauna of this area.

6. **Some families and genera are monotypic.** This means they contain only one genus or species, respectively. When these occur in the keys, couplets are not necessary and none are provided. Otherwise, these taxa are organized in the same manner as those that are polytypic (contain more than one genus or species).

ACKNOWLEDGMENTS

We are deeply indebted to the authors of previous keys for their identification of diagnostic characters. All will recognize couplets identified in their works. Numerous colleagues reviewed sections of this key, and some tested them on their students. For this help we thank Karen Anderson, Joseph Beatty, Lauren Brown, Richard Daniel, L. Lee Grismer, John B. Iverson, Richard Highton, the late Clarence J. McCoy, Michael Seidel, and Samuel S. Sweet. For many kindnesses, both direct and indirect, we are also indebted to Suzanne L. Collins, Darrel R. Frost, Elisabeth A. Hooper, Philip S. Humphrey, Kelly J. Irwin, Dean E. Metter, Brian T. Miller, Larry L. Miller, Michael E. Seidel, John S. Parmerlee, Jr., Thomas J. Sloan, Donald D. Smith, Tom Swearingen, Travis W. Taggart, Sarah Taylor, Kevin Toal, Joseph P. Ward, and Robert F. Wilkinson.

William E. Duellman, Curator, Division of Herpetology, Natural History Museum, The University of Kansas, permitted use of the herpetology collection under his care and facilitated loan requests for specimens from other institutions that were used in this project. John E. Simmons, Collection Manager, Division of Herpetology, Natural History Museum, The University of Kansas, helped access specimens, and the division's curatorial assistants retrieved and reshelved numerous jars of specimens that we requested for reference and illustration. Amy Lathrop, Randall Reiserer, Linda Trueb, and Erik Wild, all of the Division of Herpetology, Natural History Museum, The University of Kansas, made critical comments, suggestions, and provided encouragement. Walter Michener skillfully (and diplomatically) made sure we were consistent, and made numerous helpful suggestions to improve the text.

ABBREVIATIONS

The following abbreviations are used in the figure captions: BWMC = Bobby Witcher Memorial Collection, Avila College, Kansas City, Missouri; DAK = personal herpetological collection of David A. Kizirian, American Museum of Natural History, Central Park West at 79th Street, New York, New York; KU = Division of Herpetology, Natural History Museum, The University of Kansas, Lawrence; MES = personal herpetological collection of Michael E. Seidel, Department of Biological Sciences, Marshall University, Huntington, West Virginia; NCSM = North Carolina State Museum of Natural Science, Raleigh, North Carolina; UTAVC = Department of Biology, The University of Texas at Arlington, Arlington, Texas.

A map showing the area covered by this key.

KEY TO THE CLASSES AMPHIBIA AND REPTILIA

1a. Skin smooth or warty; never with epidermal scales; limbs lacking true claws .... **AMPHIBIA** (p. 5)

1b. Skin nearly always with epidermal scales; if lacking scales, skin leathery; limbs (if present) with claws .... **REPTILIA** (p. 49)

AMPHIBIA

KEY TO ORDERS OF AMPHIBIA

1a. Fore- and hindlimbs (when present) similar in structure and essentially equal in size; tail present (Fig. 1) .... **CAUDATA** (p. 5)

1b. Hindlimbs more robust than forelimbs and at least somewhat larger in size (Fig. 39); tail absent (male *Ascaphus truei* may have an everted cloaca that superficially resembles a tail) .... **ANURA** (p. 25)

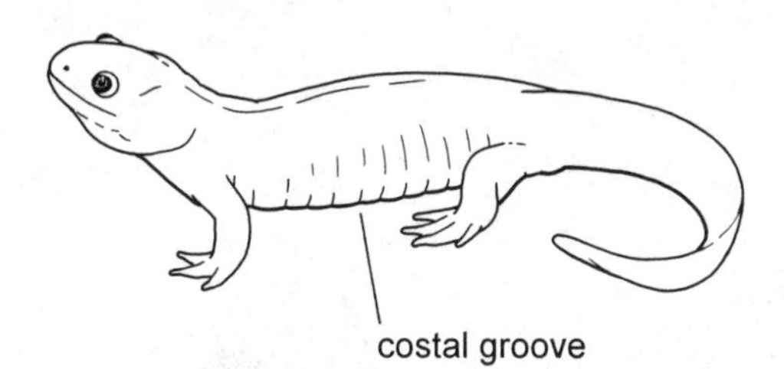

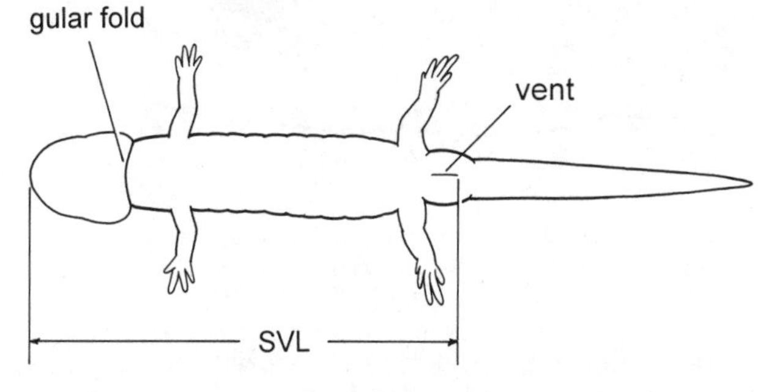

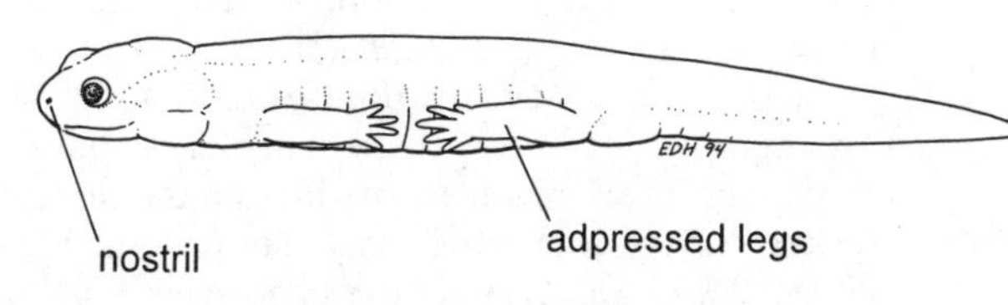

Figure 1. Generalized views of a salamander showing gross features, costal grooves, and adpressed limbs. Upper: Lateral view. Middle: Ventral view. Lower: Lateral view.

CAUDATA
(Salamanders)

KEY TO FAMILIES OF CAUDATA

1a. Body elongated and eel-like; without hindlimbs (Fig. 2) or, if hindlimbs are present, they are proportionately tiny 2

1b. Body not elongated and eel-like; if body elongated, limbs are not proportionately tiny 3

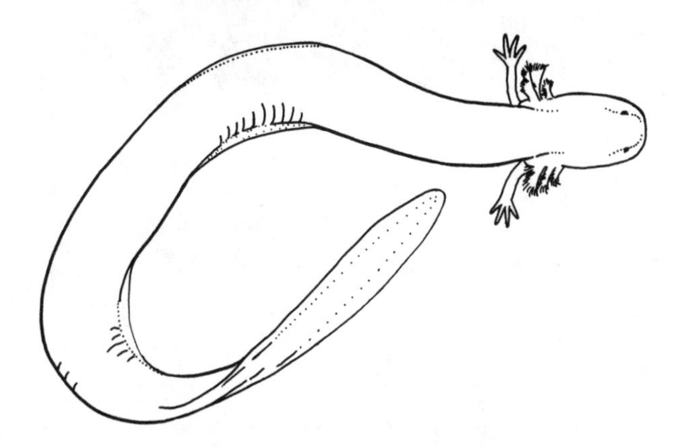

Figure 2. Dorsal view of the elongated and eel-like body typical of the salamander families Sirenidae and Amphiumidae. Adapted from Bishop (1943).

2a. One pair of appendages (no hindlimbs) (Fig. 2) .... **SIRENIDAE** (p. 9)

2b. Two pairs of proportionately tiny appendages .... **AMPHIUMIDAE** (p. 10)

3a. External gills present.. 4

3b. External gills absent 7

4a. Costal grooves indistinct; often with brown ground color and distinct reticulations, occasionally uniformly sooty tan to brown; northwestern U.S. and British Columbia .... **DICAMPTODONTIDAE** (part) (p. 7)

4b. Costal grooves distinct; pattern variable, but never brown with black reticulations 5

Figure 3. Lateral view of the external gills typical of the salamander family Proteidae.

5a. Four toes on hindlimbs; gills bushy (Fig. 3) .... **PROTEIDAE** (p. 10)

5b. Five toes on hindlimbs; if four, gills are not bushy 6

6a. Small (adult SVL less than 60 mm) and very slender;

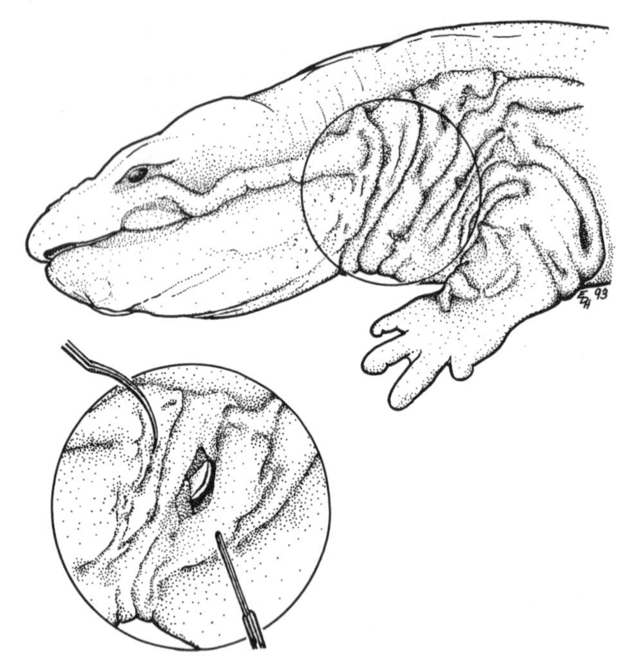

Figure 4. Lateral view of gill slits typical of the salamander family Cryptobranchidae. Drawn from a preserved specimen (KU 23283) of *Cryptobranchus alleganiensis*.

toes of adpressed limbs widely separated .... **PLETHODONTIDAE** (part) (p. 12)

6b. Large (adult SVL more than 70 mm) and more robust; toes of adpressed limbs overlap, touch, or are barely separated .... **AMBYSTOMATIDAE** (part) (p. 7)

7a. Gill slits present (Fig. 4) .... **CRYPTOBRANCHIDAE** (p. 6)

7b. Gill slits absent... 8

8a. Costal grooves present (Fig. 1) 9

8b. Costal grooves absent .... **SALAMANDRIDAE** (p. 10)

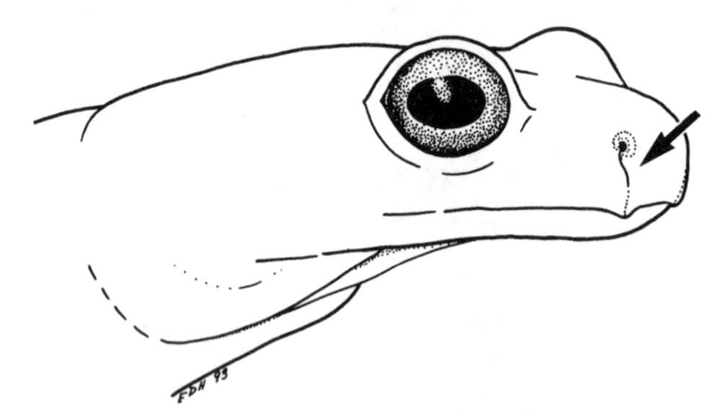

Figure 5. Lateral view of the head showing nasolabial groove, typical of the salamander family Plethodontidae. Adapted from Behler and King (1979).

9a. Nasolabial grooves present (Fig. 5) .... **PLETHODONTIDAE** (part) (p. 12)

9b. Nasolabial grooves absent.................................. 10

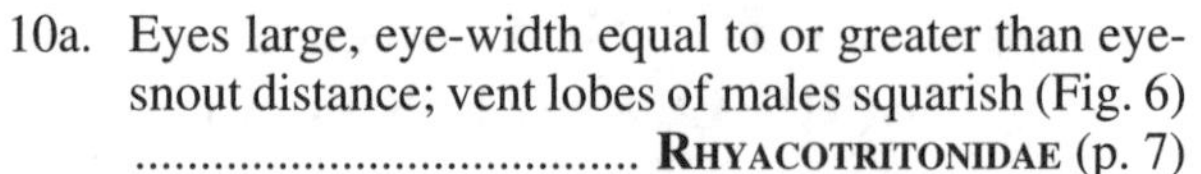

10a. Eyes large, eye-width equal to or greater than eye-snout distance; vent lobes of males squarish (Fig. 6) .... **RHYACOTRITONIDAE** (p. 7)

10b. Eyes proportionately small, eye-width less than eye-snout distance; vent lobes of males rounded (Fig. 6) 11

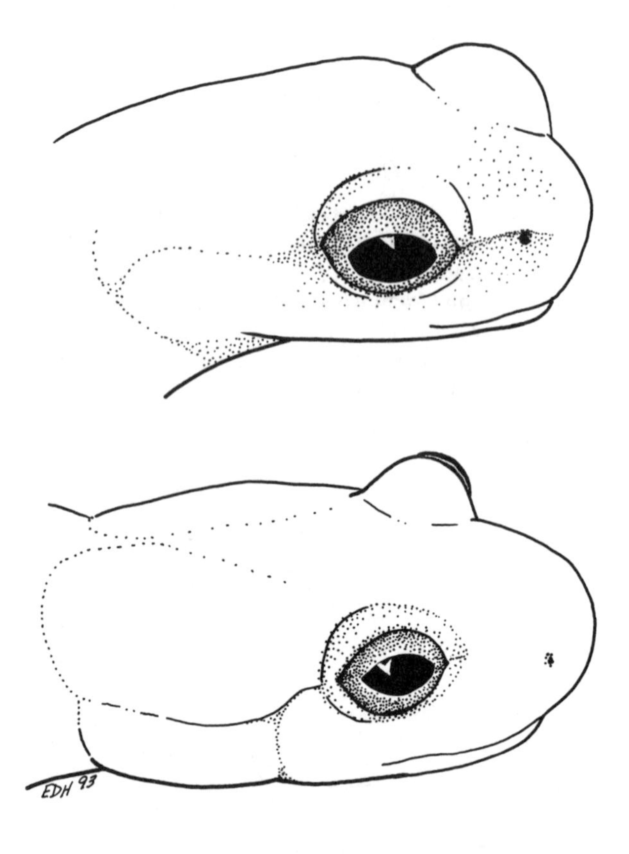

Figure 6. Lateral view of head proportions in the salamander genera *Rhyacotriton* (upper) and *Ambystoma* (lower).

11a. Costal grooves indistinct; often with brown ground color and distinct reticulations, occasionally uniformly sooty tan to brown; northwestern U.S. and British Columbia .... **DICAMPTODONTIDAE** (part) (p. 7)

11b. Costal grooves distinct; pattern variable, but never brown with black reticulations; if from the northwestern U.S. or adjacent Canada, not patterned as above .... **AMBYSTOMATIDAE** (part) (p. 7)

CRYPTOBRANCHIDAE

Cryptobranchus

CAAR 101 (1971)

Cryptobranchus alleganiensis

CC 419, pl 37 **CAAR** 101 (1971)

Some authorities have suggested that this taxon consists of two species, *C. alleganiensis* and *C. bishopi*.

Rhyacotritonidae

Rhyacotriton
CAAR 68 (1968)

Key to Species of ***Rhyacotriton***

Rhyacotriton contains species that are poorly defined morphologically and are best differentiated by geography and measures of protein divergence.

1a. At least some dark dorsal spots present 2
1b. Dorsum uniformly colored, without dark dorsal spots 3

2a. Heavily spotted dorsally; lateral spotting very pronounced, resulting in a distinct demarcation between dorsal and ventral coloration; dark preorbital stripe faint or absent; restricted to the western slopes of the Cascade Mountains from Skamania County, Washington, to northeastern Lane County, Oregon .... ***R. cascadae***
2b. Dorsal and lateral spotting generally uniform, faint to pronounced; dark preorbital stripe usually distinct; restricted to coastal ranges from Mendocino County, California, north to Polk, Tillamook, and Yamhill counties, Oregon (an isolated population on the western slopes of the Cascade Mountains in Douglas County, Oregon) .... ***R. variegatus***

3a. Demarcation between dorsal and ventral color indistinct and straight; large dark ventral spots absent; restricted to coastal ranges in northwestern Oregon and adjacent Washington .... ***R. kezeri***
3b. Demarcation between dorsal and ventral color distinct and wavy; large dark ventral spots present; restricted to the Olympic Peninsula, Washington .... ***R. olympicus***
S 4, pl 1 **CAAR** 68 (1968)

Dicamptodontidae

Dicamptodon
CAAR 76 (1969)

Key to Species of ***Dicamptodon***

Dicamptodon contains species that are poorly defined morphologically and are best differentiated by geography and measures of protein divergence.

1a. Adpressed limbs separated by 0–2 costal folds; small (total length usually less than 200 mm); head narrow (less than ⅕ SVL); metamorphosis rare in most populations .... ***D. copei***
S 1 **CAAR** 334 (1983)
1b. Adpressed limbs overlap or separated by no more than ½ costal fold); large (total length often more than 300 mm); head wider (more than ⅕ SVL); may metamorphose 2

2a. East of the Cascade Mtns (Idaho) ..... ***D. aterrimus***
2b. Coastal forests and western slopes of the Cascade Mtns (Washington, Oregon, California) 3

3a. From south of the Sonoma-Mendocino County border .... ***D. ensatus***
S 1, pl 1 **CAAR** 76 (1969) (includes *D. tenebrosus*)
3b. From Mendocino County north .... ***D. tenebrosus***

Ambystomatidae

Ambystoma
CAAR 75 (1969)

Key to Species of ***Ambystoma***

Some species of *Ambystoma* include individuals or populations that are paedomorphic and perrenibranchiate (fail to transform in nature). In many instances these specimens will be difficult to identify to species with the following key.

1a. Distinct pattern of narrow yellow-white dorsal crossbands .... ***A. annulatum***
CC 435, pl 38 **CAAR** 19 (1965)
1b. Dorsal pattern variable, but without narrow yellow-white crossbands 2

2a. White or gray dorsal crossbands combining with dorsolateral light areas to form a ladderlike pattern .... ***A. opacum***
CC 433, pl 38 **CAAR** 46 (1967)
2b. Dorsum without a ladderlike pattern 3

3a. Round yellow or orange spots in dorsolateral rows (spots do not extend onto sides) ...... ***A. maculatum***
CC 438, pl 38 **CAAR** 51 (1967)
3b. Dorsum pattern variable; if spotted, spots are not in dorsolateral rows 4

4a. Light middorsal stripe, sometimes broken into spots; extremely long toes .... ***A. macrodactylum***
S 3, pl 2 **CAAR** 4 (1963)

Some authorities have suggested that this taxon consists of two species, *A. macrodactylum* and *A. croceum*.

4b. No light middorsal stripe 5

5a. Strongly reticulated dorsal pattern with light flecks usually forming a pattern of narrow vertical bands on sides .... ***A. cingulatum***
CC 435, pl 38 **CAAR** 57 (1968)
5b. Dorsum unicolor, lichenlike, or boldly patterned (if the latter is reticulate, markings are yellow, white, or cream in color) 6

6a. Prominent paratoid glands present (Fig. 7); dorsum unmarked or (rarely) with light flecks ... ***A. gracile***
S 2, pl 2 **CAAR** 6 (1963)
6b. Paratoid glands absent; dorsum variable 7

7a. Costal grooves number 10–11; "chunky" habitus (Fig. 8); dorsum unmarked or with light flecks .... ***A. talpoideum***
CC 433, pl 38 **CAAR** 8 (1964)
7b. Costal grooves number twelve or more; more slender habitus (Fig. 8); dorsal pattern variable 8

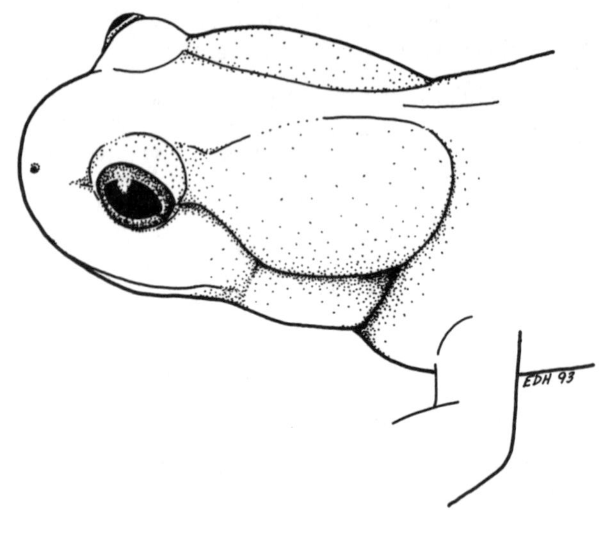

Figure 7. Lateral view of the paratoid gland of *Ambystoma gracile*. Drawn from a photograph by Suzanne L. Collins.

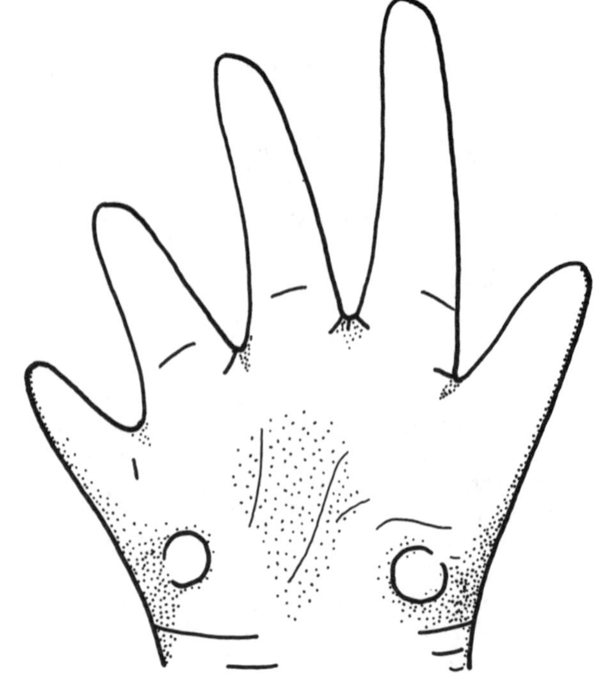

Figure 9. Palmar tubercles on the ventral surface of the foot of *Ambystoma californiense* and *Ambystoma tigrinum*. Adapted from Stebbins (1951).

8a. Palmar tubercules present (Fig. 9); often with a bold dorsal pattern of yellow, white, or cream spots, bars, or reticulations (light markings, when present, extend onto the sides) 9

8b. Palmar tubercules absent; never with a bold dorsal pattern (light flecks or lichenlike spots may be present) 11

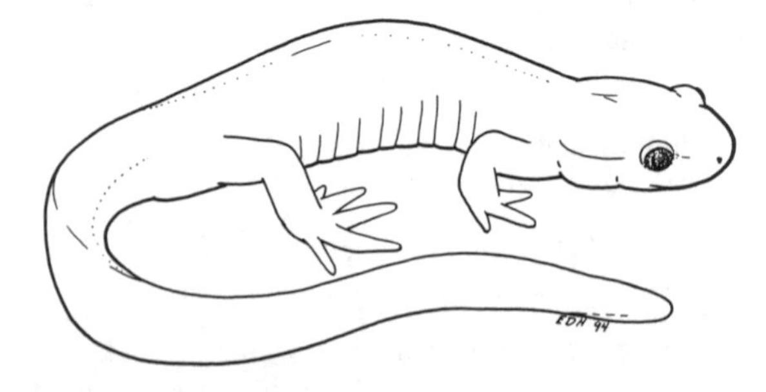

Figure 8. Lateral view of *Ambystoma talpoideum* (upper) and *A. jeffersonianum* (lower) showing differences in head and body proportions.

9a. Range limited to California; light markings often scarce or lacking middorsally ...... ***A. californiense***
S 5, pl 2
CAAR 52 (1967) (as *A. tigrinum californiense*)

9b. Found outside California; pattern variable 10

10a. Pattern of numerous light, usually yellowish, irregular spots or blotches on a dark ground color; distribution east of the Great Plains .... ***A. tigrinum***
CC 441, pl 38 **CAAR** 52 (1967) (as *A. t. tigrinum*)

10b. Pattern variable, but typically one of the following descriptions applies: (a) large light blotches, often forming vertical light bars or separated by vertical dark bars on sides; (b) dark ground color restricted to a network surrounding the light blotches; (c) dorsal ground color light to dark with scattered, small dark brown to black spots; or (d) numerous light (usually yellowish) irregular spots or blotches on a dark ground color; in all instances, the range includes the Great Plains or desert/montane areas to the west... .... ***A. mavortium***
CC 441, pl 38 **S** 5 **CAAR** 52 (1967)
(as *A. t. mavortium*, *A. t. diaboli*,
A. t. melanostictum, *A. t. nebulosum*
or *A. t. stebbinsi*)

11a. Costal grooves number thirteen or more; tongue with a longitudinal median furrow (Fig. 10) 12

11b. Costal grooves number twelve; tongue without a median furrow (a shallow central depression may be present) (Fig. 11) 14

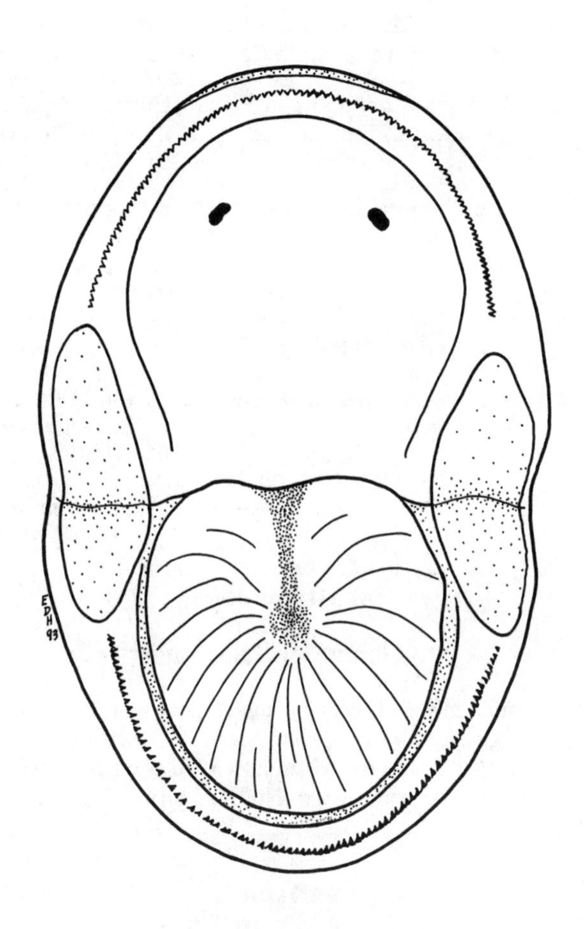

Figure 10. Interior view of the mouth of *Ambystoma mabeei* showing a median furrow on the tongue and single row of marginal teeth on the upper and lower jaws. Drawn from a preserved specimen (KU 68049).

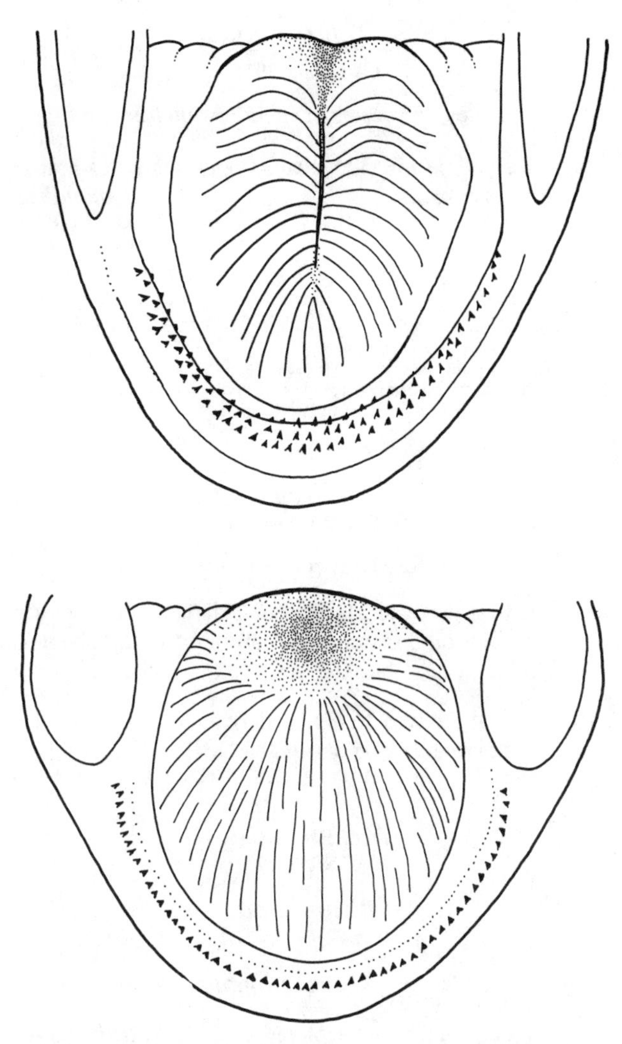

Figure 11. Interior view of the lower jaw of an ambystomatid salamander (upper) showing more than one row of marginal teeth and a median furrow on the tongue. Interior view of the lower jaw of *Ambystoma laterale* (lower) showing a single row of marginal teeth and the lack of a median furrow on the tongue. Adapted from Pfingsten and Downs (1989).

12a. Costal grooves number thirteen; teeth on margin of jaw in a single row (Figs. 10, 11); head broad; mouth large .... ***A. mabeei***
CC 437, pl 38 **CAAR** 81 (1970)

12b. Usually fourteen or more costal grooves; teeth on margin of jaw in more than one row (Fig. 11); head narrow; mouth small 13

13a. Maxillary and premaxillary teeth with short spatulate lingual cusps; breeds in streams; range restricted to portions of the Ohio River drainage in Indiana, Kentucky, Ohio, and West Virginia .... ***A. barbouri***
CC 437 **CAAR** 621 (1996)

13b. At least some maxillary and premaxillary teeth with long lingual cusps; within the range of *A. barbouri*, breeds in ponds .... ***A. texanum***
CC 437, pl 38 **CAAR** 37 (1967)

Ambystoma barbouri and *A. texanum* cannot be distinguished on the basis of external features; if tooth characters cannot be resolved and the breeding sites are unknown, specimens from within the range of *A. barbouri* cannot be accurately identified.

14a. Numerous pale spots and flecks on body; area around vent black .... ***A. laterale***
CC 438, pl 38 **CAAR** 48 (1967)

14b. Pale spots and flecks absent or small and scattered; area around vent gray .... ***A. jeffersonianum***
CC 438, pl 38 **CAAR** 47 (1967)

Hybrids of *A. laterale* and *A. jeffersonianum*, often difficult to distinguish, are found in large portions of the ranges of these two species (**CC** 438). These hybrids were once thought to represent the parthenogenetic species *A. platineum* (**CAAR** 49, 1967) and *A. tremblayi* (**CAAR** 50, 1967).

SIRENIDAE
CAAR 151 (1974)

Key to Genera of Sirenidae

1a. Three toes on each foot ... ***Pseudobranchus*** (p. 10)
1b. Four toes on each foot .... ***Siren*** (p. 10)

Pseudobranchus

CAAR 118 (1972)

Key to Species of *Pseudobranchus*

1a. Karyotype $n = 32$; inhabits open marsh or prairie ponds .... ***P. axanthus***
CC 431, pl 39, p 431 **CAAR** 118 (1972) (as *P. striatus axanthus*)

1b. Karyotype $n = 24$; inhabits cypress ponds in acid pine flatwoods .... ***P. striatus***
CC 431, pl 39, p 431 **CAAR** 118 (1972)

Pseudobranchus axanthus and *P. striatus* cannot be identified solely on the basis of morphological characters, and are best distinguished on the basis of karyological studies and habitat preferences.

Siren

CAAR 152 (1974)

Key to Species of *Siren*

1a. Costal grooves 31–38; small dark spots generally present on dorsum; tip of tail pointed; tail about ⅓ SVL .... ***S. intermedia***
CC 428, pl 37 **CAAR** 127 (1973)

1b. Costal grooves 36–39; no dark spots on dorsum; tip of tail rounded; tail less than ⅓ SVL .... ***S. lacertina***
CC 428, pl 37 **CAAR** 128 (1973)

AMPHIUMIDAE

CAAR 147 (1973)

Amphiuma

CAAR 147 (1973)

Key to Species of *Amphiuma*

1a. Three toes; body bicolored .... ***A. tridactylum***
CC 426, pl 37 **CAAR** 149 (1973)

1b. Fewer than three toes; body not bicolored 2

2a. One toe .... ***A. pholeter***
CC 426, pl 37 **CAAR** 622 (1996)

2b. Two toes .... ***A. means***
CC 425, pl 37 **CAAR** 148 (1973)

PROTEIDAE

Necturus

Key to Species of *Necturus*

1a. Venter pigmented (not white), often with darker spots (Fig. 12) 2

1b. At least a portion of the venter white and without spots (Fig. 12) 4

2a. Large (to more than 400 mm total length); dorsum darker than venter (larvae have dorsal stripes) .... ***N. maculosus***
CC 420, pl 37, p 421

2b. Small (less than 300 mm total length); dorsal and ventral ground color essentially the same 3

3a. Few large dark spots on dorsum; Neuse and Tar River drainage of North Carolina .... ***N. lewisi***
CC 423, p 422 **CAAR** 456 (1990)

3b. Many small dark spots on dorsum; Gulf Coast drainage .... ***N. beyeri***
CC 423, pl 37, p 421

4a. Large (to more than 400 mm total length); white, unmarked portion of venter limited to a narrow median strip (Fig. 12) .... ***N. louisianensis***
CC 420, pl 37, p 421 (as *N. maculosus louisianensis*)

4b. Small (less than 225 mm total length); nearly the entire venter white (Fig. 12) 5

5a. Dorsum without dark blotches (rarely with scattered small spots) .... ***N. punctatus***
CC 424, pl 37, p 421

5b. Dark blotches or spots on dorsum .... ***N. alabamensis***
CC 424, p 421

SALAMANDRIDAE

Key to Genera of Salamandridae

1a. Uniformly dark dorsum; top of head without distinct ridges; from western North America .... ***Taricha*** (p. 10)

1b. Striped or spotted dorsum; two ridges on top of head (Fig. 13); from eastern North America .... ***Notophthalmus*** (p. 10)

Taricha

CAAR 271 (1981)

Key to Species of *Taricha*

1a. Lower eyelid dark (Fig. 14) .... ***T. granulosa***
S 6, pl 1 **CAAR** 272 (1981)

1b. Lower eyelid at least partly light (Fig. 14) 2

2a. Dorsal color sharply demarcated from ventral color; iris dark; usually with a dark band across the vent .... ***T. rivularis***
S 9, pl 1 **CAAR** 9 (1964)

2b. Dorsal color blends into ventral color; iris with light areas; no dark band across the vent .... ***T. torosa***
S 8, pl 1 **CAAR** 273 (1981)

Some authorities have suggested that this taxon consists of two species, *T. torosa* and *T. sierrae*.

Notophthalmus

Key to Species of *Notophthalmus*

1a. At least some dark spots on body and tail are distinctly bordered and as large as eye (Fig. 15) .... ***N. meridionalis***
CC 444, pl 39 **CAAR** 74 (1968)

1b. Spots on body and tail smaller than eye or, if as large as eye, with very indistinct borders 2

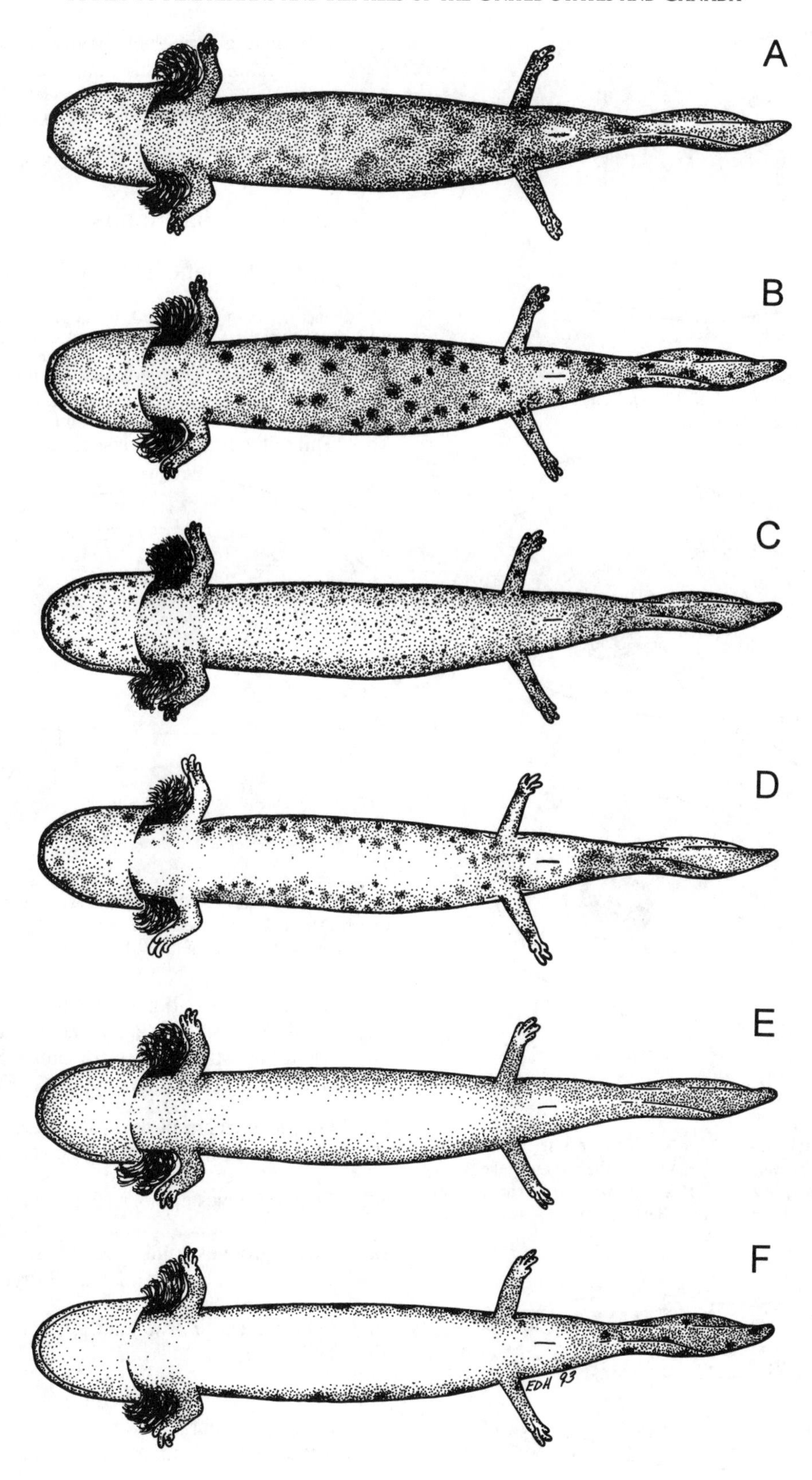

Figure 12. Venters of six species of the salamander genus *Necturus*. A. *N. maculosus*, B. *N. lewisi*, C. *N. beyeri*, D. *N. louisianensis*, E. *N. punctatus*, F. *N. alabamensis*. Adapted from Bishop (1943).

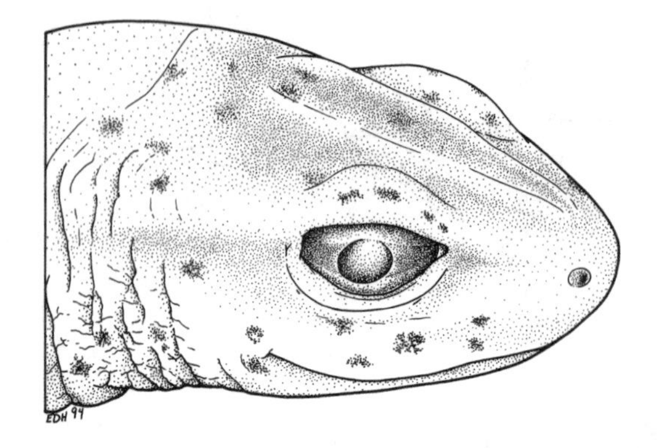

Figure 13. Dorsolateral view showing ridges on the head, typical of the salamander genus *Notophthalmus*. Drawn from a preserved specimen (KU 213262) of *Notophthalmus viridescens*.

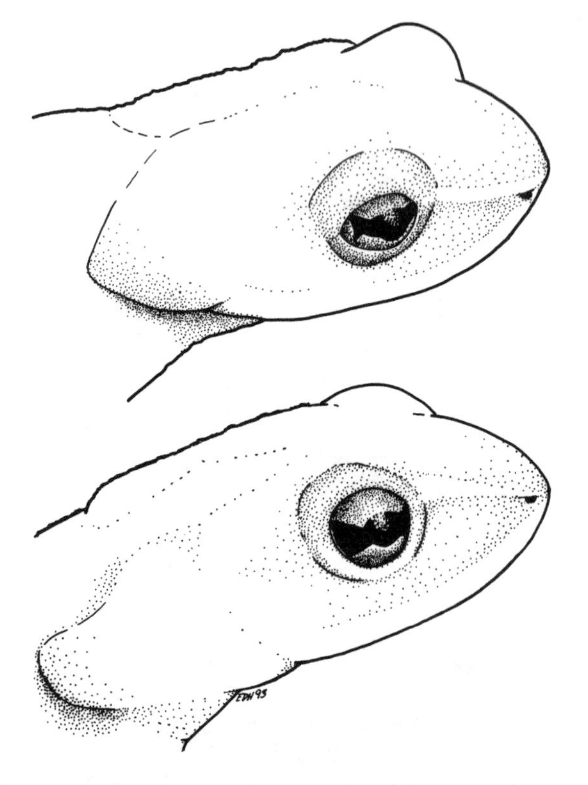

Figure 14. Lateral views of two species of the salamander genus *Taricha*. Upper: *Taricha granulosa* showing dark lower eyelid. Drawn from a photograph by Suzanne L. Collins. Lower: *Taricha torosa* showing light lower eyelid. Adapted from Behler and King (1979).

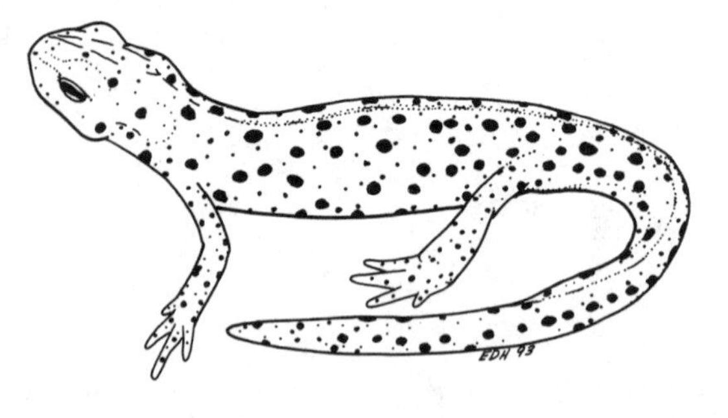

Figure 15. Lateral view of *Notophthalmus meridionalis* showing large dark spots on head, body, and tail.

2a. Dorsolateral stripe not heavily black-bordered .... ***N. perstriatus***
CC 444, pl 39 **CAAR** 38 (1967)

2b. Dorsolateral stripe absent or heavily black-bordered .... ***N. viridescens***
CC 443, pl 39 **CAAR** 53 (1967)

PLETHODONTIDAE

Key to Genera of Plethodontidae

Some genera of the Plethodontidae include individuals or populations that are paedomorphic and perrenibranchiate (fail to transform in nature). In many instances these specimens will be difficult to identify to species with the following key.

1a. Tail with a basal constriction (Fig. 16) 2
1b. Tail without a basal constriction 3

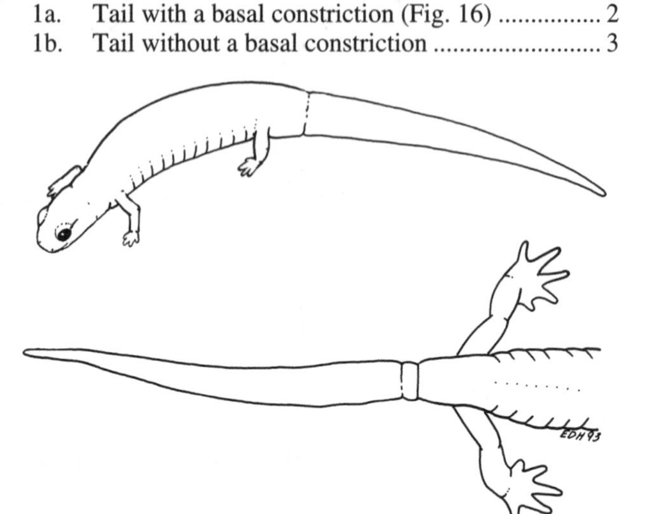

Figure 16. Views of two salamanders showing basal constriction of tail. Upper: Dorsolateral view of *Hemidactylium scutatum*. Lower: Dorsal view of the posterior half of the body and tail of *Ensatina eschscholtzii*.

2a. Venter with small black spots; eastern and central North America .... ***Hemidactylium*** (p. 14)
2b. Venter not spotted; western United States .... ***Ensatina*** (p. 14)

3a. Four toes on hindlimb 4
3b. Five toes on hindlimb.. 5

4b. Costal grooves number 16–22; western U.S. .... ***Batrachoseps*** (p. 16)
4a. Costal grooves number 15–16; eastern U.S. .... ***Eurycea*** (part) (p. 20)

5a. Gills present 6
5b. Gills absent.. 11

6a. Body pigment greatly reduced or lacking; eyes reduced or absent 7
6b. Not as above 10

7a. With a distinct canthus rostralis (Fig. 17) .... ***Gyrinophilus*** (part) (p. 15)
7b. No distinct canthus rostralis 8

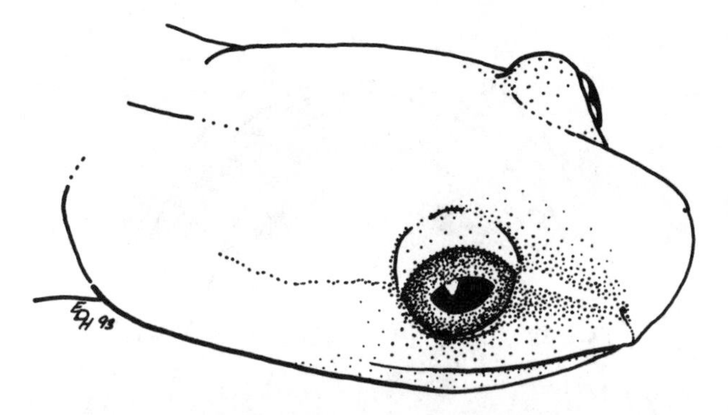

Figure 17. Dorsolateral view of the head of *Gyrinophilus porphyriticus* showing the distinct canthus rostralis.

8a. Thin limbs; eyes absent or greatly reduced 9

8b. Normal limbs; eyes not greatly reduced .... ***Eurycea*** (part) (p. 20)

9a. Sides of head nearly parallel (Fig. 18); neck narrower than body; from Georgia .... ***Haideotriton*** (p. 15)

9b. Sides of head taper toward snout (Fig. 18); neck not narrower than body; from central Texas .... ***Typhlomolge*** (p. 16)

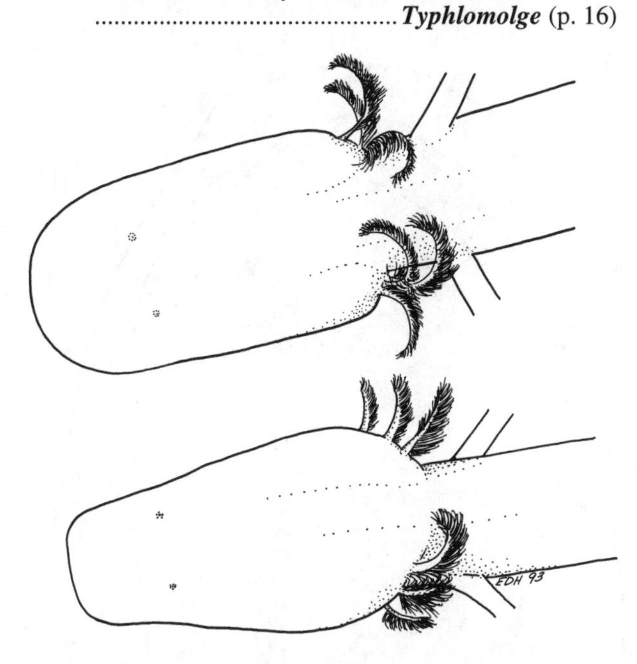

Figure 18. Dorsal views of the heads of *Haideotriton wallacei* (upper) and *Typhlomolge rathbuni* (lower).

10a. With a distinct canthus rostralis (Fig. 17) .... ***Gyrinophilus*** (part) (p. 15)

10b. No distinct canthus rostralis .... ***Eurycea*** (part) (p. 20)

11a. Neck distinctly wider than head; neck bent (head and body are not on the same plane in lateral view)..... 12

11b. Neck narrower to barely wider than head; neck straight (head and body are on the same plane) 13

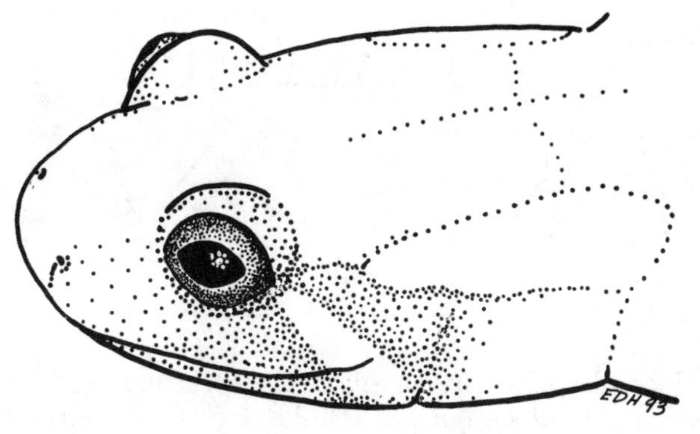

Figure 19. Dorsolateral view of the head showing light line from eye to angle of jaw in *Desmognathus*.

12a. Costal grooves number 20–22 .... ***Phaeognathus*** (p. 16)

12b. Costal grooves number 13–15; often with light line from eye to angle of jaw (Fig. 19) .... ***Desmognathus*** (p. 17)

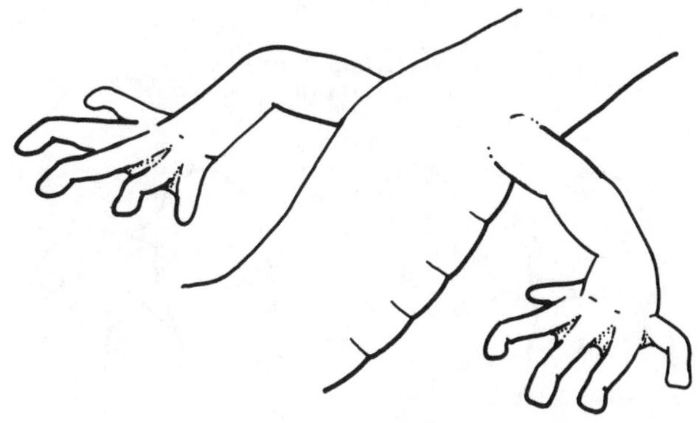

Figure 20. Dorsal view of rear limbs of *Aneides ferreus* showing square-tipped digits typical of the genus.

13a. Digits square-tipped (Fig. 20); may have protruding teeth .... ***Aneides*** (p. 15)

13b. Tips of digits rounded 14

14a. Eyes reduced; pigment greatly reduced .... ***Typhlotriton*** (p. 15)

14b. Eyes normal; body pigmented 15

15a. Alternating dark and light longitudinal lines on sides .... ***Stereochilus*** (p. 16)

15b. Pattern variable, but never with alternating dark and light longitudinal lines on sides 16

16a. Toes webbed at least half the length of digits (Fig. 21) .... ***Hydromantes*** (p. 16)

16b. Toes barely webbed or not at all 17

17a. Tongue attached in front (Fig. 22) .... ***Plethodon*** (p. 21)

17b. Tongue free all around (Fig. 22) 18

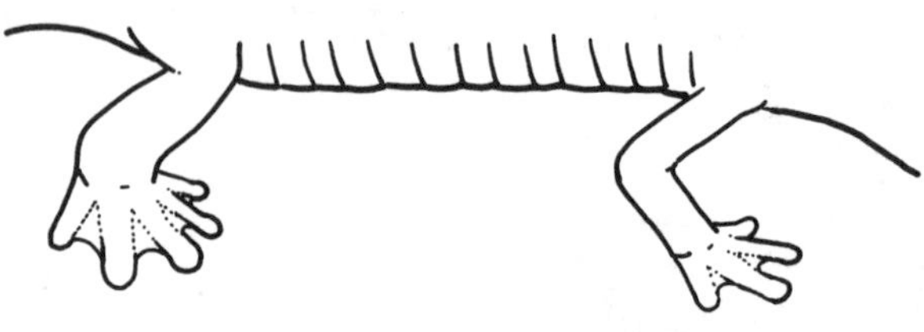

Figure 21. Lateral view of the lower body and limbs of *Hydromantes platycephalus* showing the webbed feet typical of the genus.

18a. Vomerine and parasphenoid teeth continuous (Fig. 23); stocky or robust habitus or with a distinct canthus rostralis 19
18b. Vomerine and parasphenoid teeth not continuous (Fig. 23); slender habitus ...... ***Eurycea*** (part) (p. 20)

19a. Distinct canthus rostralis (Fig. 17) .... ***Gyrinophilus*** (part) (p. 15)
19b. No distinct canthus rostralis .... ***Pseudotriton*** (p. 15)

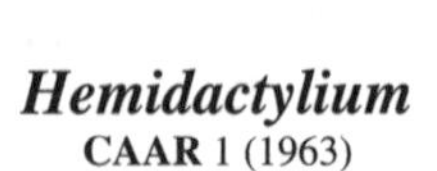

Hemidactylium

CAAR 1 (1963)

Hemidactylium scutatum

CC 478, pl 43 **CAAR** 2 (1963)

Figure 22. Lateral view of a diagram of tongue attachment in the mouths of two salamanders. Left: Tongue completely free around margins, sitting on a pedestal. Right: Tongue almost fully attached in front and rear.

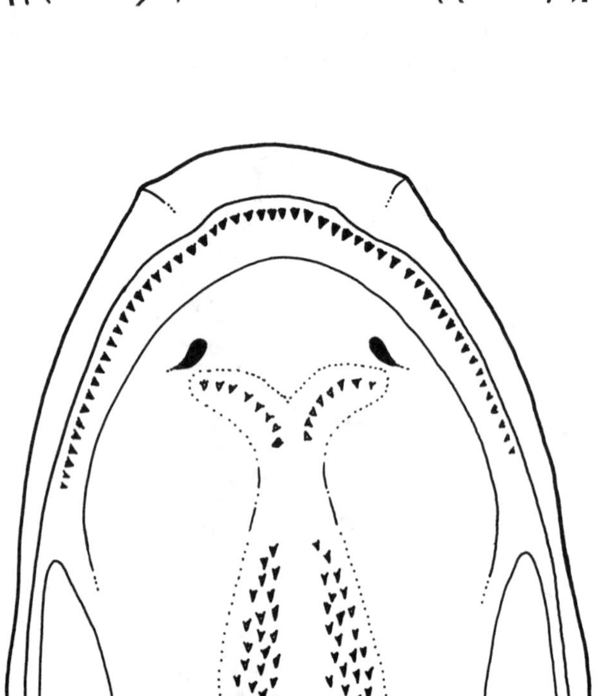

Figure 23. Interior view of upper jaws of two salamander genera. Upper: Vomerine and parasphenoid teeth continuous, as in the genus *Gyrinophilus*. Lower: Vomerine and parasphenoid teeth not continuous, as in the genus *Eurycea*. Adapted from Pfingsten and Downs (1989).

Ensatina

Ensatina eschscholtzii

S 10, pl 3

Some authorities have suggested that this taxon consists of several species.

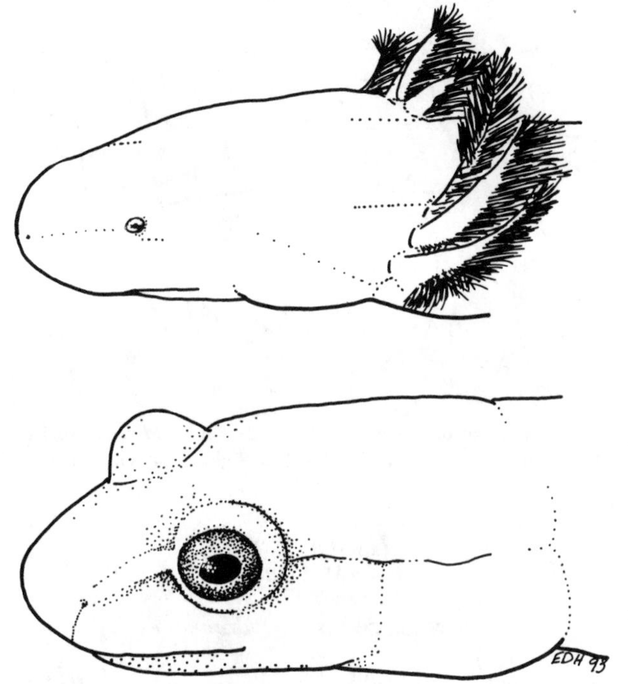

Figure 24. Lateral views of heads of the salamander genus *Gyrinophilus*. Upper: *G. palleucus* showing small, poorly developed eye. Lower: *G. porphyriticus* showing large, well developed eye.

Haideotriton
CAAR 39 (1967)

Haideotriton wallacei
CC 498, p 499 **CAAR** 39 (1967)

Gyrinophilus
CAAR 31 (1967)

Key to Species of ***Gyrinophilus***

1a. Eye small (Fig. 24), eye-snout distance 4–5 times eye diameter 2

1b. Eye large (Fig. 24), eye-snout distance no more than 3½ times eye diameter .... ***G. porphyriticus***
CC 482, pl 42 **CAAR** 33 (1967)

2a. Gray network on pale dorsum; West Virginia .... ***G. subterraneus***
CC 482

2b. Dorsum uniformly colored or with spots 3

3a. With a dark stripe or blotch on forward half of throat; from caves in eastern Tennessee..... ***G. gulolineatus***
CC 484 **CAAR** 32 (1967)
(as *G. palleucus gulolineatus*)

3b. Throat without a dark stripe or blotch; from caves in central and south-central Tennessee and northern Alabama .... ***G. palleucus***
CC 484, pl 39 **CAAR** 32 (1967)

Pseudotriton
CAAR 165 (1975)

Key to Species of ***Pseudotriton***

1a. Snout long, at least 1½ times eye diameter (Fig. 25); a dark- bordered line from eye to nostril; dorsum with numerous, irregular spots, which often run together .... ***P. ruber***
CC 487, pl 42 **CAAR** 167 (1975)

1b. Snout short, no more than 1½ times eye diameter (Fig. 25); no light line from eye to nostril; dorsum with nearly circular, well-separated spots 2

2a. Dorsal spots numerous; venter unmarked or with dark spots or flecks; from southern New Jersey through the Carolinas to northern Florida and along the Gulf Coast to extreme eastern Louisiana .... ***P. montanus***
CC 484, pl 42 **CAAR** 166 (1975)

2b. Dorsal spots few in number and widely scattered; venter unmarked or occasionally with a dark line on rim of lower jaw (individuals in proximate, but not sympatric, populations of *P. montanus* have spotted or flecked venters); from southern Ohio and West Virginia through much of Kentucky to southern Tennessee .... ***P. diastictus***
CC 484 **CAAR** 166 (1975) (as *P. montanus diastictus*)

Aneides
CAAR 157 (1974)

Key to Species of ***Aneides***

1a. Eastern U.S. ....***A. aeneus***
CC 479, pl 43 **CAAR** 30 (1967)

1b. Western U.S. 2

2a. Venter and dorsum black or slate......................... .... ***A. flavipunctatus***
S 22, pl 6, 7 **CAAR** 158 (1974)

2b. Venter light, dorsum dark 3

3a. Tips of digits expanded (Fig. 26) 4

3b. Tips of digits not expanded (Fig. 26)...... ***A. hardii***
S 21, pl 7 **CAAR** 17 (1965)

4a. Venter unmarked; dorsum often with yellow spots .... ***A. lugubris***
S 20, pl 6, 7 **CAAR** 159 (1974)

4b. Venter dusky and lightly speckled; no yellow dorsal spots .... ***A. ferreus***
S 19, pl 7 **CAAR** 16 (1965)

Typhlotriton
CAAR 84 (1970)

Typhlotriton spelaeus
CC 498, pl 39 **CAAR** 84 (1970)

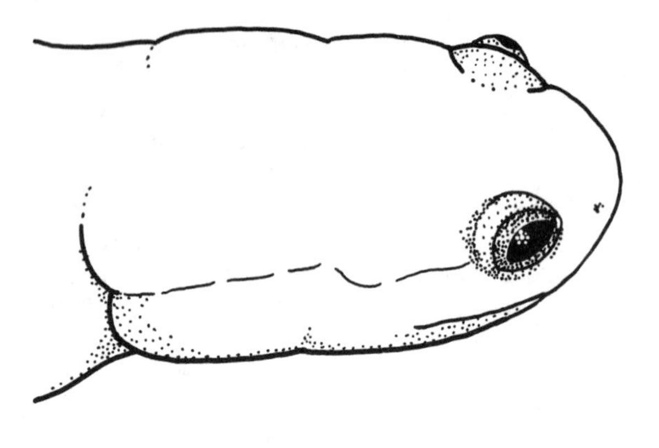

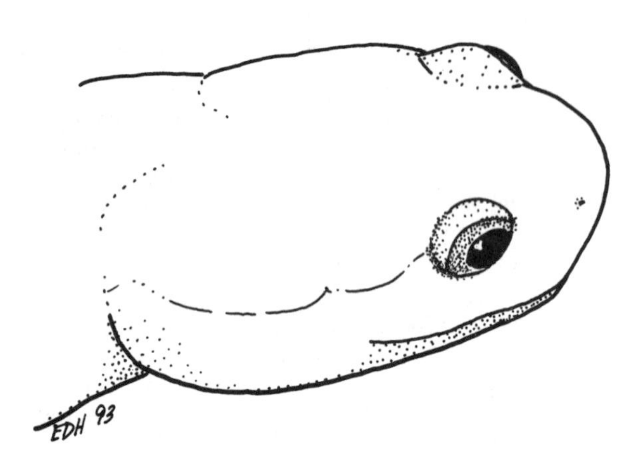

Figure 25. Lateral views of the heads of the salamander genus *Pseudotriton*. Upper: *Pseudotriton montanus* showing anterior position of eye on snout. Lower: *P. ruber* showing more posterior position of eye on snout. Adapted from Pfingsten and Downs (1989).

Phaeognathus
CAAR 26 (1966)

Phaeognathus hubrichti
CC 461, pl 41 **CAAR** 26 (1966)

Typhlomolge

Key to Species of ***Typhlomolge***

1a. Robust habitus; adpressed limbs overlap by one costal fold .... ***T. robusta***
CC 496
1b. Slender habitus; adpressed limbs overlap by 5–9 costal folds .... ***T. rathbuni***
CC 496, pl 39, p 496

Stereochilus
CAAR 25 (1966)

Stereochilus marginatus
CC 478, pl 43 **CAAR** 25 (1966)

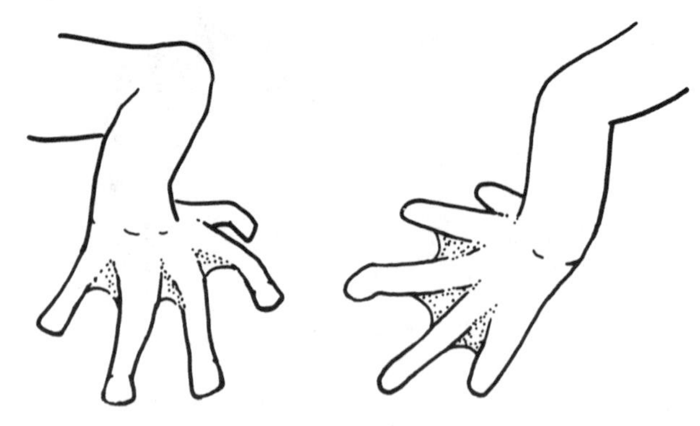

Figure 26. Dorsolateral views of feet of salamander genus *Aneides*. Left: Expanded digits in *Aneides lugubris*. Right: Digits not expanded in *Aneides hardii*.

Hydromantes
CAAR 10 (1964)

Key to Species of ***Hydromantes***

1a. Dorsum uniformly pigmented .... ***H. brunus***
S 7, pl 6 **CAAR** 11 (1964)
1b. Dorsum mottled 2

2a. Venter dark with light flecks; adpressed limbs not overlapping .... ***H. platycephalus***
S 7, pl 6 **CAAR** 11 (1964)
2b. Venter with light blotches; adpressed limbs overlap .... ***H. shastae***
S 7, pl 6 **CAAR** 11 (1964)

Batrachoseps

Key to Species of ***Batrachoseps***

1a. No longitudinal dorsal stripe (but may have pairs of stripes over the shoulders and the pelvic region or a pair of dorsolateral stripes slightly lighter than the ground color) 2
1b. Longitudinal dorsal stripe present (may be indistinct in some preserved specimens) 4

2a. With 20–21 costal grooves .... ***B. simatus*** (part)
S 17, pl 9
2b. With 16–19 costal grooves 3

3a. Dark above and below, without light highlights (some individuals may have faint dorsolateral stripes); tail short, about ¾ of SVL; 16–18 costal grooves; from the Inyo Mountains, Inyo County, California .... ***B. campi***
S 17, pl 8
3b. Dark above and below, but with light highlights, often including a pair of light stripes above the shoulders and pelvic region; tail longer than SVL; 18–19 costal grooves; from San Gabriel Mountain, Los Angeles County, California .... ***B. gabrieli***

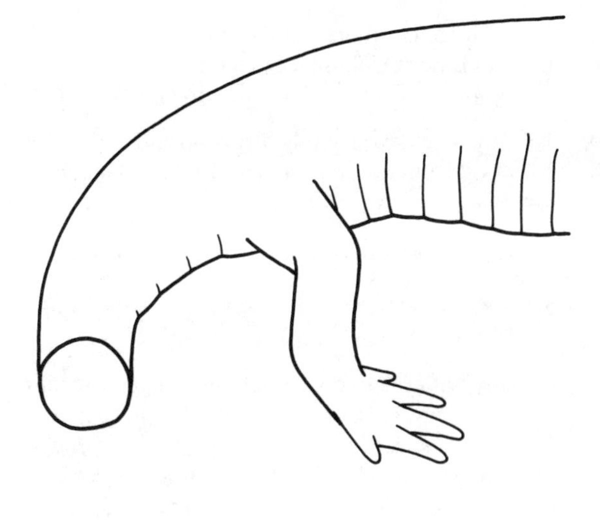

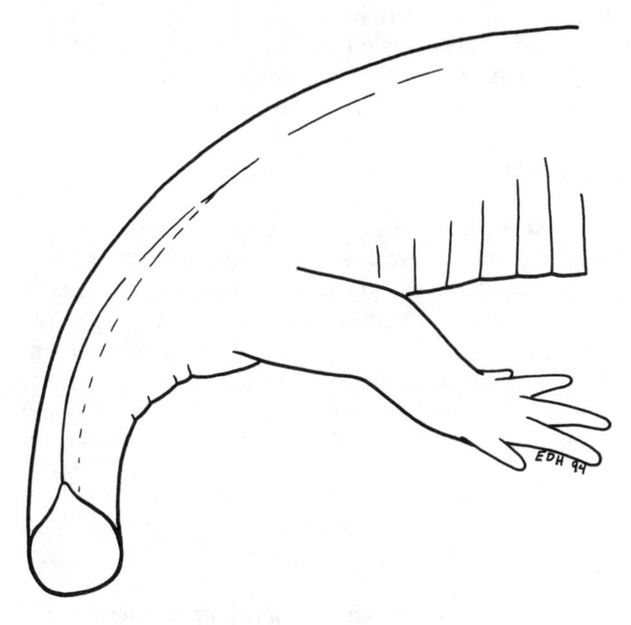

Figure 27. Cross-sections of tails of the salamander genus *Desmognathus*. Upper: Tail round as in *Desmognathus wrighti*. Lower: Tail triangular and often with dorsal keel as in *Desmognathus monticola*.

4a. Large white patches on sides of venter .... ***B. wrightorum***
S 18, pl 7, 8 **CAAR** 506 (1991) (as *B. wrighti*)
4b. Small white dots on sides of venter 5

5a. Pale venter with dark spots 6
5b. Dark venter with light spots 7

6a. Costal folds number 9–12 between adpressed limbs; head no wider than body .... ***B. major***
S 17, pl 9 (as *B. pacificus major*)
6b. Costal folds number 5½–8 between adpressed limbs; head wider than body .... ***B. pacificus***
S 17, pl 7

7a. Underside of tail distinctly paler than venter .... ***B. aridus***
S 17, pl 8
7b. Underside of tail not or barely paler than venter 8

8a. Very short limbs (SVL > 6 times hindlimb length) 9
8b. Limbs longer (SVL 4 ½–6 times hindlimb . length) 10

9a. Venter medium to dark gray .... ***B. attenuatus***
S 17, pl 6, 7, 9
9b. Venter black .... ***B. nigriventris***
S 17, pl 9

10a. Only 4–5 costal folds overlapped by adpressed hindlimb .... ***B. relictus***
S 17, pl 9 (as *B. pacificus relictus*)
10b. At least 5–6½ costal folds overlapped by adpressed hindlimb 11

11a. Costal grooves number 20–21 ......***B. simatus*** (part)
S 17, pl 9
11b. Costal grooves number 18–19 .... ***B. stebbinsi***
S 17, pl 8

Desmognathus

Key to Species of *Desmognathus*

1a. Tail near base oval or round in cross section, not keeled above (Fig. 27).. 2
1b. Tail near base triangular in cross section, often keeled above (Fig. 27).. 9

2a. Dorsal chevrons very distinct; tail usually less than ½ total length; venter white (Fig. 28) ... ***D. wrighti***
CC 459, pl 42, p 458, 459
2b. Dorsal pattern variable (some very indistinct chevron-like markings may be present on some individuals; tail at least ½ total length; venter variable, white or darker ground color 3

3a. Often with continuous dorsolateral dark lines; large (to more than 60 mm total length) 4
3b. No continuous dorsolateral dark lines; small (to 30 mm total length) .... ***D. aeneus***
CC 459, p 458, 459 **CAAR** 534 (1992)

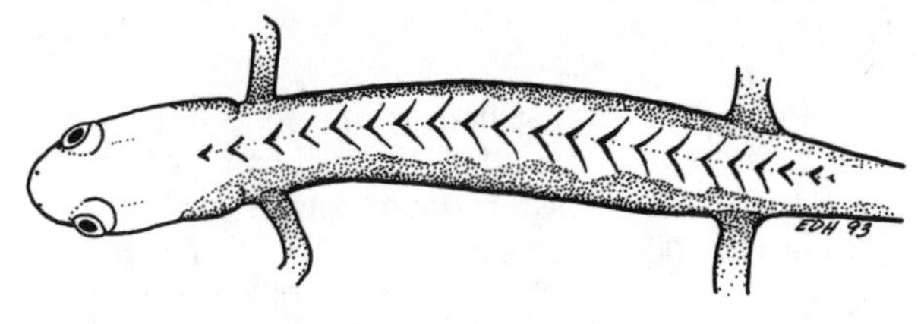

Figure 28. Dorsal view of the body of *Desmognathus wrighti* showing the distinct chevron pattern typical for this species. Adapted from Bishop (1943).

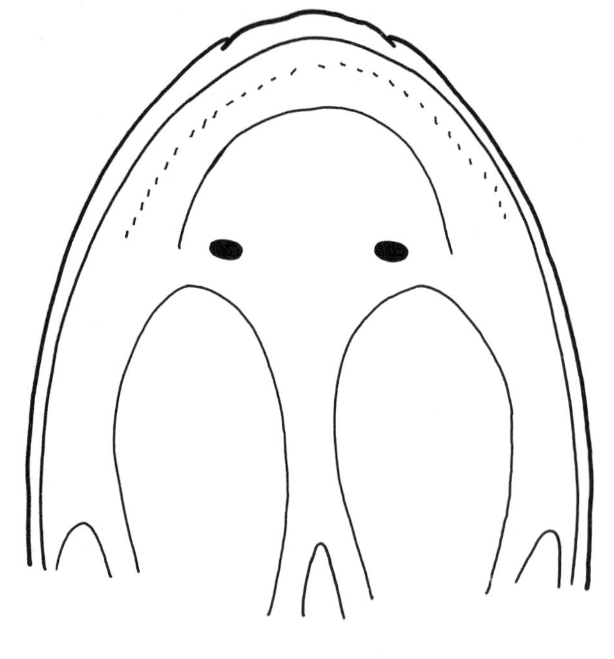

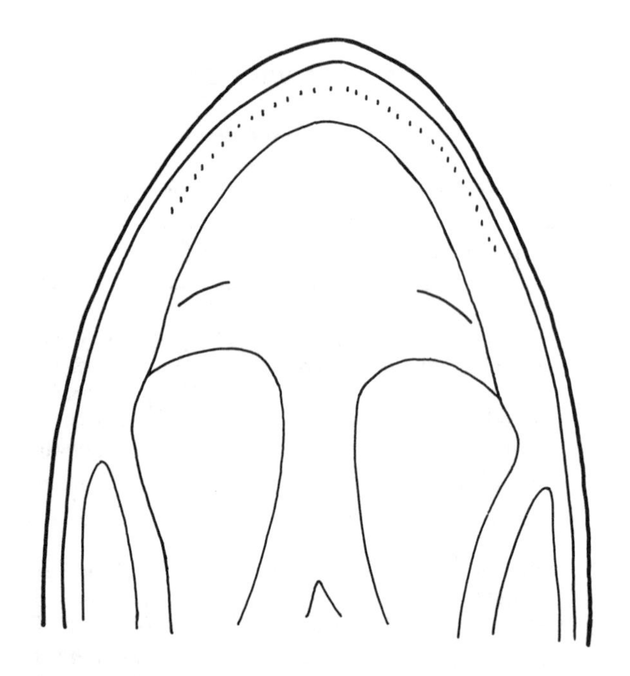

Figure 29. Interior view of the upper jaw of other species of the genus *Desmognathus* (upper) and *D. marmoratus* (lower) showing position and appearance of choanae.

4a. Reddish cheek patches (in life); line from eye to angle of jaw often dark (instead of light) and indistinct; dorsal pattern (if present) distinct, but lacking a middorsal stripe .... ***D. imitator***
CC 457, pl 40 **CAAR** 359 (1985)

4b. No red cheek patches; line from eye to angle of jaw light and usually distinct; dorsum variable, if distinctly patterned, often with a middorsal stripe or a series of blotches merging to form a line 5

5a. Dorsal blotches, if present, indistinct; from south of the Fall Line at Columbus, Georgia .... ***D. apalachicolae***
CC 457

5b. Usually with a moderately distinct pattern of dorsal blotches; from north of the Fall Line at Columbus, Georgia 6

The following four species constitute the *Desmognathus ochrophaeus* complex, species of which are morphologically cryptic and are best differentiated by geography and measures of protein divergence.

6a. Usually with relatively straight dorsolateral stripes and often with a row of middorsal dark patches; from New York and extreme southern Quebec through eastern Ohio, much of Pennsylvania and West Virginia, western Virginia and eastern Kentucky to central Tennessee .... ***D. ochrophaeus***
CC 455, pl 42, p 447, 449, 455 **CAAR** 129 (1973)

6b. Dorsolateral stripes straight or wavy, occasionally absent; rarely with middorsal dark patches; from southern Virginia through western North Carolina and extreme eastern Tennessee to northeastern Georgia and an isolated population in northeastern Alabama 7

7a. Most individuals with wavy dorsolateral stripes (except in the Great Smoky Mountains where specimens have straight stripes); from extreme southwestern North Carolina and southeastern Tennessee into northwestern South Carolina and northeastern Georgia, and a disjunct population in the northeastern corner of Alabama .... ***D. ocoee***
CC 455, pl 42, p 449, 455 (as *D. ochrophaeus,* southern type)
CAAR 129 (1973)

7b. Dorsolateral stripes straight or wavy; from southern Virginia into northwestern and west-central North Carolina and adjacent Tennessee 8

8a. From southern Virginia into northwestern North Carolina and extreme northeastern Tennessee .... ***D. orestes***
CC 455, pl 42, p 449, 455 (as *D. ochrophaeus,* southern type)
CAAR 129 (1973)

8b. From extreme west-central North Carolina and adjacent eastern Tennessee .... ***D. carolinensis***
CC 455, pl 42, p 449, 455 (as *D. ochrophaeus,* southern type)
CAAR 129 (1973)

9a. Head distinctly flattened, sloping down toward snout from behind small eyes; choanae slitlike (Fig. 29), often obscure; light line from eye to angle of jaw frequently indistinct or absent .... ***D. marmoratus***
CC 461 , pl 41, p 460 **CAAR** 3 (1963)
(as *Leurognathus marmoratus*)

9b. Head not distinctly flattened; slope down to snout begins in front of eyes; choanae round (Fig. 29); light (rarely dark) line from eye to angle of jaw usually distinct 10

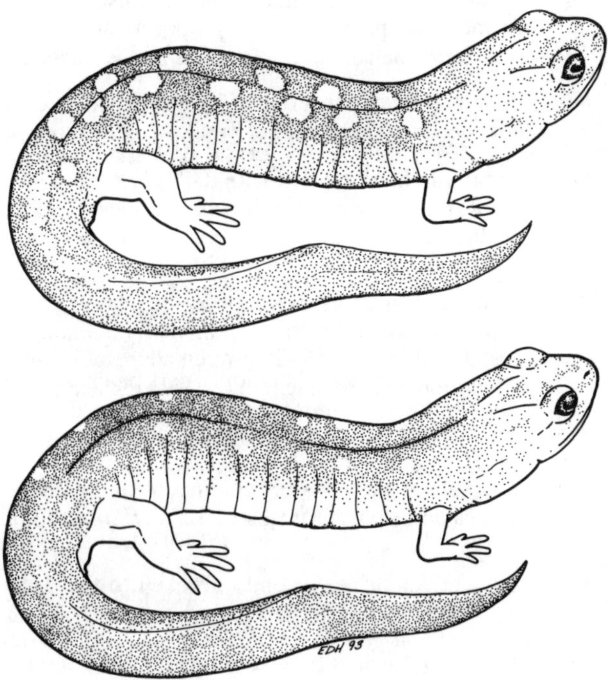

Figure 30. Dorsolateral views of the salamander genus *Desmognathus*. Upper: *D. santeetlah* showing large, narrowly-spaced spots on the dorsum. Drawn from a photograph by Suzanne L. Collins. Lower: *D. auriculatus* showing small, widely-spaced spots on dorsum. Drawn from a preserved specimen (KU 22773).

10a. Venter mottled 11
10b. Venter uniformly pigmented 17

11a. Venter distinctly dark with light mottling .... ***D. quadramaculatus*** (part)
CC 455, pl 41 **CAAR** 153 (1974)
11b. Venter light, mottling distinct or not 12

12a. Thirteen (rarely 14) costal grooves; venter lightly mottled .... ***D. monticola*** (part)
CC 453, pl 41, p 449
12b. Fourteen (rarely 15) costal grooves; ventral mottling variable, but usually distinct 13

13a. Large, light, narrowly-spaced spots on dorsum (Fig. 30), occasionally fused into a line; usually 2–3 (rarely 4) costal folds between adpressed limbs 14
13b. Small, light, widely-spaced spots on dorsum (Fig. 30); 4½–5½ costal folds between adpressed limbs in specimens with SVL more than 45 mm, at least 3½ in smaller specimens 16

14a. Body robust; venter lightly mottled or, if extensively mottled, dorsolateral stripes and dorsal pattern distinct 15
14b. Body less robust; venter extensively mottled; dorsolateral stripes narrow and indistinct; dorsal pattern subdued .... ***D. santeetlah***
CC 450

15a. Usually with a distinct dorsal pattern of light spots fused into a band; at least a portion of the venter without mottling 19
15b. Dorsum mottled, usually lacking a light-colored band; entire venter mottled .... ***D. welteri***
CC 452

16a. Light-colored dorsal spots number 7–9 between limbs .... ***D. auriculatus***
CC 450, p 451
16b. Pale dorsal spots between limbs number 11–14, or dorsum uniformly brown; west of Mississippi River .... ***D. brimleyorum***
CC 452, pl 42, p 452

17a. Venter black or dark brown ....
D. quadramaculatus (part)
CC 455, pl 41 **CAAR** 153 (1974)
17b. Venter light 18

18a. With distinct dorsal markings ... ***D. monticola*** (part)
CC 453, pl 41, p 449
18b. Dorsum with a distinct light band sometimes broken into spots; dark markings (if present) are indistinct 19

19a. Light dorsal spots usually fused into a band (Fig. 31); from southern New Brunswick and Quebec to southeastern Indiana and the Carolinas (often disjunct through the southern Appalachian Mountains) .... ***D. fuscus***
CC 448, pl 41, p 447, 449
19b. Light dorsal spots usually distinct, not fused; from extreme southern Illinois, western Kentucky and Tennessee south to the Gulf of Mexico and east to the Florida panhandle and the Fall Line at Columbus, Georgia; also from northeastern Arkansas and extreme southern Arkansas and northcentral Louisiana .... ***D. conanti***
CC 448, pl 42, p 447 (as *D. fuscus conanti*)

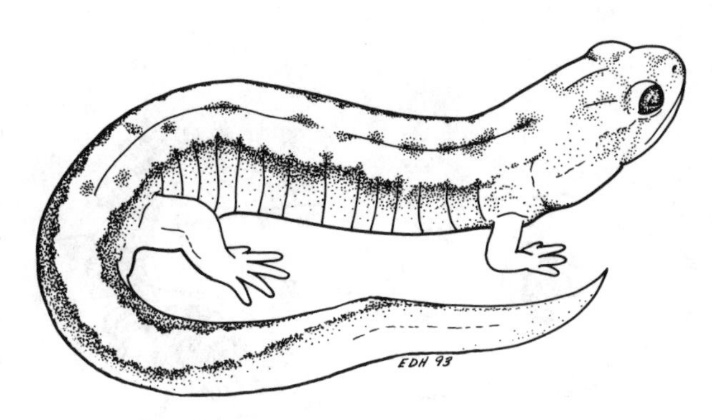

Figure 31. Dorsolateral view of *Desmognathus fuscus* showing a dorsal pattern of rows of spots fused into a single medial band. Drawn from photographs by Suzanne L. Collins and in Martof et al. (1980).

Eurycea

Key to Species of *Eurycea*

1a. Four toes on each hind foot .... ***E. quadridigitata***
CC 494, pl 43 **CAAR** 44 (1967) (as *Manculus quadridigitatus*)
1b. Five toes on each hind foot 2

2a. Costal grooves number nineteen or more 3
2b. Costal grooves number seventeen or less 5

3a. External gills present 4
3b. Without external gills .... ***E. multiplicata*** (part)
CC 491, pl 43 **CAAR** 21 (1965)

4a. Venter uniformly light ....***E. tynerensis***
CC 494, pl 39 **CAAR** 22 (1965)
4b. Venter with dark granules ......***E. multiplicata*** (part)
CC 491, pl 43 **CAAR** 21 (1965)

Distinguishing between *Eurycea multiplicata* and *E. tynerensis* is difficult; recent research indicates a different relationship in this complex than the traditional arrangement presented here.

5a. Adults with gills; central Texas 6
5b. Adults without gills; not in central Texas 9

6a. Eyes small (diameter no more than ⅓ interorbital distance), often at least partly covered by skin (Fig. 32); with a *shovelnose* 7
6b. Eyes large (diameter at least ½ interorbital distance) (Fig. 32) 8

7a. *Shovelnose* distinct; dorsal pattern variable, but usually without pale patches; limbs proportionately short .... ***E. tridentifera***
CC 495, p 496 **CAAR** 199 (1977)
7b. *Shovelnose* not pronounced; dorsum usually with distinct pale patches; limbs proportionately long; eyes very small .... ***E. sosorum***

8a. Dorsum brown with a row of light spots down each side; dark ring at margin of pupil .... ***E. nana***
CC 495, pl 39 **CAAR** 35 (1967)
8b. Dorsum lighter, mottled with dark pigment, with a double row of light flecks down each side (at least in smaller individuals) .... ***E. neotenes***
CC 496, pl 39, p 496 **CAAR** 36 (1967)

9a. Costal folds number 2–5 between adpressed limbs; tail about ½ total length 10
9b. Costal folds number 0–2 between adpressed limbs; tail more than ½ total length 13

10a. Dorsolateral stripes essentially unbroken on body and extending onto tail 11
10b. Dorsolateral stripes rarely continuous; tail with numerous dark spots or flecks .... ***E. junaluska***
CC 490 **CAAR** 321 (1983)

11a. Dorsolateral stripes extend unbroken to the distal fourth or to the tip of the tail .... ***E. cirrigera***
CC 488 **CAAR** 45 (1967) (as *E. bislineata cirrigera*)
11b. Dorsolateral stripes rarely extend unbroken past the midpoint of the tail 12

12a. Costal grooves 13–14 (rarely 15); dorsolateral stripes black .... ***E. wilderae***
CC 488, pl 43 **CAAR** 45 (1967) (as *E. bislineata wilderae*)
12b. Costal grooves 15–16; dorsolateral stripes brown .... ***E. bislineata***
CC 488, pl 43 **CAAR** 45 (1967)

13a. No vertical bars on tail; pattern usually of scattered black spots .... ***E. lucifuga***
CC 492, pl 43 **CAAR** 24 (1966)
13b. Dark vertical bars on sides of tail (Fig. 33) (occasionally fused into a solid line); pattern variable 14

14a. Middorsal and dorsolateral lines form distinct three-lined pattern .... ***E. guttolineata***
CC 492, pl 43 **CAAR** 221 (1979)
(as *E. longicauda guttolineata*)
14a. Dorsum without middorsal line or series of spots (spots and flecks often present, but not as middorsal line) .... ***E. longicauda***
CC 492, pl 43 **CAAR** 221 (1979)

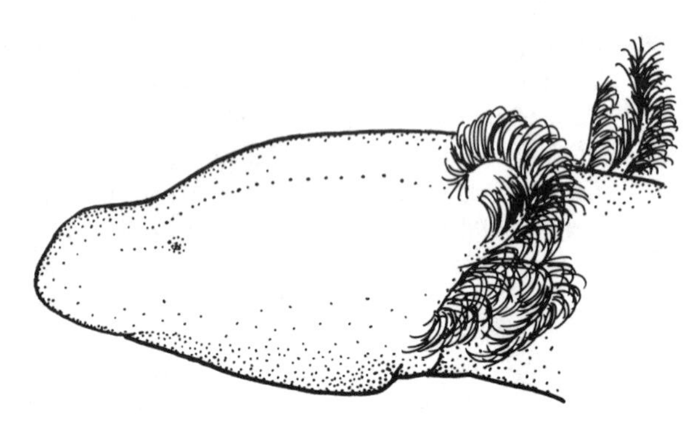

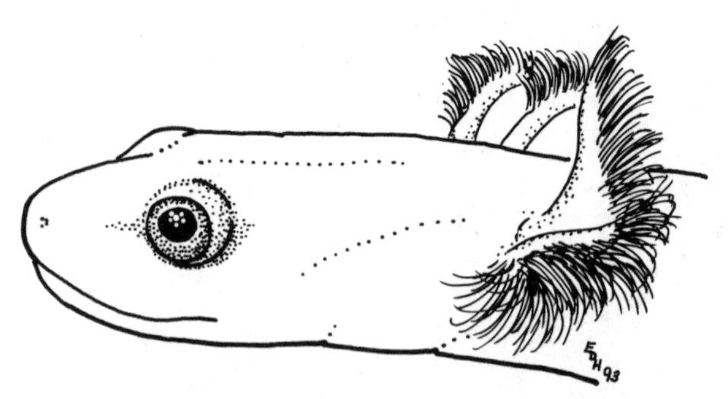

Figure 32. Lateral views of heads of completely aquatic species of the salamander genus *Eurycea*. Upper: *Eurycea tridentifera* showing small, poorly developed eye. Lower: Other completely aquatic species of the genus.

Figure 33. Lateral view of *Eurycea longicauda* showing vertical bars on the tail. Adapted from Johnson (1977).

Plethodon

Key to Species of *Plethodon*

Due to the nature of species complexes within this genus, having very accurate locality data for the specimen to be identified is important; this is often the only conclusive means of determining the identity of a species in this genus.

1a. Found west of the 100th meridian 2
1b. Found east of the 100th meridian......................... 9

2a. Fifth toe absent or with only one segment 3
2b. Fifth toe with two segments 4

3a. Costal grooves number 18–20; New Mexico......... .... ***P. neomexicanus***
S 15, pl 5 **CAAR** 131 (1973)
3b. Costal grooves number 14–16; Oregon or Washington .... ***P. larselli***
S 14, pl 5 **CAAR** 13 (1964)

4a. Fourteen (rarely 15) costal grooves; paratoid glands distinct (especially in larger adults) (Fig. 34) 5
4b. Fifteen or more costal grooves (very rarely 14); paratoid glands absent 6

5a. Even-edged light middorsal stripe with few or no dark flecks or spots, extends to the tip of the tail (Fig. 35); from western Washington............ ***P. vandykei***
S 16, pl 5 **CAAR** 91 (1970)
5b. Light middorsal stripe with dark flecks or spots, rarely extending unbroken to tip of tail (Fig. 35); from Idaho and Montana .... ***P. idahoensis***
S 16, pl 5 **CAAR** 91 (1970) (as *P. vandykei idahoensis*)

6a. Usually 17–18 costal grooves; short toes............ 7
6b. Usually 15–16 costal grooves; long toes 8

7a. Dark above and below; usually eighteen costal grooves; six or more costal folds between adpressed limbs .... ***P. elongatus***
S 13, pl 5 **CAAR** 102 (1971)
7b. Light brown above and below; usually seventeen costal grooves; less than six costal folds between adpressed limbs .... ***P. stormi***
S 13, pl 5 (as *P. elongatus stormi*) **CAAR** 103 (1971)

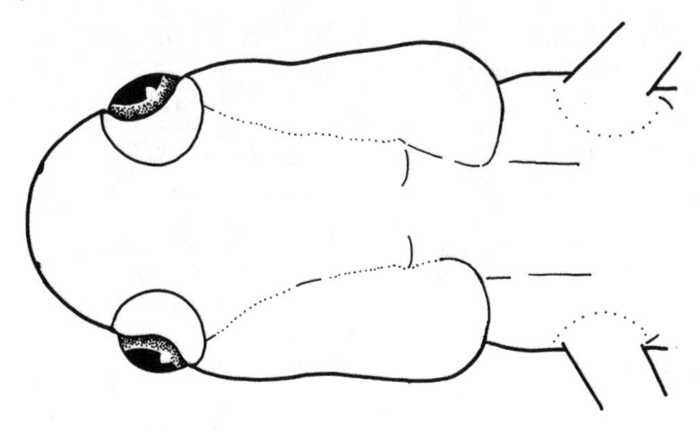

Figure 34. Dorsal view of the head of *Plethodon vandykei* showing paratoid glands typical of some species of the genus. Adapted from Bishop (1943).

Figure 35. Dorsal views of *Plethodon vandykei* (upper) showing even-edged middorsal stripe that extends to tip of tail and contains few or no dark flecks, and *P. idahoensis* (lower) showing uneven-edged middorsal stripe that does not extend to tip of tail and contains numerous dark flecks.

8a. Usually fifteen costal grooves; dorsal stripe (if present) greenish-yellow to tan in life and extending onto tail, but not to tip ....***P. dunni***
S 12, pl 5 **CAAR** 82 (1970)
8b. Usually sixteen costal grooves; dorsal stripe (if present) yellow, orange, brown, or red in life and extending to tip of tail .... ***P. vehiculum***
S 11, pl 5 **CAAR** 83 (1970)

9a. Found west of the Mississippi River.................. 10
9b. Found east of the Mississippi River................... 20

10a. Seventeen or fewer costal grooves; three or fewer costal folds between adpressed limbs 11

These and the following characters identify western members of the *P. glutinosus* complex, which contains species that are morphologically poorly defined and are best differentiated by geography and measures of protein divergence.

10b. Seventeen or more costal grooves; six or more costal folds between adpressed limbs........................... 43

11a. Chin dark.. 12
11b. Chin light 14

12a. Chin and venter uniformly dark; from southern Missouri, northern and western Arkansas, and eastern Oklahoma (exclusive of the range of *P. kiamichi* and *P. sequoyah*)............................... ***P. albagula*** (part)
CC 472
12b. Chin slightly lighter than venter 13

13a. From Beaver's Bend State Park, McCurtain County, Oklahoma .... ***P. sequoyah*** (part)
CC 472

13b. From central Louisiana through southern Arkansas .... ***P. kisatchie*** (part)
CC 472

14a. Light color extending onto chest; from the Caddo Mountains of Montgomery and Polk counties, Arkansas .... ***P. caddoensis***
CC 475, pl 40 **CAAR** 14 (1964)

14b. Light color restricted to chin 15

15a. Chin only slightly lighter than venter 16

15b. Chin distinctly lighter than venter....................... 17

16a. From Beaver's Bend State Park, McCurtain County, Oklahoma .... ***P. sequoyah*** (part)
CC 472

16b. From central Louisiana through southern Arkansas .... ***P. kisatchie*** (part)
CC 472

17a. From south-central Texas .... ***P. albagula*** (part)
CC 472

17b. From Arkansas or Oklahoma 18

18a. Only from Round and Kiamichi Mountains of Polk County, Arkansas and Leflore County, Oklahoma .... ***P. kiamichi***
CC 472

18b. Not from Round and Kiamichi Mountains 19

19a. Dorsum usually with two longitudinal rows of large light spots (Fig. 36); from Fourche and Irons Fork Mountains, Polk and Scott counties, Arkansas .... ***P. fourchensis***
CC 475, pl 40, p 476 **CAAR** 391 (1986)

19b. Dorsum usually with small light spots (or none), not in two rows, often with chestnut markings (in life) .... ***P. ouachitae***
CC 475, pl 40 **CAAR** 40 (1967)

20a. Four or fewer costal folds between adpressed limbs; seventeen or fewer costal grooves (except 18 in *P. punctatus*); 21

20b. Five or more costal folds between adpressed limbs; seventeen or more costal grooves 43

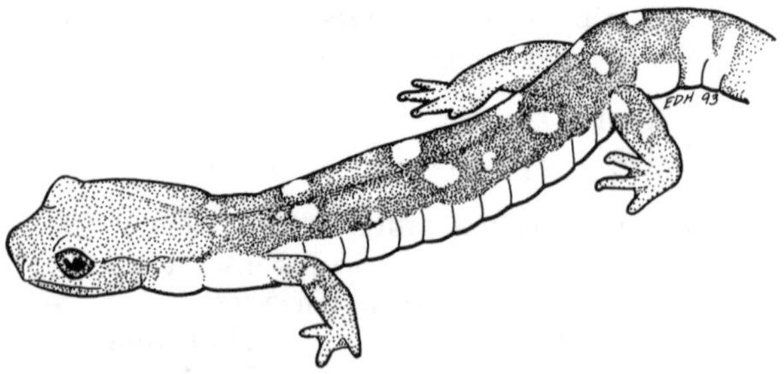

Figure 36. Dorsolateral view of *Plethodon fourchensis* showing two longitudinal rows of large, light spots. Drawn from a photograph by Suzanne L. Collins.

21a. Usually sixteen costal grooves; chin dark or, if light, never white or pinkish (in life); web between 3rd and 4th digits of hindlimb absent or reduced; adpressed limbs overlap or are separated by 0–3 (occasionally 4 in *P. welleri*) costal folds 23

21b. Usually 17–18 costal grooves; chin white or very light (pinkish in life); web present between 3rd and 4th digits of hindlimb; 2–4 costal folds between adpressed limbs 22

22a. Dorsum with bronze (in life) mottling or small light flecks, faint spots (if present) reddish in life; seventeen (rarely 16) costal grooves .... ***P. wehrlei***
CC 469, pl 41 **CAAR** 402 (1987)

22b. Dorsum with large light spots; 17–18 costal grooves .... ***P. punctatus***
CC 469, p 476 **CAAR** 414 (1988)

23a. Costal folds number 3–4 between adpressed limbs; slender (SVL = 6 or more head widths); golden or silvery (in life) blotches above .... ***P. welleri***
CC 468, pl 40 **CAAR** 12 (1964)

23b. Costal folds number 0–3 between adpressed limbs (digits may overlap in some specimens); more robust (SVL less than 6 head widths); dorsal pattern variable 24

These and the following characters identify members of the *P. jordani* complex and of the eastern *P. glutinosus* complex, both of which contain species that are morphologically cryptic and are best differentiated by geography and measures of protein divergence.

24a. Dorsum uniformly colored 25

24b. Dorsum with light flecks or spots 29

25a. Dorsum lighter than head or tail, invested with chestnut color in life; sides white or with many large white blotches (latter often appear to be gray); large (to more than 200 mm) .... ***P. yonahlossee*** (part)
CC 475, pl 40 **CAAR** 15 (1965)

25b. Dorsum not lighter than head or tail 26

26a. Venter and chin dark, barely lighter than dorsum, if at all .... ***P. glutinosus*** (part)
CC 472, pl 41

26b. Venter or at least chin lighter than dorsum 27

27a. From the southern Coastal Plain of South Carolina and adjacent southeastern Georgia......................... .... ***P. variolatus*** (part)
CC 472

27b. Not from the southern Coastal Plain of South Carolina and adjacent southeastern Georgia.............. 28

28a. Few if any light lateral spots; venter often distinctly lighter than dorsum .... ***P. jordani*** (part)
CC 477, pl 40 **CAAR** 130 (1973)

28b. Abundant lateral spotting; from the Chattahoochee National Forest in northern Georgia and adjacent Cherokee County, North Carolina .... ***P. chattahoochee*** (part)
CC 472

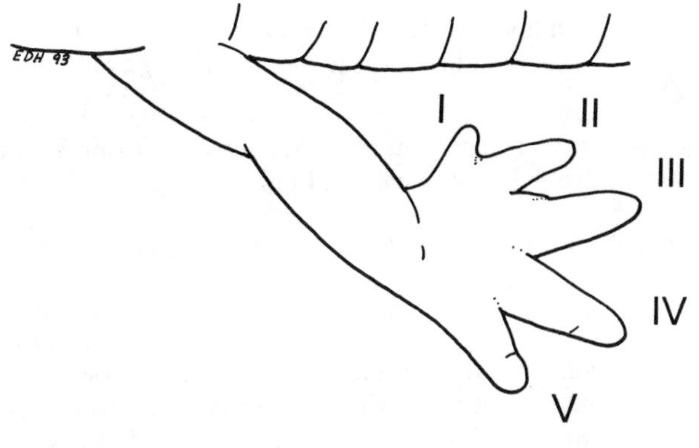

Figure 37. Dorsolateral view of blunt toe tips on rear foot of *Plethodon petraeus*. When adpressed, the 5th toe extends beyond the 2nd joint of the 4th toe.

29a. Dorsum with numerous light spots and chestnut (in life) blotches; sides white or with many large white blotches (latter often appear to be gray); large (to more than 200 mm) .... ***P. yonahlossee*** (part)
CC 475, pl 40 **CAAR** 15 (1965)
29b. Dorsum lacking numerous light spots and chestnut (in life) blotches; sides dark 30

30a. Dorsum with reddish-brown color (in life) extending onto head and tail; toe tips blunt (Fig. 37); 5th toe on hindlimb extending beyond the 2nd joint of the adjacent digit; from Walker County, Georgia........ .... ***P. petraeus***
CC 475, pl 41
30b. Dorsum never reddish-brown in life (may have small red spots); toe tips rounded; 5th toe on hindlimb extends to, but not beyond 2nd joint of the adjacent digit 31

31a. Dorsum with scattered small light flecks, sides may have larger light spots 32
31b. Larger light spots usually present on dorsum 38

32a. Few if any light lateral spots; venter often distinctly lighter than dorsum .... ***P. jordani*** (part)
CC 477, pl 40 **CAAR** 130 (1973)
32b. Lateral spots larger and more abundant than on dorsum 33

33a. Venter as dark as dorsum (chin lighter); from central Georgia .... ***P. ocmulgee***
CC 472
33b. Venter at least slightly lighter than dorsum 34

34a. From the Cumberland Plateau of eastern Kentucky and adjacent regions .... ***P. kentucki***
CC 475 **CAAR** 382 (1986)
34b. Not from the Cumberland Plateau of eastern Kentucky and adjacent regions 35

35a. From the Chattahoochee National Forest in northern Georgia and adjacent Cherokee County, North Carolina .... ***P. chattahoochee*** (part)
CC 472
35b. Not from the Chattahoochee National Forest in northern Georgia and adjacent North Carolina 36

36a. From southwestern North Carolina and adjacent regions .... ***P. oconaluftee***
CC 475 **CAAR** 401 (1987) (as *P. teyahalee*)
36b. Not from southwestern North Carolina and adjacent regions 37

37a. From the Piedmont and Coastal Plain of central and northern South Carolina ....***P. chlorobryonis***
CC 472
37b. From the southern Coastal Plain of South Carolina and adjacent southeastern Georgia........................ .... ***P. variolatus*** (part)
CC 472

38a. Venter as dark as dorsum (chin lighter) 39
38b. Venter at least slightly lighter than dorsum 40

39a. From southern Alabama and Georgia to central Florida .... ***P. grobmani***
CC 472
39b. From southeastern Louisiana through Mississippi and western Alabama to western Kentucky and Tennessee ....***P. mississippi***
CC 472

40a. From Burke, Jefferson, and Richmond counties, Georgia; small (SVL to 69 mm); venter with a few light spots .... ***P. savannah***
CC 472
40b. Distribution not as above; usually larger; venter variable, but usually lacking spots 41

41a. Venter dark, barely lighter than dorsum; chin barely lighter than venter; from Connecticut west to Illinois and south to Georgia .... ***P. glutinosus*** (part)
CC 472, pl 41
41b. Venter lighter than dorsum 42

42a. From eastern Monroe County, Tennessee and adjacent regions; brassy pigment in dorsal spots (in life) .... ***P. aureolus***
CC 473 **CAAR** 381 (1986)
42b. From the Piedmont and Blue Ridge of Virginia south to the northern Piedmont of South Carolina; large white dorsal spots interspersed with smaller spots ....***P. cylindraceus***
CC 472

43a. Venter light with extensive dark mottling 44
43b. Venter dark, with or without light mottling 48

These and the following characters identify members of the *Plethodon cinereus* complex, which contains species that are morphologically cryptic and are best differentiated by geography and measures of protein divergence.

44a. Middorsal stripe (if present) very narrow with straight edges or wide with wavy or scalloped edges, if wide, broader (or not narrower) at base of tail (Fig. 38); 17–19 costal grooves; 6–7 costal folds between adpressed limbs; often with some red pigment on venter (in life) 45

44b. Middorsal stripe (if present) wide with straight or saw-toothed edges, narrowing slightly at base of tail (Fig. 38); 17–22 costal grooves; 7–10 costal folds between adpressed limbs; no red pigment on venter 47

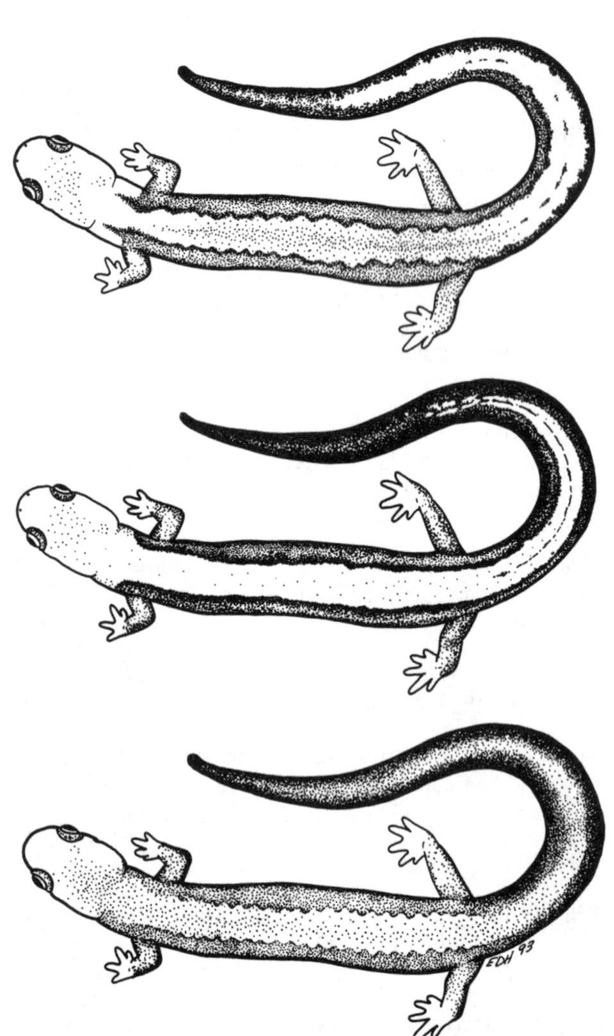

Figure 38. Dorsal views showing middorsal stripes of three species of eastern small *Plethodon*. Upper: *P. dorsalis*. Middle: *P. cinereus*. Lower: *P. serratus*. All drawn from photographs by Suzanne L. Collins.

45a. From northwestern Arkansas and adjacent areas of Missouri and Oklahoma .... ***P. angusticlavius***
CC 468, p 463 **CAAR** 29 (1966)
(as *P. dorsalis angusticlavius*)

45b. From Indiana south to Alabama, or from South Carolina west to eastern Louisiana 46

46a. From Indiana south to Alabama; when in sympatry with *P. websteri*, middorsal stripe is absent .... ***P. dorsalis***
CC 468, pl 40, p 463 **CAAR** 29 (1966)

46b. From South Carolina west to eastern Louisiana; when in sympatry with *P. dorsalis*, middorsal stripe is present .... ***P. websteri***
CC 469 **CAAR** 384 (1986)

47a. Stripe (if present) straight-edged (Fig. 38); from southern Canada south to North Carolina and Indiana .... ***P. cinereus***
CC 465, pl 40, p 463, 466 **CAAR** 5 (1963)

47b. Stripe (if present) with saw-toothed edges (Fig. 38); from southeastern Missouri, southwestern Arkansas, and adjacent Oklahoma, or from northwestern Georgia and adjacent regions in neighboring states .... ***P. serratus***
CC 465, p 463 **CAAR** 394 (1986)

48a. Five to seven costal folds between adpressed limbs .... ***P. nettingi***
CC 465, pl 40 **CAAR** 383 (1986)

48b. Nine to ten costal folds between adpressed limbs 49

49a. Chin mostly light .... ***P. hoffmani***
CC 467, p 466 **CAAR** 392 (1986)

49b. Chin dark, may be lightly mottled 50

50a. Usually 20–23 costal grooves; very slender; from western Pennsylvania and southeastern Indiana south to northeastern Tennessee and northwestern North Carolina .... ***P. richmondi***
CC 467, pl 41, p 466

50b. Usually 18–20 costal grooves; less slender; from Virginia 51

51a. Usually eighteen costal grooves; from Shenandoah National Park, Virginia .... ***P. shenandoah***
CC 465 **CAAR** 413 (1988)

51b. Usually nineteen costal grooves; from Botetourt and Bedford counties, Virginia .... ***P. hubrichti***
CC 465 **CAAR** 393 (1986)

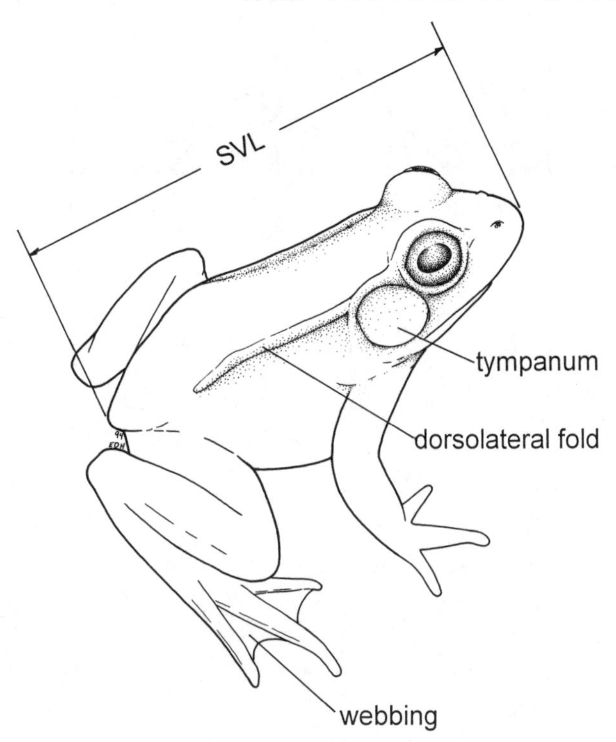

Figure 39. Generalized view of a frog, showing gross features, toe webbing, method of measuring the length from snout to vent (SVL), and location of the tympanum and dorsolateral fold.

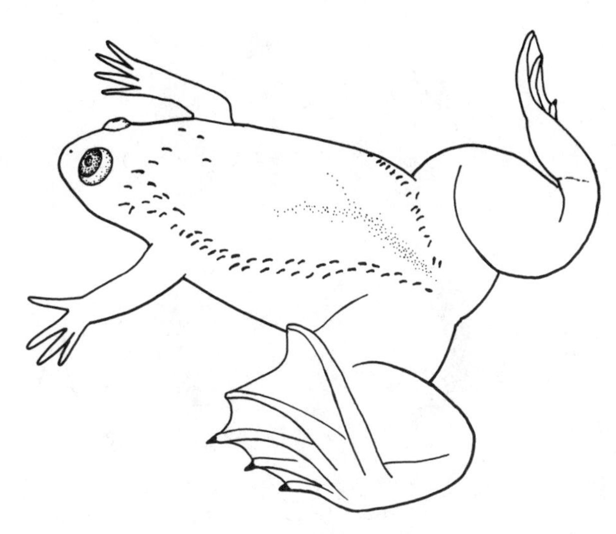

Figure 40. Generalized view of *Xenopus laevis,* showing the "claws" on the inner three toes. Adapted from a photograph in Passmore and Carruthers (1979).

ANURA
(Frogs and Toads)

KEY TO FAMILIES OF ANURA

1a. Inner three toes with cornified tips (= "claws"); toes fully webbed (Fig. 40); tongue absent .... **PIPIDAE** (p. 26)
1b. Toes lacking "claws;" webbing variable; tongue present 2

2a. Paratoid glands very distinct (Fig. 41); skin glandular (often warty in appearance) .... **BUFONIDAE** (p. 28)
2b. Paratoid glands absent or, if present, very indistinct (barely noticeable); skin smooth to glandular 3

3a. Tongue bicornuate behind (Fig. 42) .... **RANIDAE** (p. 42)
3b. Tongue not bicornuate behind 4

4a. Tongue free in front (Fig. 43); very stocky habitus with a cone-shaped nose (Fig. 44); pupil of eye vertical (Fig. 46)............. **RHINOPHRYNIDAE** (p. 26)
4b. Tongue attached in front; habitus not as above; if stocky, with a fold of skin across head behind eyes and nose not cone-shaped; pupil of eye vertical or horizontal (Fig. 46) 5

5a. Tympanum absent 6
5b. Tympanum present (Fig. 39) 7

6a. With a prominant fold of skin across the head behind the eyes (Fig. 45); pupil of eye horizontal (Fig. 46); very stocky habitus with small head; toes essentially equal in diameter .... **MICROHYLIDAE** (p. 33)
6b. No prominent fold of skin across the head behind the eyes; pupil of eye vertical (Fig. 46); head not disproportionately small; outermost toe thicker than others; with an everted cloaca ("tail") in males (Fig. 47) .... **ASCAPHIDAE** (p. 26)

7a. Dark, sharp-edged metatarsal tubercle ("spade") present (Fig. 48); pupil of eye vertical (Fig. 46).... .... **PELOBATIDAE** (p. 27)
7b. Without a dark, sharp-edged metatarsal tubercle; pupil of eye horizontal 8

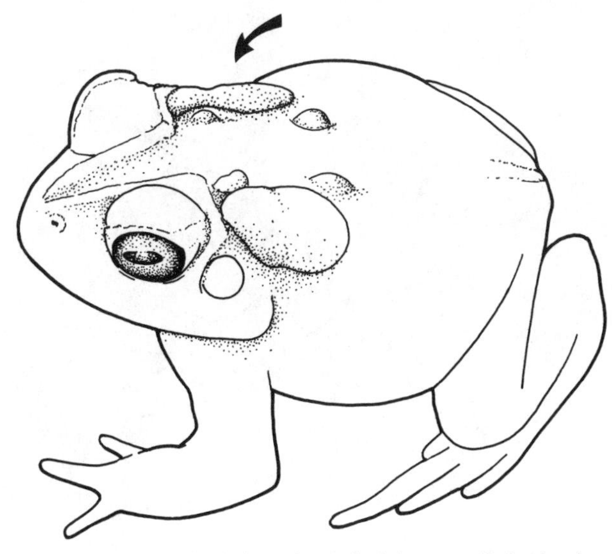

Figure 41. Generalized view of a toad of the genus *Bufo,* showing paratoid glands typical of the family Bufonidae.

Figure 42. Frontal view of the open mouth of *Rana pretiosa,* showing bicornuate tongue typical of the family Ranidae. Drawn from a preserved specimen (KU 29386).

8a. With at least short webs between toes; intercalary cartilage present (Fig. 49) .... **HYLIDAE** (p. 34)
8b. Webs between toes greatly reduced; no intercalary cartilage (Fig. 49) .... **LEPTODACTYLIDAE** (p. 33)

PIPIDAE

Xenopus

***Xenopus laevis* (A)**
S 56, pl 14

ASCAPHIDAE

Ascaphus

CAAR 69 (1968)

Ascaphus truei
S 27, pl 16 **CAAR** 69 (1968)

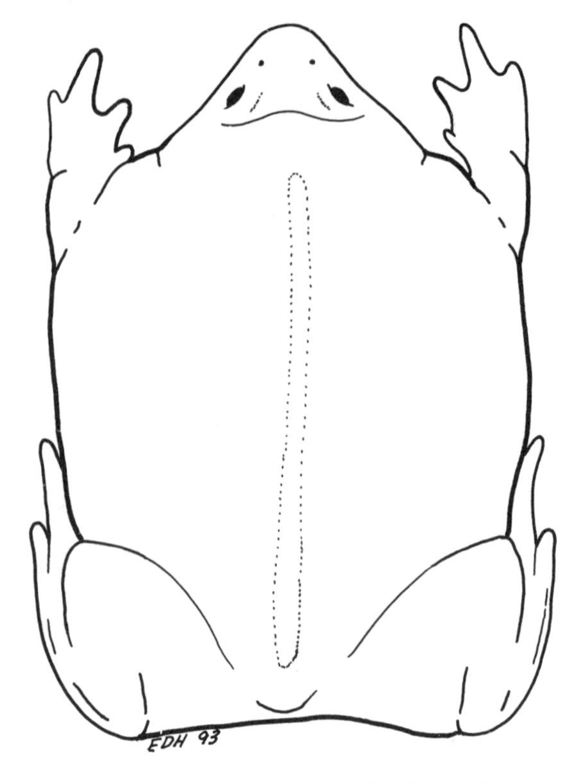

Figure 44. Dorsal view of *Rhinophrynus dorsalis,* showing habitus. Drawn from a preserved specimen (KU 70983).

RHINOPHRYNIDAE

CAAR 78 (1969)

Rhinophrynus

CAAR 78 (1969)

Rhinophrynus dorsalis
CC 501, pl 45 **CAAR** 78 (1969)

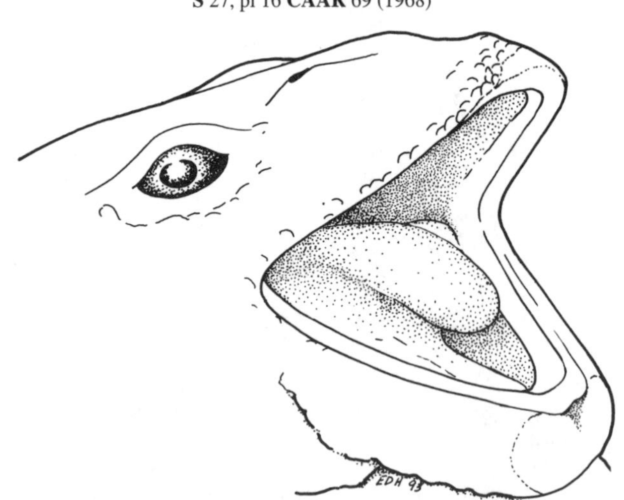

Figure 43. Lateral view of the head of *Rhinophrynus dorsalis,* showing tongue free in front. Drawn from preserved specimens (KU 70976 & 86647).

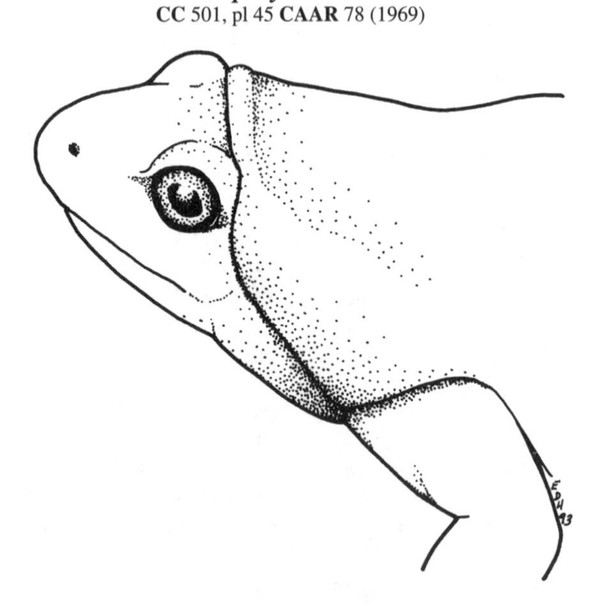

Figure 45. Lateral view of the head of *Gastrophryne olivacea,* showing fold of skin across head, typical of frogs of the family Microhylidae. Drawn from a photograph by Suzanne L. Collins.

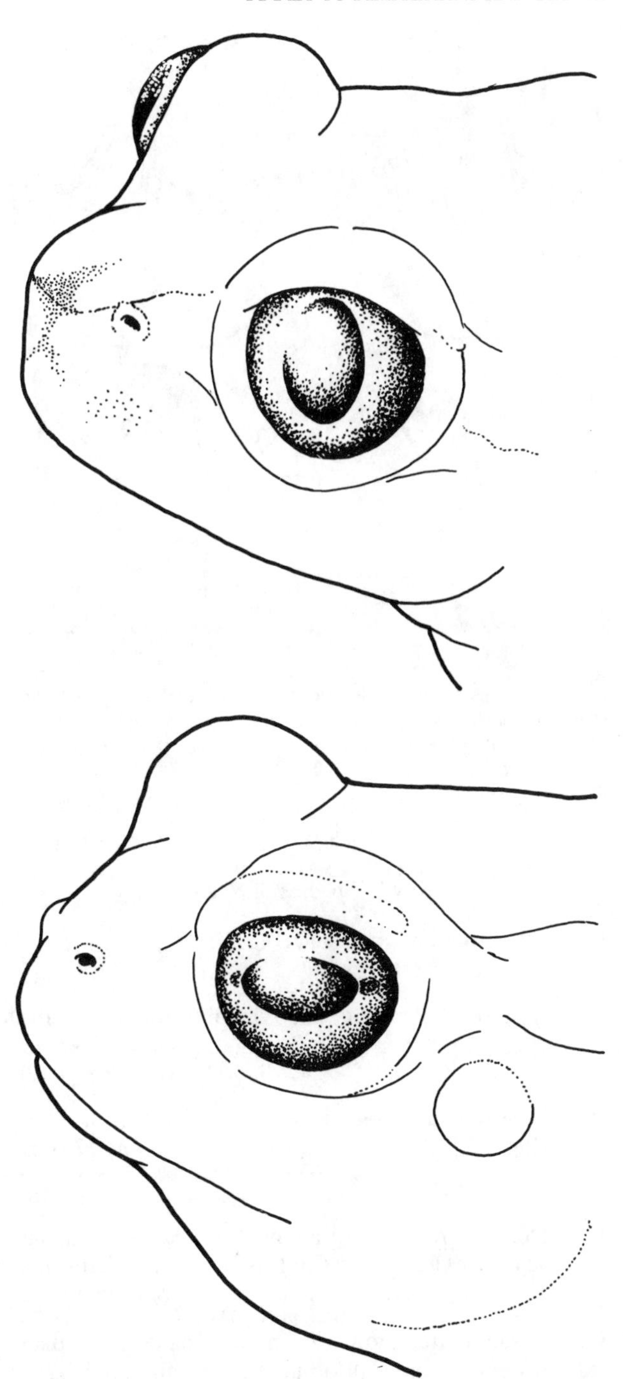

Figure 46. The pupils of frogs. Upper: Vertical pupil of *Spea intermontana*, typical of the family Pelobatidae. Drawn from a photograph in Baxter and Stone (1980) Lower: Horizontal pupil of *Bufo microscaphus*, typical of other frog families found in the area covered by this key. Drawn from a photograph in Behler and King (1979).

PELOBATIDAE

Key to Genera of Pelobatidae

1a. Spade on hind foot elongated, sickle-shaped (Fig. 50) .... ***Scaphiopus*** (p. 27)

1b. Spade short, rounded, wedge-shaped (Fig. 50) .... ***Spea*** (p. 28)

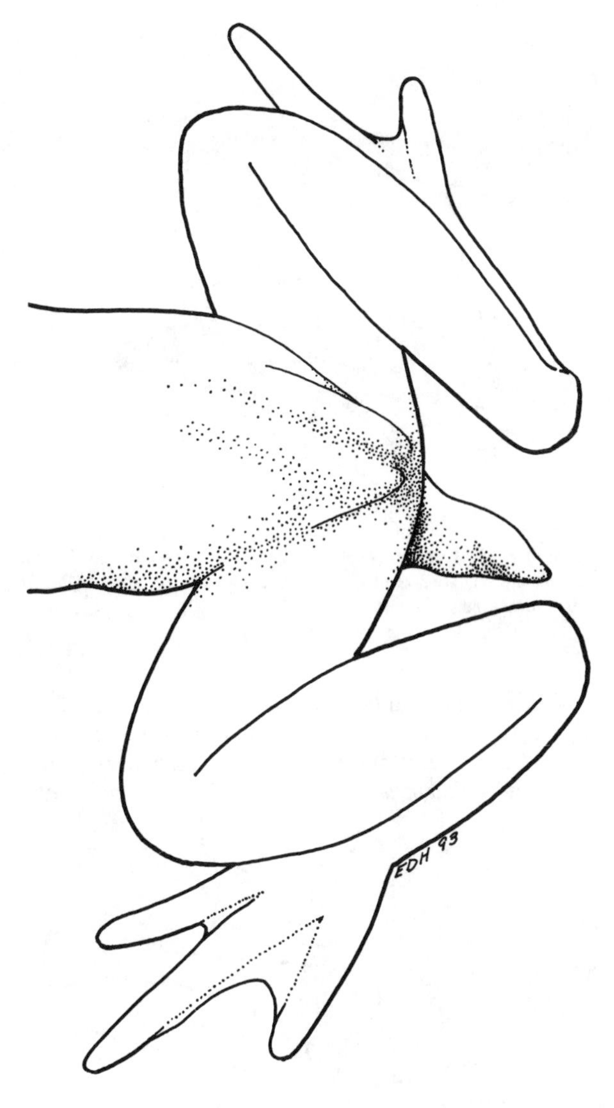

Figure 47. Dorsal view of the posterior half of the body of a male *Ascaphus truei*, showing the everted cloaca, or "tail," typical of the family Ascaphidae. Drawn from a 35mm color slide (KU).

Scaphiopus

Key to Species of *Scaphiopus*

1a. Dark edge of spade no more than twice as long as wide; fingers webbed; pectoral glands absent........ .... ***S. couchii***
CC 504, pl 44, p 502 **S** 26, pl 10 **CAAR** 85 (1970)

1b. Dark edge of spade three times as long as wide; fingers essentially without webs; pectoral glands present (Fig. 51) 2

2a. Raised boss between or slightly behind the eyes; from central and western Arkansas, and northwestern Louisiana through eastern Oklahoma and Texas .... ***S. hurterii***
CC 504, pl 44, p 503 **CAAR** 70 (1968)
(as *S. holbrookii hurterii*)

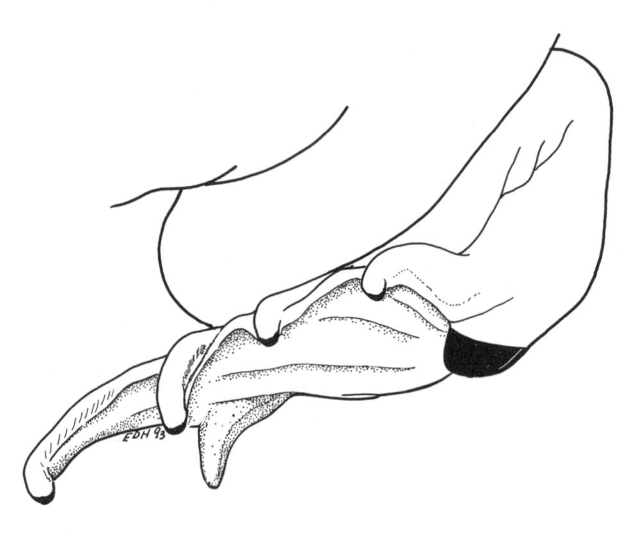

Figure 48. View of the hind foot of *Spea intermontana*, showing the location and relative size of the metatarsal tubercle, or "spade." Drawn from a preserved specimen (KU 204557).

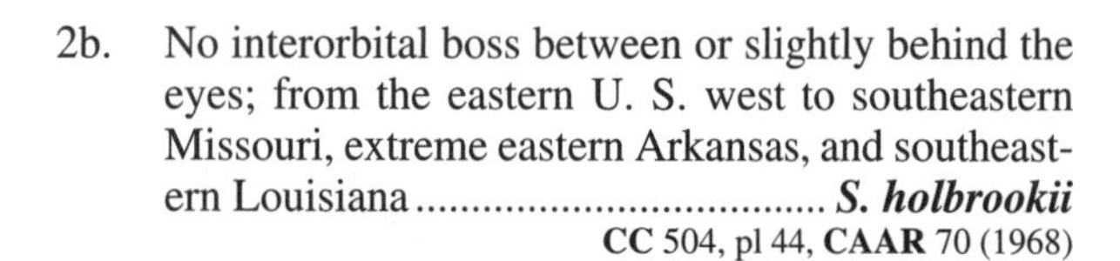

2b. No interorbital boss between or slightly behind the eyes; from the eastern U. S. west to southeastern Missouri, extreme eastern Arkansas, and southeastern Louisiana .... ***S. holbrookii***
CC 504, pl 44, **CAAR** 70 (1968)

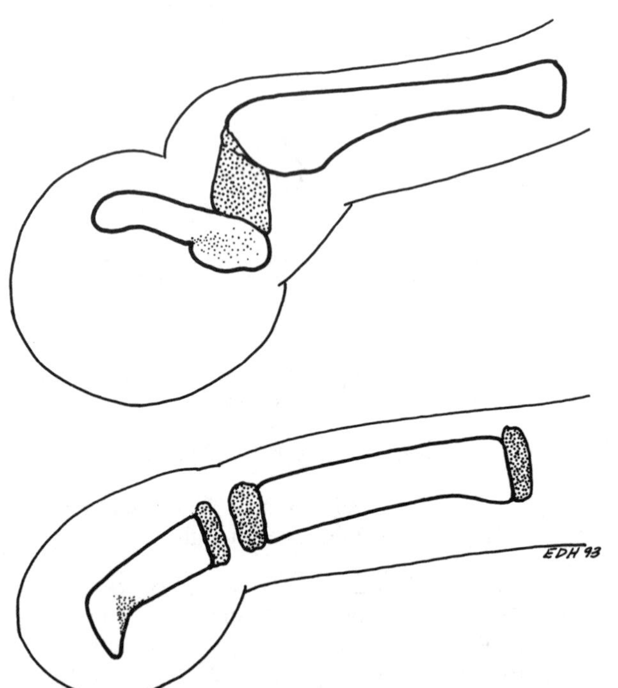

Figure 49. Lateral views of the toes of frogs, showing intercalary cartilage. Upper: Intercalary cartilage present, typical of the family Hylidae. Drawn from a preserved specimen (KU 146900) of *Hyla chrysoscelis*. Lower: Intercalary cartilage absent, typical of the family Leptodactylidae. Drawn from a preserved specimen (KU 203413) of *Eleutherodactylus planirostris*.

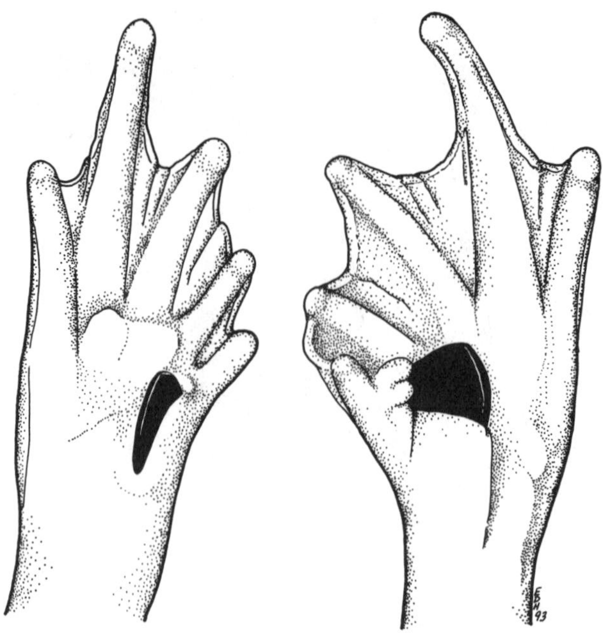

Figure 50. Ventral views of the hind feet of frogs of the family Pelobatidae. Left: Sickle-shaped spade, typical of the genus *Scaphiopus*. Right: Wedge-shaped spade, typical of the genus *Spea*. Both drawn from preserved specimens.

Spea

Key to Species of ***Spea***

1a. Interorbital boss present (Fig. 52) 2
1b. No interorbital boss 3

2a. Boss low, not bony, not extending forward of the eyes (Fig. 52) .... ***S. intermontana***
S 23, pl 10 **CAAR** 650 (1998)
(as *Scaphiopus intermontanus*)

2b. Boss high and bony, extending forward of the eyes (Fig. 52) .... ***S. bombifrons***
CC 507, pl 44, p 502, 503 **S** 24, pl 10
(as *Scaphiopus bombifrons*)

3a. Dorsum greenish or gray (in life); spade short, about as wide as long; from California and Baja California .... ***S. hammondii***
S 25, pl 10 (as *Scaphiopus hammondii*)

3b. Dorsum often brownish (in life); spade longer than wide; not from California or Baja California......... .... ***S. multiplicata***
CC 507, pl 44, p 502 **S** 25
(as *Scaphiopus multiplicatus*)

BUFONIDAE

Bufo

Key to Species of ***Bufo***

1a. Paratoid glands as long as head, tapering posteriorly (Fig. 53) 2
1b. Paratoid glands not as long as head; if large, not tapering posteriorly 4

Figure 51. Ventral view of *Scaphiopus holbrookii*, showing the size and location of the pectoral glands. Drawn from a preserved specimen (KU 211216).

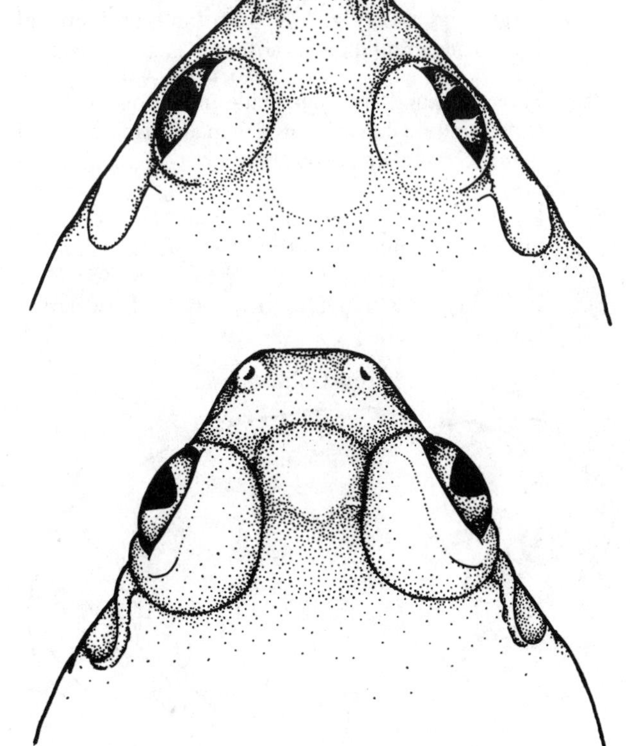

Figure 52. Dorsal views of heads of two examples of the genus *Spea*, showing size and location of interorbital bosses. Upper: *S. intermontana*. Drawn from a preserved specimen (KU 204557). Lower: *S. bombifrons*. Drawn from a preserved specimen (KU 218750).

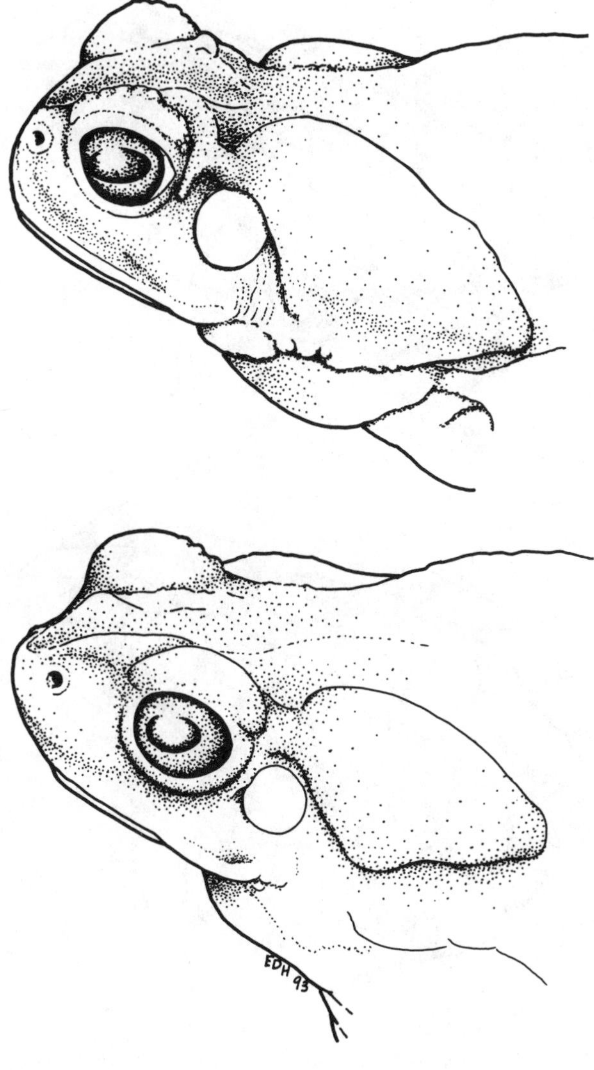

Figure 53. Lateral views of the heads of two examples of the genus *Bufo*, showing the size and location of the paratoid glands. Upper: *B. marinus*. Lower. *B. retiformis*. Both adapted from photographs in Behler and King (1979) and from a 35mm color slide (KU).

2a. Cranial crests prominent; large (to more than 60 mm SVL); brownish in life .... ***B. marinus***
(A) (Florida populations)
CC 526, pl 44 **CAAR** 395 (1986)

2b. Cranial crests low; small (SVL less than 60 mm); greenish in life 3

3a. Dorsal pattern of small separated spots or lines (Fig. 54); from western Kansas, Oklahoma, and Texas west to southeastern Arizona and northeastern Sonora and farther south in México .... ***B. debilis***
CC 525, pl 46 **S** 37, pl 10

3b. Dark lines on dorsum form a continuous reticulum (Fig. 54); from south-central Arizona into adjacent regions of Sonora .... ***B. retiformis***
S 37, pl 10 **CAAR** 207 (1978)

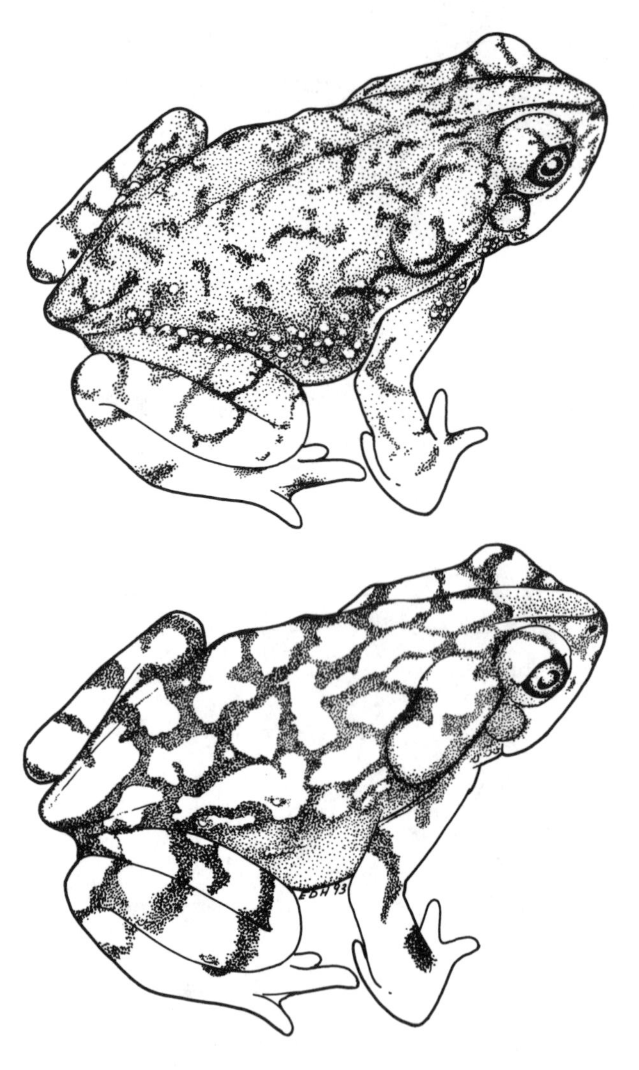

Figure 54. Dorsolateral views of two species of the genus *Bufo*, showing dorsal patterns. Upper: *B. debilis*. Lower: *B. retiformis*. Both adapted from photographs in Leviton (1971).

4a. Paratoid glands round (or transversely oval) (Fig. 55); no cranial crests .... ***B. punctatus***
CC 524, pl 44, 515 **S** 39, pl 12

4b. Paratoid glands variable in shape, but always longitudinally elongated (never round or transversely oval); cranial crests may be present or not........... 5

5a. Small (adults less than 35 mm); paratoid glands large, nearly as long as head; conspicuous light middorsal line; from southeastern U.S. .... ***B. quercicus***
CC 522, pl 46, p 512 **CAAR** 222 (1979)

5b. Larger (adults more than 35 mm); paratoid glands smaller; middorsal line present or not................. 6

6a. Thigh with 1–3 greatly enlarged dorsal glands (Fig. 56) .... ***B. alvarius***
S 38, pl 11 **CAAR** 93 (1970)

6b. Thigh without or with only few slightly enlarged glands 7

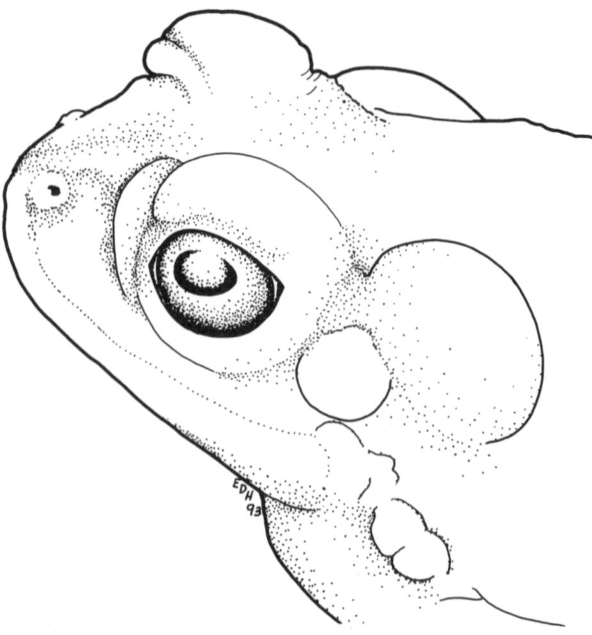

Figure 55. Dorsolateral view of the head of *Bufo punctatus*, showing size and location of round paratoid gland. Drawn from a 35mm color slide (KU).

7a. Paratoid glands flat, wide, and close together (interparatoid distance about equal to the width of a paratoid gland) (Fig. 57); from the high Sierra of California; cranial crests absent .... ***B. canorus***
S 31, pl 11 **CAAR** 132 (1973)

7b. Paratoid glands not flat, wide, and close together (interparatoid distance more than width of a paratoid gland); cranial crests present or not 8

8a. Paratoid glands triangular; with a deep valley between the cranial crests (Fig. 58) .... ***B. valliceps***
CC 521, pl 44, p 515 **CAAR** 94 (1970)

8b. Paratoid glands elongate; no deep valley between the cranial crests 9

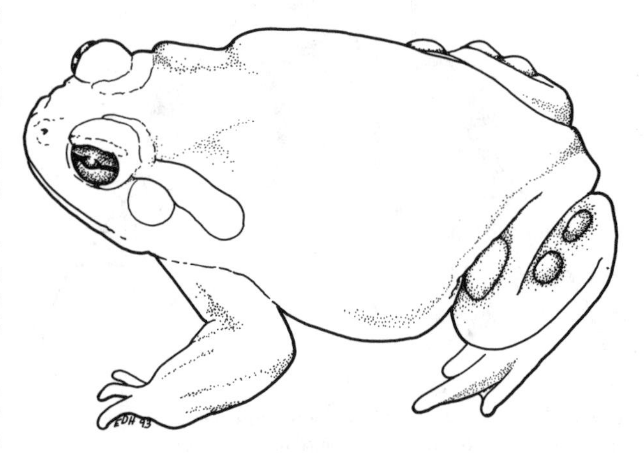

Figure 56. Dorsolateral view of *Bufo alvarius*, showing size and location of thigh glands. Drawn from a 35mm color slide (KU Color Slide 45).

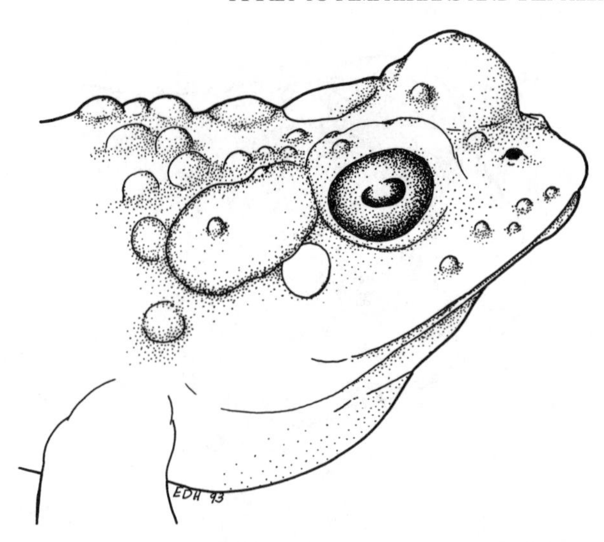

Figure 57. Dorsolateral view of the head of a female *Bufo canorus*, showing size and location of paratoid gland. Adapted from a photograph in Leviton (1971).

9a. Parietal cranial crests form a conspicuous knob (Fig. 59) .... ***B. terrestris***
CC 514, pl 44, p 515 **CAAR** 223 (1979)

9b. Parietal cranial crests do not form a conspicuous knob 10

10a. Interorbital cranial crests unite at some point to form a boss (Fig. 59) 11

10b. No boss present 13

11a. Boss at anterior border of eyes; paratoid glands strongly divergent (Fig. 59); dark dorsal markings with light outlines .... ***B. cognatus***
CC 522, pl 44, p 512, 515 **S** 35, pl 11 **CAAR** 457 (1990)

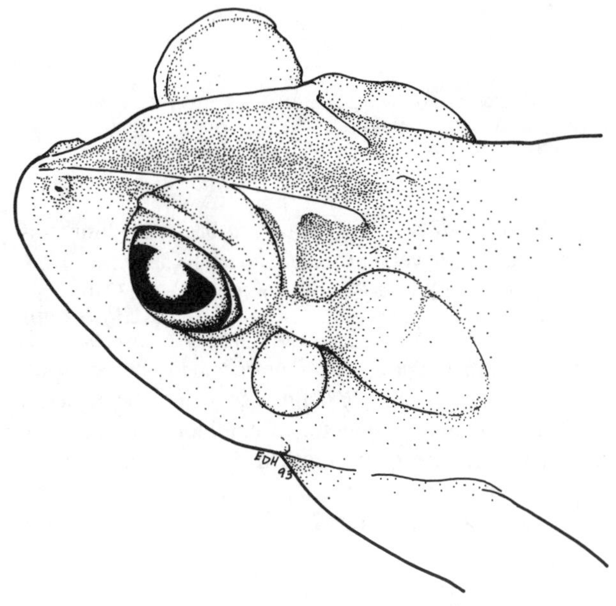

Figure 58. Dorsolateral view of the head of *Bufo valliceps*, showing size and location of paratoid gland and cranial crests. Drawn from a 35mm color slide (KU).

11b. Boss between eyes; paratoid glands essentially parallel (Fig. 59); dark spots on dorsum without distinct light outlines 12

12a. From western and northwestern Minnesota, northwestern South Dakota, and southeastern Manitoba through much of North Dakota and Saskatchewan to northern Montana, the eastern half of Alberta, and extreme southcentral Northwest Territories .... ***B. hemiophrys***
CC 516, pl 44, p 503, 515 **S** 34, pl 12

12b. From Wyoming .... ***B. baxteri***
CC 516 (as western subspecies) **S** 34 (as *B. hemiophrys*)

13a. Cranial crests distinct 14

13b. Cranial crests absent or very indistinct 17

14a. Postorbital cranial crests thicker than interorbitals (Fig. 59); from southeastern Texas .... ***B. houstonensis***
CC 516, p 515 **CAAR** 133 (1973)

14b. Postorbital cranial crests not thicker than interorbitals 15

15a. Parotoid glands reniform, nearly always separated from cranial crests (Fig. 60); light middorsal stripe may be present or absent; warts on shank often larger than those on thigh; dark dorsal spots usually encircle 1–3 warts; venter usually with dark spots, chest often heavily spotted; range largely east of the Great Plains .... ***B. americanus***
CC 514, pl 44, p 515

15b. Paratoid glands oval, usually in contact with cranial crests (Fig. 60); light middorsal stripe nearly always present; warts on shank not larger than those on thigh; dark dorsal spots may contain up to six warts; venter unmarked or with few spots limited to chest 16

16a. Dorsal spots relatively indistinct (appearing almost "washed-out") and encircling one to several warts; range largely restricted to the Great Plains and arid southwestern U.S. .... ***B. woodhousii***
CC 519, pl 44 **S** 33, pl 12

16b. Dorsal spots usually large and well-defined, frequently six in number, and encircling three or more warts; range largely east of the Great Plains .... ***B. fowleri***
CC 519, pl 44, 515 **S** 33 (as *B. woodhousii fowleri*)

17a. Vertebral line distinct; enlarged gland on shank 18

17b. Vertebral line absent or indistinct; no enlarged gland on shank 19

18a. Very dark; dorsal mottling separated by wavy light lines; heavy black blotches on venter; skin between warts smooth .... ***B. exsul***
S 30

18b. Pattern not as above; skin between warts tuberculate 20

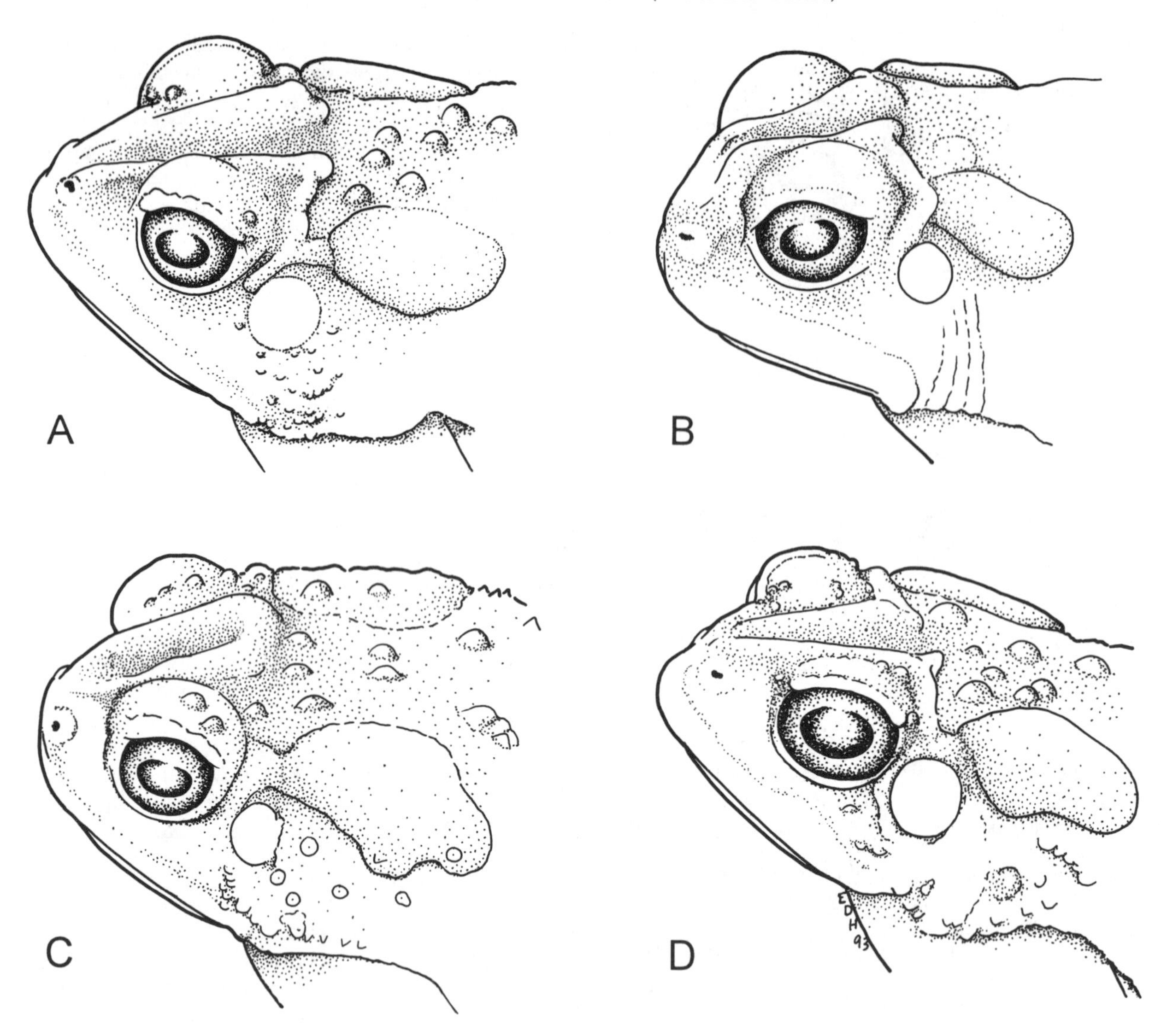

Figure 59. Dorsolateral views of four species of the genus *Bufo*, showing the size and location of cranial crests and bosses. (A) *B. terrestris* (KU 19149); (B) *B. cognatus* (KU 45727); (C) *B. hemiophrys* (KU 96249); (D) *B. houstonensis* (KU 190154).

19a. Limbs long (elbow and knee touch when laid straight alongside body); broad head with blunt snout .... ***B. boreas***
S 30, pl 11

19b. Limbs short (elbow and knee do not touch when laid straight alongside of body); narrow head with long snout; restricted to Amargosa River Valley, Nye County, Nevada .... ***B. nelsoni***
S 30

20a. Large inner metatarsal tubercle elongate, sharp-edged (Fig. 61); paratoid gland uniformly colored .... ***B. speciosus***
CC 524, pl 44, p 512 **S** 36 pl 12

20b. Large inner metatarsal tubercle rounded, not sharp-edged (Fig. 61); anterior end of paratoid gland often light colored 21

21a. With distinct dark spotting on back and limb surfaces; skin very granular; coastal California from San Luis Obispo County south into Baja California .... ***B. californicus***
S 32 pl 12 (as *B. microscaphus californicus*)

21b. Dorsum and upper limb surfaces without dark spotting (a few, vaguely defined spots may be present); skin relatively smooth; discontinuous range from southwestern Utah, southern Nevada, and extreme eastern California through central Arizona and western New Mexico into México ..... ***B. microscaphus***
S 32 pl 12

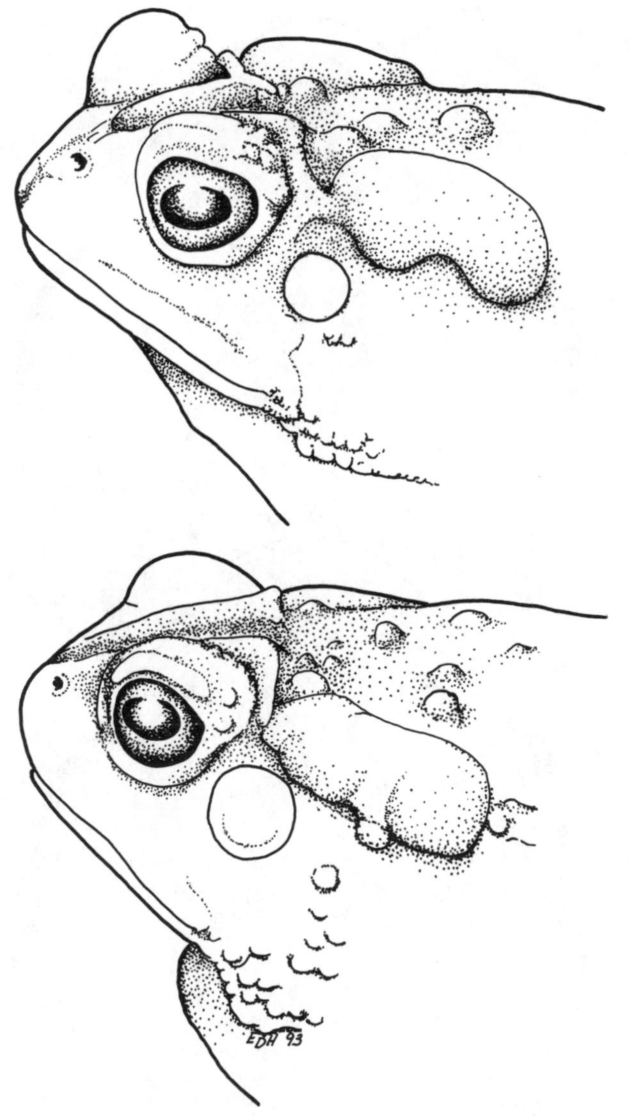

Figure 60. Dorsolateral views of two species of the genus *Bufo*, showing the size, shape, and location of paratoid glands in relation to cranial crests. Upper: *B. americanus*. Lower: *B. woodhousii*. Both drawn from photographs by Suzanne L. Collins.

MICROHYLIDAE

Key to Genera of Microhylidae

1a. With two metatarsal tubercles on each hindlimb; narrow, but distinct vertebral line present .... ***Hypopachus*** (p. 33)

1b. With one metatarsal tubercle on each hindlimb; vertebral line absent .... ***Gastrophryne*** (p. 33)

Hypopachus

Hypopachus variolosus
CC 554, pl 45, p 553

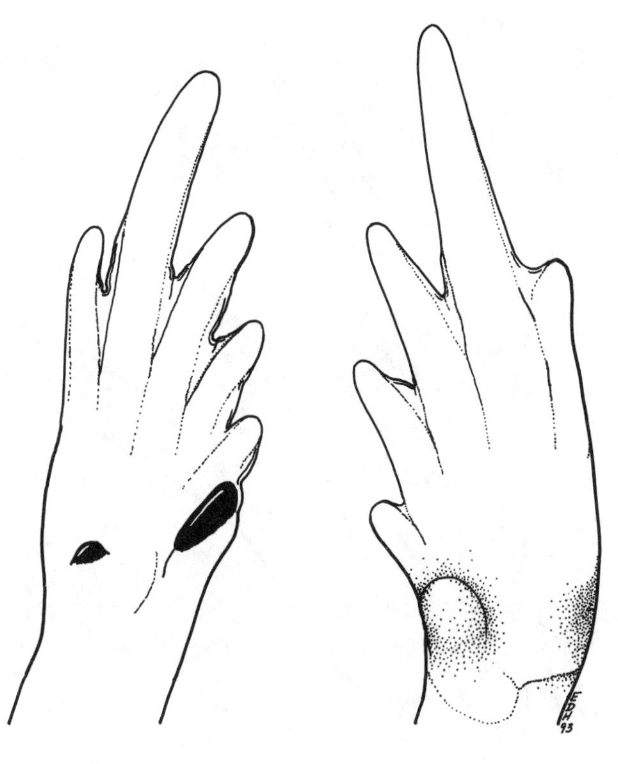

Figure 61. Ventral views of the hind feet of two species of the genus *Bufo*, showing the size and location of metatarsal tubercles. Left: *B. speciosus* (KU 145451). Right: *B. microscaphus* (KU 60424).

Gastrophryne

CAAR 134 (1973)

Key to Species of *Gastrophryne*

1a. Venter heavily mottled .... ***G. carolinensis***
CC 552, pl 45, p 553 **CAAR** 120 (1972)

1b. Venter not mottled .... ***G. olivacea***
CC 552, pl 45, p 553 **S** 29, pl 16 **CAAR** 122 (1972)

LEPTODACTYLIDAE

Key to Genera of Leptodactylidae

1a. First front toe longer than second by one phalanx (Fig. 62); white stripe on upper lip .... ***Leptodactylus*** (p. 33)

1b. First front toe no longer than second or barely longer; white stripe on upper lip absent or indistinct .... ***Eleutherodactylus*** (p. 33)

Leptodactylus

Leptodactylus labialis
CC 508, pl 45, p 507 **CAAR** 104 (1971)

Eleutherodactylus

Key to Species of *Eleutherodactylus*

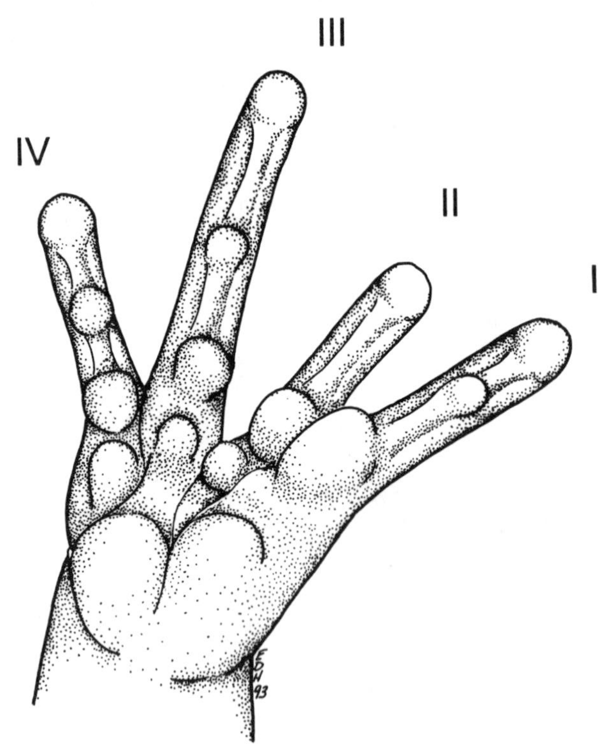

Figure 62. Ventral view of front foot of *Leptodactylus labialis*, showing first toe (I) longer than second toe (II) by one phalanx. Drawn from a preserved specimen (KU 265824).

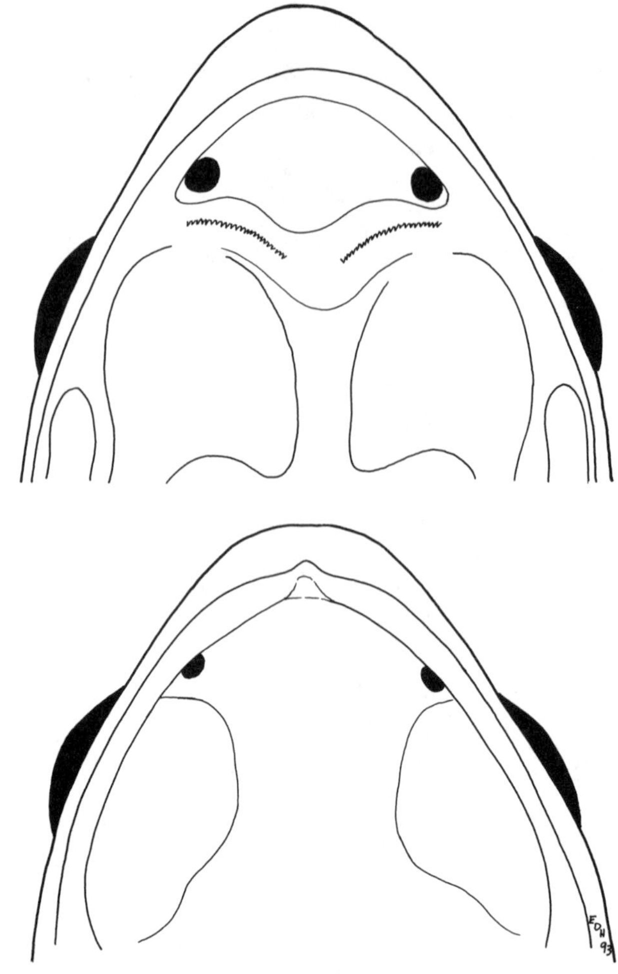

Figure 63. Views of the roofs of the mouths of two species of the genus *Eleutherodactylus*. Upper: *E. planirostris* (drawn from a preserved specimen, KU 68827), showing presence of vomerine teeth (also present in *E. coqui* and *E. augusti*). Lower: *E. marnockii* (drawn from a preserved specimen, KU 18440), showing absence of vomerine teeth (also absent in *E. cystignathoides* and *E. guttilatus*).

1a. Vomerine teeth present (Fig. 63); from Florida or, if not from Florida, with dorsolateral skin folds 2

1b. Vomerine teeth absent (Fig. 63); from Texas; no dorsolateral skin folds 4

2a. Dorsolateral skin folds present (Fig. 64) .... ***E. augusti***
CC 508, pl 45, p 507
S 28, pl 12 **CAAR** 41 (1967) (as *Hylactophryne augusti*)

2b. Dorsolateral skin folds absent 3

3a. Terminal discs barely wider than toes (Fig. 65); vomerine teeth in two broad fasciculi across palate .... ***E. planirostris* (A)**
CC 509, pl 45 **CAAR** 154 (1974)

3b. Terminal discs distinctly wider than toes (Fig. 65); vomerine teeth in two small clumps posterior and medial to choanae .... ***E. coqui* (A)**
CC 510, pl 45

4a. Usually with a dark bar between eyes (Fig. 66); dorsal pattern often with a network of dark reticulation; from the Big Bend region of Texas (or southeastern Coahuila to Guanajuato .... ***E. guttilatus***
CC 511, pl 45 (as *Syrrhophus guttilatus*)

4b. No dark bar between eyes; dorsal pattern of scattered spots, lines, or blotches; range not as above 5

5a. Skin granular; tips of digits barely dilated, rounded (Fig. 67); from the lower Rio Grande Valley south into México .... ***E. cystignathoides***
CC 511, pl 45 (as *Syrrhophus cystignathoides*)

5b. Skin smooth; tips of digits distinctly dilated, blunt (Fig. 67); from south-central Texas .... ***E. marnockii***
CC 510, pl 45 (as *Syrrhophus marnockii*)

HYLIDAE

Key to Genera of Hylidae

1a. Webs on feet significantly reduced, not over ¼ length of longest toe (Fig. 68), or, if longer, with an X on the middle of the back (Fig. 72); terminal discs no wider or barely wider than toes, or, if wider, with an X on the back 2

Figure 64. Lateral view of *Eleutherodactylus augusti*, showing the arrangement of the skin folds. Drawn from a 35mm color slide (KU).

1b. Webs on feet at least ½ length of longest toe (Fig. 68); terminal discs distinctly wider than toes; without an X on the middle of the back 3

2a. A distinct ridge across back of head (Fig. 69) .... ***Pternohyla*** (p. 36)

2b. No ridge across back of head .... ***Pseudacris*** (p. 36)

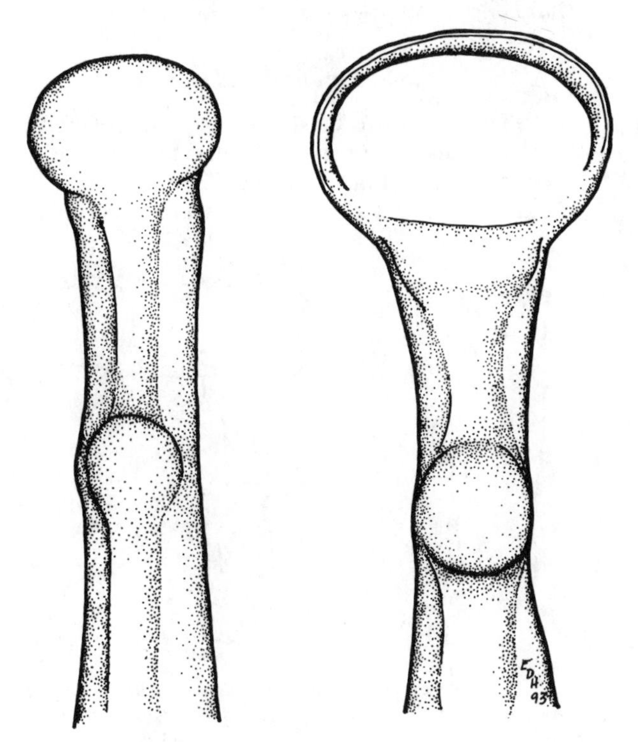

Figure 65. Ventral view of the toe tips of two species of the genus *Eleutherodactylus*. Left: Terminal disc in *E. planirostris* (KU 92653). Right: Terminal disc in *E. coqui* (KU 276676).

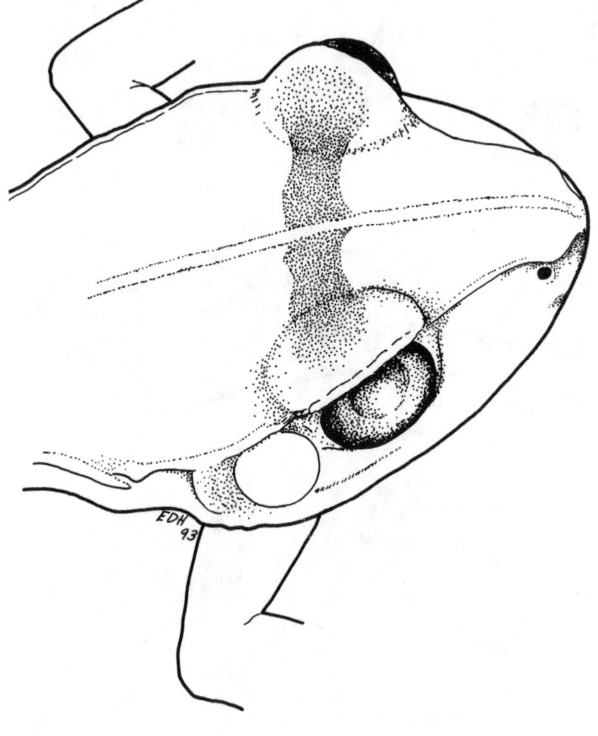

Figure 66. Dorsal view of the head of *Eleutherodactylus guttilatus*, showing bar between eyes. Drawn from a preserved specimen (KU 193270).

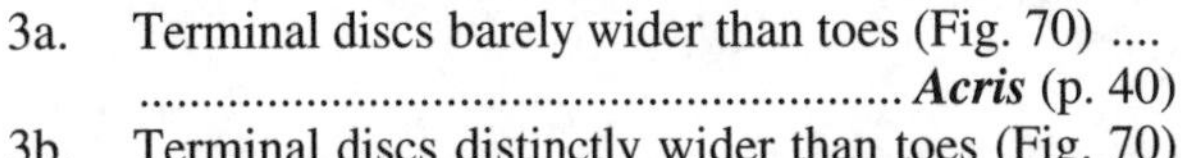

3a. Terminal discs barely wider than toes (Fig. 70) .... ***Acris*** (p. 40)

3b. Terminal discs distinctly wider than toes (Fig. 70) 4

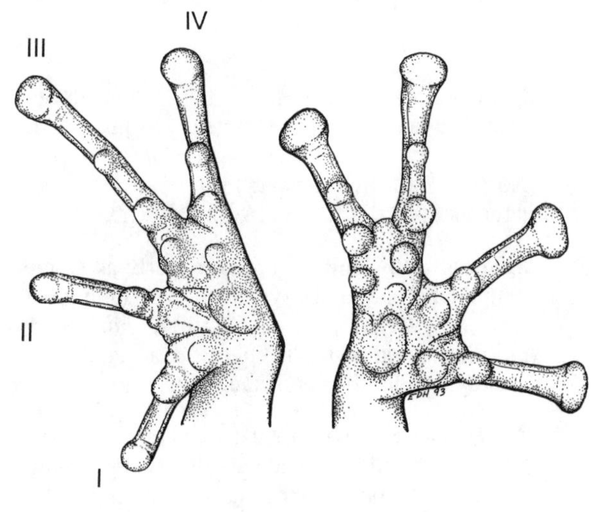

Figure 67. Ventral views of front feet of two species of the genus *Eleutherodactylus*. Left: *E. cystignathoides*, showing reduced size of tips of fingers. Drawn from a preserved specimen (KU 8135). Right: *E. marnockii*, showing enlarged size of tips of fingers. Drawn from a preserved specimen (KU 18440).

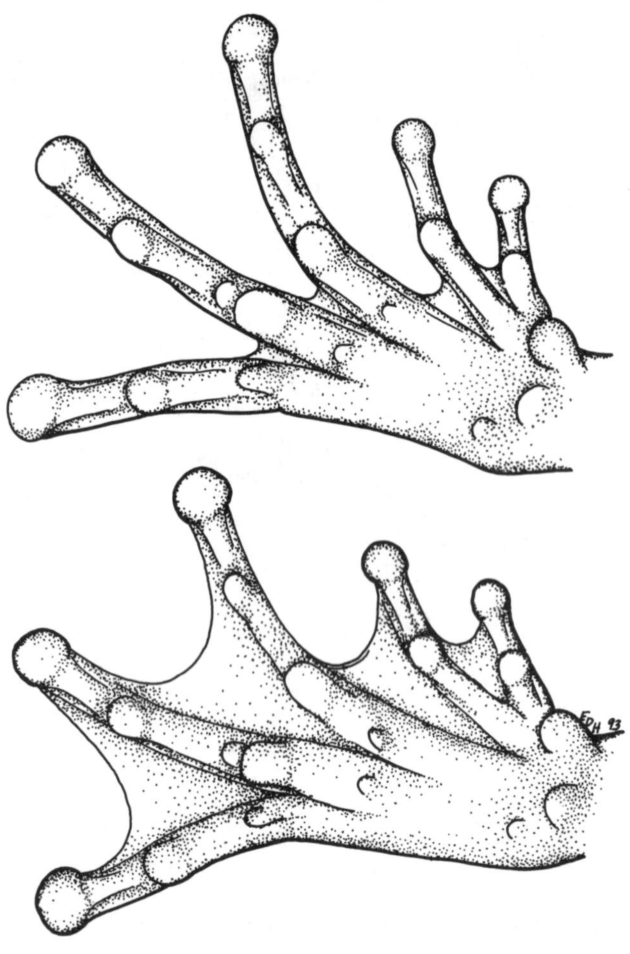

Figure 68. Ventral views of hind feet of species of the genera *Hyla* and *Pseudacris* (drawing adapted from a preserved specimen of *Hyla regilla*, KU 204576). Upper: Webbing reduced, not over one half the length of longest toe. Lower: Webbing not reduced, at least one half the length of longest toe.

4a. With broad dark lines from tip of snout through nostrils and eyes; tympanum nearly as large as eye .... ***Smilisca*** (p. 40)

4b. No such dark line; tympanum no more than ¾ as large as eye 5

5a. Skin fused to skull; finger discs nearly as large as tympanum (may not be evident in small individuals) .... ***Osteopilus*** (p. 40)

5b. Skin not fused to skull; finger discs variable, but usually smaller than tympanum 6

6a. Appearance extremely robust; size large (to 100 mm SVL); with a thick glandular supratympanic ridge extending to the shoulder (Fig. 71); finger discs nearly as large as tympanum .... ***Litoria*** (p. 40)

6b. Appearance usually slender; size small to moderate (≤ 70 mm SVL); no thick supratympanic ridge; finger discs distinctly smaller than tympanum....... .... ***Hyla*** (p. 40)

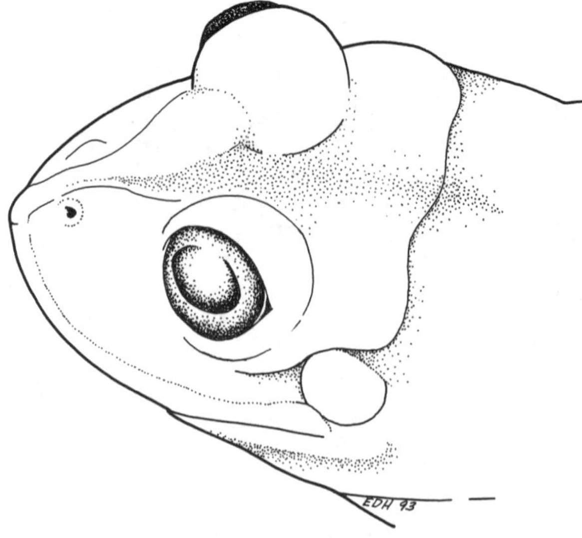

Figure 69. Dorsolateral view of head of *Pternohyla fodiens*, showing location of fold of skin behind eyes. Drawn from a 35mm color slide (KU).

Pternohyla

CAAR 77 (1969)

Pternohyla fodiens

S 41, pl 16 **CAAR** 77 (1969)

Pseudacris

Recent research indicates that the composition of the genus *Pseudacris* may be different than the arrangement shown here.

Key to Species of *Pseudacris*

1a. Webs on feet not reduced, at least ½ length of digits; terminal discs distinctly wider than toes; with an X on the middle of the back (Fig. 72) .... ***Pseudacris crucifer*** **CC** 541, pl 46

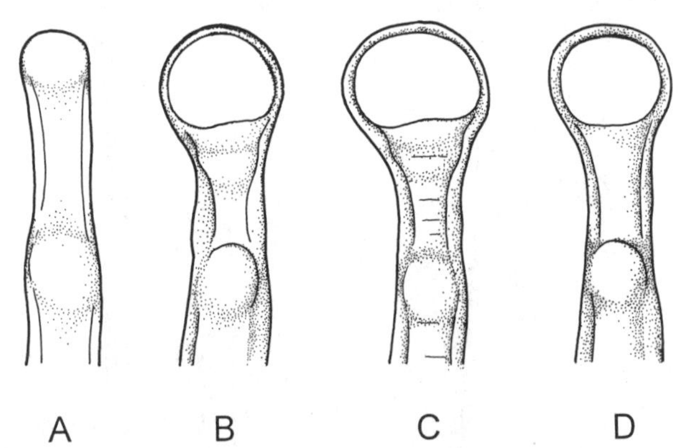

Figure 70. Ventral views of the fingers of four genera of frogs, showing the size and shape of the terminal discs. (A) *Acris crepitans*. Drawn from a preserved specimen (KU 197384); (B) *Smilisca baudinii*. Drawn from a preserved specimen (KU 184817); (C) *Osteopilus septentrionalis*. Drawn from a preserved specimen (KU 204263); (D) *Hyla femoralis*. Drawn from a preserved specimen (KU 60291).

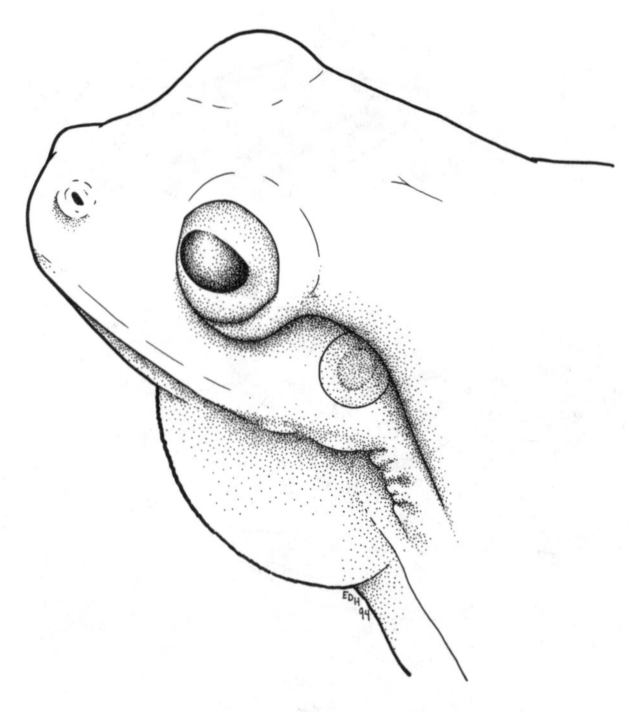

Figure 71. Dorsolateral view of the head of *Litoria caerulea*, showing the location and configuration of the supratympanic ridge. Drawn from a 35mm color slide (KU 6158).

1b. Webs on feet reduced, not over ¼ length of digits; terminal discs barely as wide or no wider than toes; with or without an X on the back 2

2a. A dark stripe through each eye and onto the sides of the body (Fig. 73) and an unspotted chest; very small (adults less than 18 mm); lacking palatal teeth .... ***P. ocularis***
CC 550, pl 46
CAAR 209 (1978) (as *Limnaoedus ocularis*)

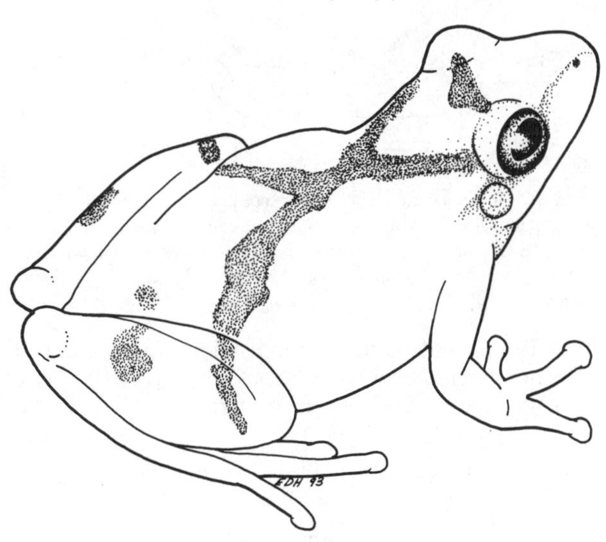

Figure 72. Dorsolateral view of *Pseudacris crucifer*, showing size and location of X-shaped mark on back. Adapted from a photograph in Johnson (1977).

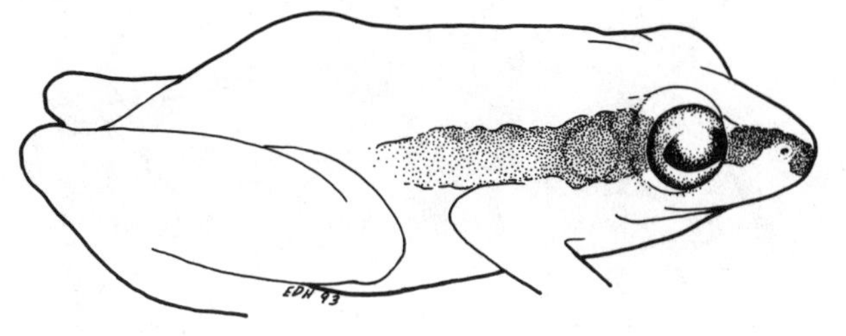

Figure 73. Lateral view of *Pseudacris ocularis*, showing location of dark lateral stripe from snout through eye onto side of body. Drawn from a photograph by Suzanne L. Collins.

2b. No dark lines through eyes or, if present, spots on the chest; larger (adults more than 18 mm); with palatal teeth (Fig. 74) 3

3a. With two broad dorsal stripes that approach or meet each other in the middle of the back like reversed parentheses (Fig. 75); terminal discs wider than toes .... ***P. brachyphona***
CC 548, pl 46, p 542 **CAAR** 234 (1980)

3b. Dorsal pattern variable, but not with reversed parentheses; terminal discs barely as wide or no wider than toes 4

4a. Body stout; head wider than long 5

4b. Body slender; head narrower or not wider than long 7

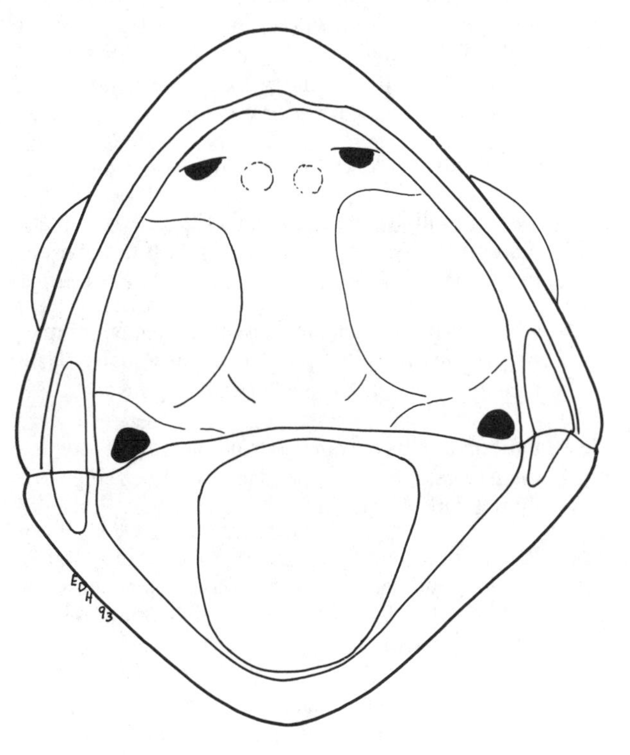

Figure 74. Frontal view of the open mouth of *Pseudacris brachyphona*, showing palatal teeth. Drawn from a preserved specimen (KU 143682).

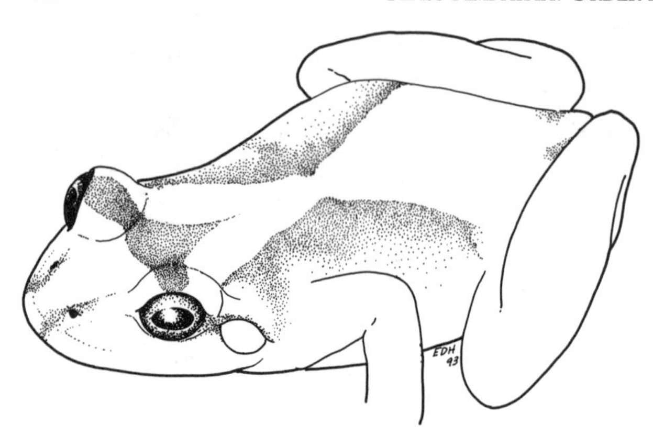

Figure 75. Dorsolateral view of *Pseudacris brachyphona*, showing pattern on head and back. Drawn from a 35mm color slide (KU).

5a. Single dark spot on lip below eye; flanks and groin lacking bold black spots 6

5b. No dark spot on lip; flanks and groin with bold black spots .... ***P. ornata***
CC 548, pl 46

6a. Dark stripe through eye and dark spots along side of body usually in strong contrast with pale ground color; from southcentral Kansas and all but western Oklahoma south through eastern Texas to the Gulf of Mexico, east into western Arkansas and extreme northwestern Louisiana ....***P. streckeri***
CC 550, pl 46 **CAAR** 27 (1966)

6a. Dark stripe through eye and dark spots along side of body poorly developed and not in strong contrast with pale ground color; in west-central Illinois and from southern Illinois through extreme southeastern Missouri to adjacent regions of Arkansas ....
.... ***P. illinoensis***
CC 550 **CAAR** 27 (1966) (as *P. streckeri illinoensis*)

7a. Chest usually spotted; middorsal stripes often paler than outer two stripes; lacking a dark triangle between eyes .... ***P. brimleyi***
CC 547, pl 46 **CAAR** 311 (1983)

7b. Chest unspotted; dorsum lacking stripes (spots may be present) or, if striped, all about equally dark; usually with a dark triangle between eyes 8

8a. Pattern of spots or (rarely) stripes not in five longitudinal rows; spots or stripes darkly outlined (Fig. 76); lip not dark-edged .... ***P. clarkii***
CC 547, pl 46 **CAAR** 458 (1990)

8b. Pattern of stripes (sometimes broken into spots) in five longitudinal rows (Fig. 76); stripes or spots not darkly outlined; upper lips with a light line and dark-edged, or mostly dark 9

9a. Dark tibial bands broad, with narrow light interspaces (Fig. 77); dorsal markings very dark; snout pointed (Fig. 78); from the southeastern U.S. ....
.... ***P. nigrita***
CC 545, pl 46, p 542

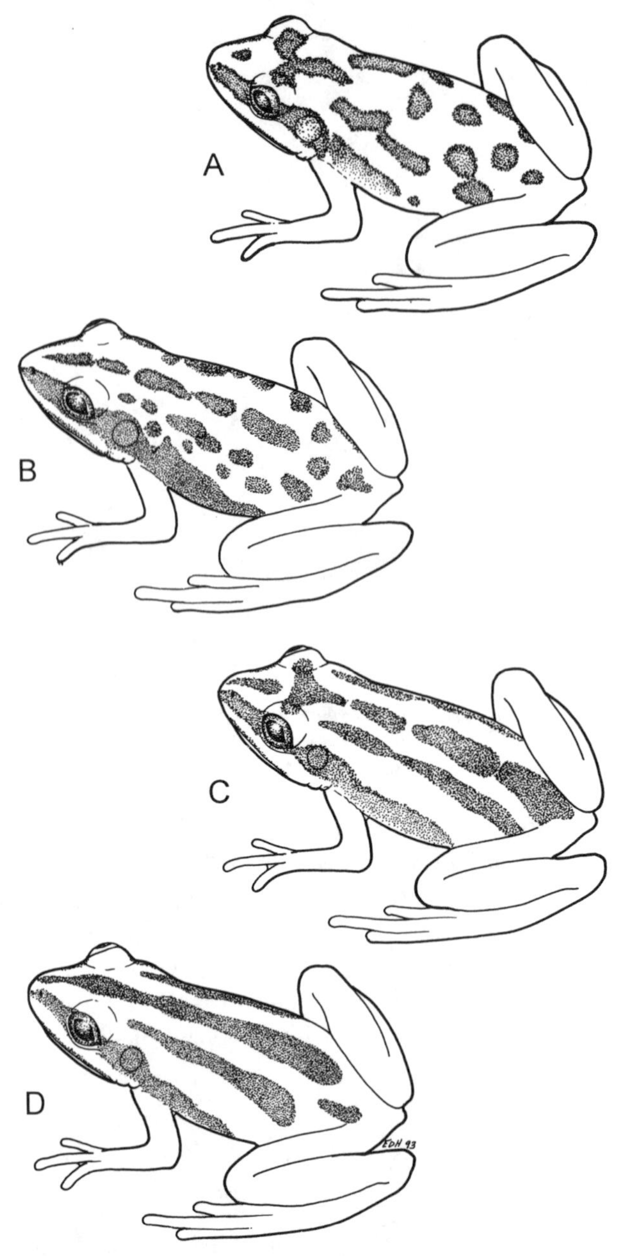

Figure 76. Dorsolateral views of four species of the genus *Pseudacris*, showing dorsal patterns of spots, stripes, or a combination of both. (A) *P. clarkii*. (B) *P. nigrita*. (C) *P. feriarum*. (D) *P. triseriata*. Drawn from photographs in Martof et al. (1980), Garrett and Barker (1987), and by Suzanne L. Collins.

9b. Tibial bands narrow or indistinct (Fig. 77); dorsal markings not very dark; snout less pointed (Fig. 78); not restricted to the southeastern U.S. 10

These and the following species of *Pseudacris* are exceedingly variable; geography may be the most effective means of identification.

10a. Dorsal stripes thin and usually broken into rows of spots (Fig. 76); from northern New Jersey south to the Florida panhandle west to eastern Texas and southeastern Oklahoma .... ***P. feriarum***
CC 543, pl 46, p 542 (as *P. triseriata feriarum*)

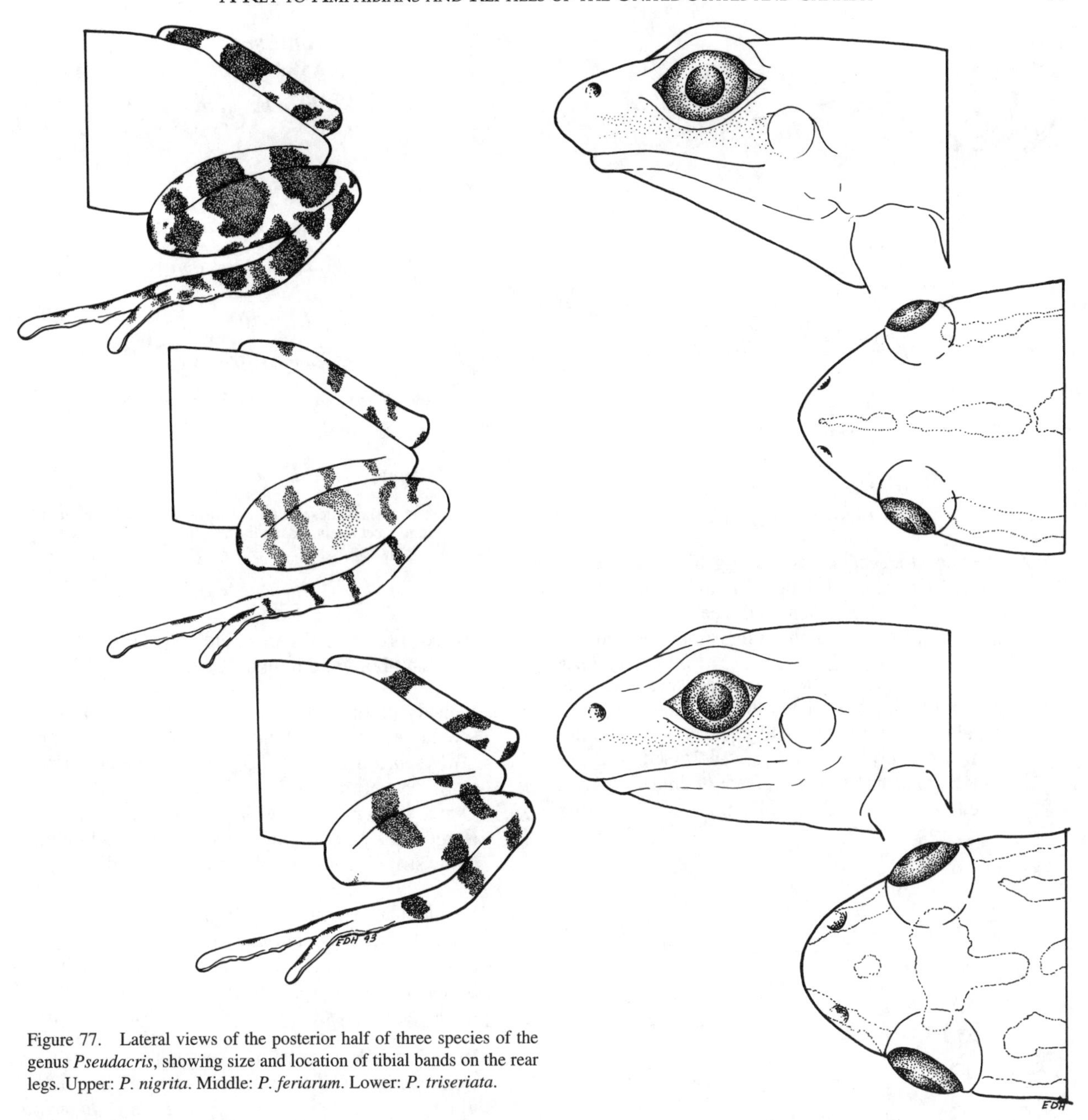

Figure 77. Lateral views of the posterior half of three species of the genus *Pseudacris*, showing size and location of tibial bands on the rear legs. Upper: *P. nigrita*. Middle: *P. feriarum*. Lower: *P. triseriata*.

Figure 78. Views of the heads of two species of the genus *Pseudacris*. Upper two images: *P. nigrita* (KU 20091), showing a more flattened (left) and pointed (right) snout. Lower two images: *P. feriarum* (KU 218715), showing a less flattened (left) and less pointed (right) snout. Both drawn from the preserved specimens indicated plus images in Mount (1975).

10b. Dorsal stripes broader and usually unbroken (Fig. 76) or, if broken, usually restricted to the middle stripe 11

11a. Slender habitus; tibia not notably shortened (Fig. 79); stripes usually unbroken (Fig. 76); from extreme southern Quebec and adjacent New York southwest to Kansas and Oklahoma, a disjunct area in New Mexico and Arizona, and an isolate in southwestern Oklahoma .... ***P. triseriata***
CC 543, pl 46, p 542 **S** 40, pl 16

11b. Stocky habitus; tibia notably shortened or not; middorsal stripe broken or not; range not as above 12

12a. Tibia notably shortened (Fig. 79); dorsal stripes (especially the middorsal stripe) often broken; from northern Ontario to the vicinity of Great Bear Lake in northwestern Canada south to Utah and northern New Mexico .... ***P. maculata***
CC 543, pl 46 **S** 40, pl 16 (as *P. triseriata maculata*)

12b. Tibia not shortened (hindlimbs longer) (Fig. 79); dorsal stripes broad, distinct, and rarely broken; from along the coastal plain of the Delmarva Peninsula north to Staten Island, New York ......***P. kalmi***
CC 543 (as *P. triseriata kalmi*)

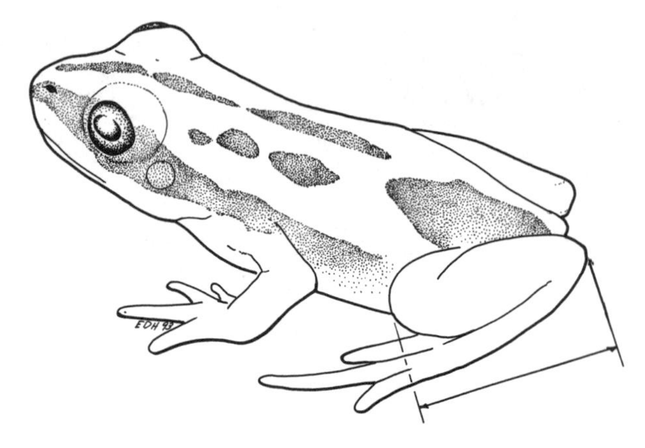

Figure 79. Lateral view of *Pseudacris maculata*, showing habitus and method of measuring tibial length.

Acris

Key to Species of *Acris*

1a. Innermost toe webbed to disc (Fig. 80); 1½ or fewer (rarely 2) terminal phalanges of 4th hind toe free of web; with hindlimb adpressed, heel does not reach snout; stripe on rear of thigh indistinctly edged..... ....***A. crepitans***
CC 529, pl 46, p 527, 529 **S** 42, pl 16

1b. Innermost toe ½ webbed (Fig. 80); two or more (usually 3 or more) terminal phalanges of 4th hind toe free of web; with hindlimb adpressed, heel extends beyond snout; stripe on rear of thigh distinctly edged .... ***A. gryllus***
CC 529, pl 46, p 527, 529

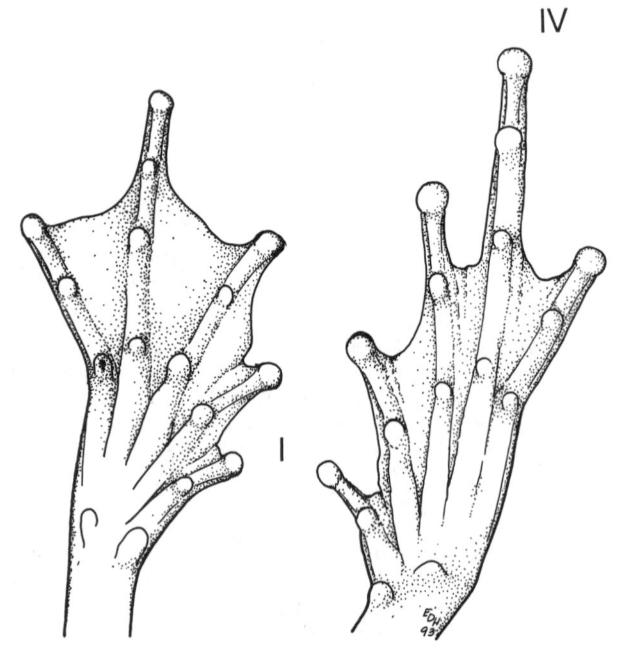

Figure 80. Ventral view of hind feet of both species of the genus *Acris*, showing location of webbing on innermost toe (I) and fourth toe (IV). Left: *A. crepitans* (KU 197366). Right: *A. gryllus* (KU 176192). Both drawn from the preserved specimens indicated.

Smilisca

CAAR 58 (1968)

Smilisca baudinii
CC 539, pl 47 **CAAR** 59 (1968)

Osteopilus

***Osteopilus septentrionalis* (A)**
CC 539, pl 47
CAAR 92 (1970) (as *Hyla septentrionalis*)

Litoria

***Litoria caerulea* (A)**

This recently introduced species is currently restricted to southern Florida.

Hyla

Recent research indicates that *Hyla cadaverina* and *H. regilla* may be more accurately placed in the genus *Pseudacris*, as currently defined.

Key to Species of *Hyla*

1a. Black-bordered light area below eye (Fig. 81) 2
1b. No black-bordered light area below eye 6

2a. Dorsal pattern usually consisting of small dark spots (no larger and usually smaller than eye); almost always lacking canthal or supratympanic stripes; from west of the 100th meridian 3
2b. Dorsal pattern usually with large dark blotches (larger than eye); often with a canthal or supratympanic stripe; from east of the 100th meridian 4

3a. Last phalanx of 4th toe fully free of webbing; from southern Utah and all but extreme western Arizona east into Colorado and New Mexico, and south into Sonora and western Chihuahua; isolated populations in western Texas .... ***Hyla arenicolor***
CC 538, pl 47

3b. Webbing at least touching last phalanx of 4th toe; from southern California and northern Baja California .... ***Hyla cadaverina***
S 45, pl 16 **CAAR** 225 (1979)

4a. Rear of thigh mottled, green, white, or yellow (in life); relatively small (usually less than 44 mm SVL, never more than 55 mm SVL); call a ringing bird-like whistle .... ***H. avivoca***
CC 538, pl 47 **CAAR** 28 (1966)

4b. Rear of thigh with or without a dark bar, yellow-orange (in life); slightly larger than *H. avivoca* (frequently to 50 mm SVL); call a trill 5

5a. Rear of thigh often with a dark bar; call a slow trill; karyotype $n = 24$ .... ***H. versicolor***
CC 536, pl 47

5b. Rear of thigh usually without a dark bar; call a rapid trill; karyotype $n = 12$ .... ***H. chrysoscelis***
CC 536, pl 47

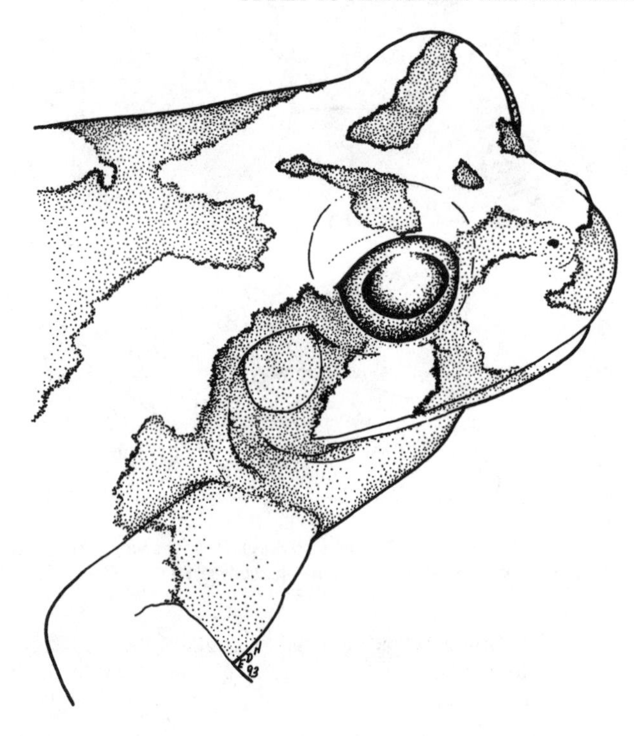

Figure 81. Dorsolateral view of the head of *Hyla chrysoscelis*, showing black-bordered light area below eye. Drawn from a photograph in Johnson (1977).

Preserved specimens of *Hyla avivoca, H. chrysoscelis* and *H. versicolor* are very difficult to identify. Live *H. avivoca* can be distinguished by the color of the hidden surfaces of the thigh, and all three species can be distinguished on the basis of call type. *H. chrysoscelis* and *H. versicolor* may also be differentiated by karyotype.

6a. Rear of thigh dark with light spots (Fig. 82) .... ***H. femoralis***
CC 535, pl 47, p 527

6b. Rear of thigh mottled or light with scattered dark spots 7

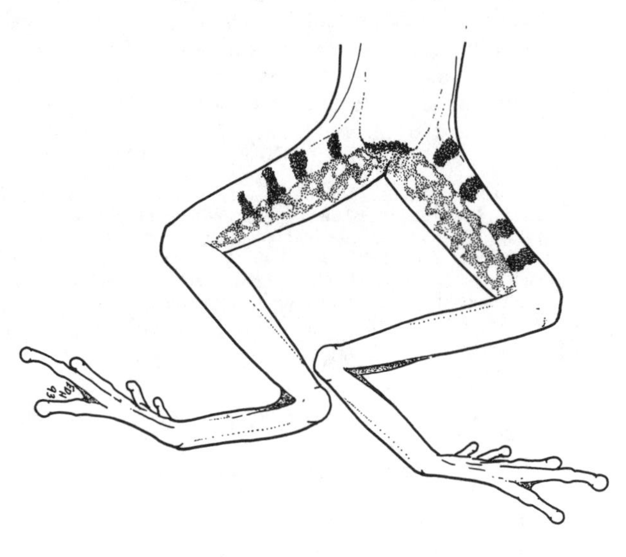

Figure 82. Dorsal view of posterior half of *Hyla femoralis*, showing pattern on rear of thigh. Drawn from a preserved specimen (KU 740).

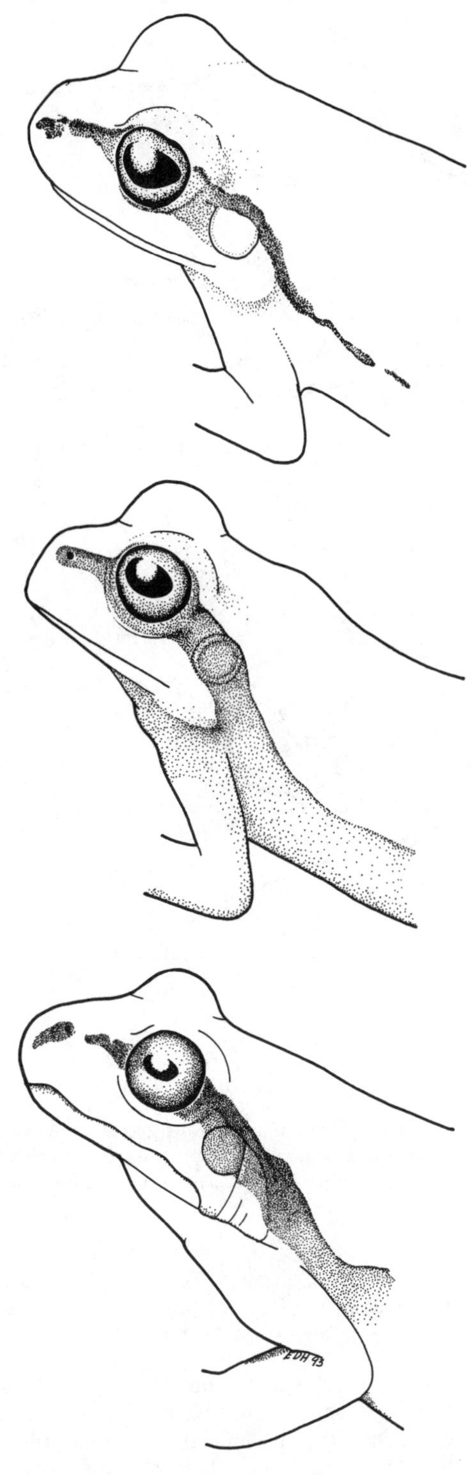

Figure 83. Lateral views of head and anterior part of body of three species of the genus *Hyla*, showing location of dark stripe from snout through eye. Upper: *H. femoralis*. Middle: *H. andersonii*. Lower: *H. eximia*.

7a. Dark stripe through eye (Fig. 83) 8

7b. No dark stripe through eye 10

8a. Dark stripe extends back only to shoulder, not separated from dorsal ground color by a thin light line;

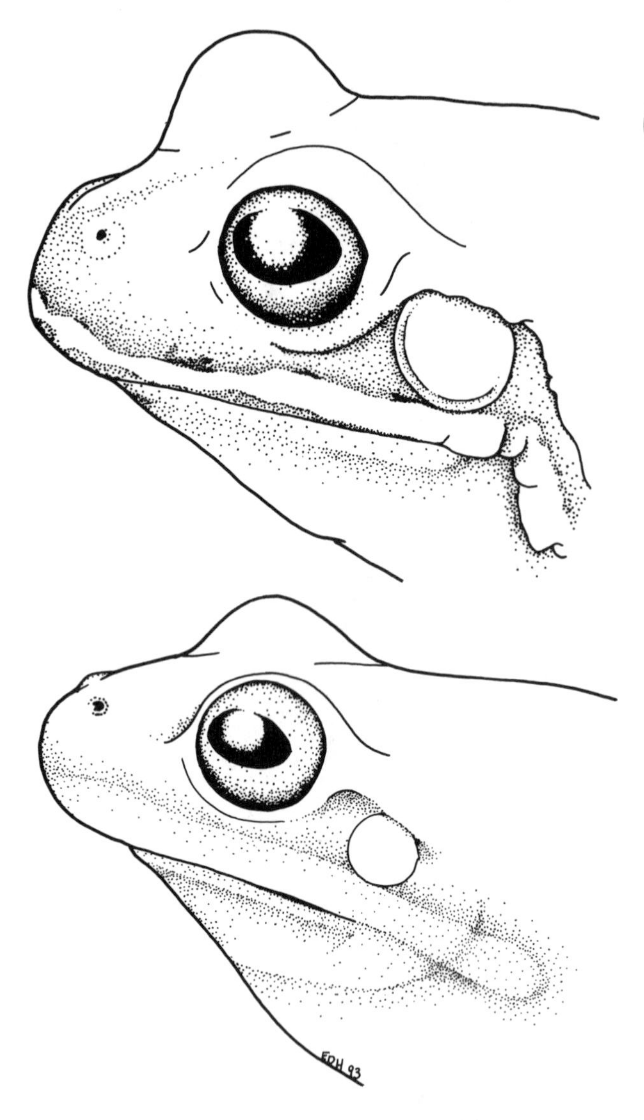

Figure 84. Lateral views of heads of two species of the genus *Hyla*. Upper: *H. gratiosa*, showing distinct light streak on upper lip. Lower: *H. squirella*, showing vague light streak on upper lip. Both drawn from 35mm color slides (KU).

from British Columbia south into Baja California and west of the Rocky Mountains, barely entering extreme northwestern Arizona .... ***H. regilla***
S 43, pl 16

8b. Dark stripe, separated from dorsal ground color by a thin light line, extends back onto body at least to groin (line may be broken into spots in the groin) (Fig. 83); from central or southern Arizona and western New Mexico, eastern Sonora and western Chihuahua south in the Sierra Madre Occidental, or the eastern U. S. 9

9a. Webbing extends less than ½ the length of the toes; from mountains of central Arizona and western New Mexico, extreme southern Arizona, and south from Sonora and Chihuahua in the Sierra Madre Occidental .... ***H. eximia***
S 44, pl 16

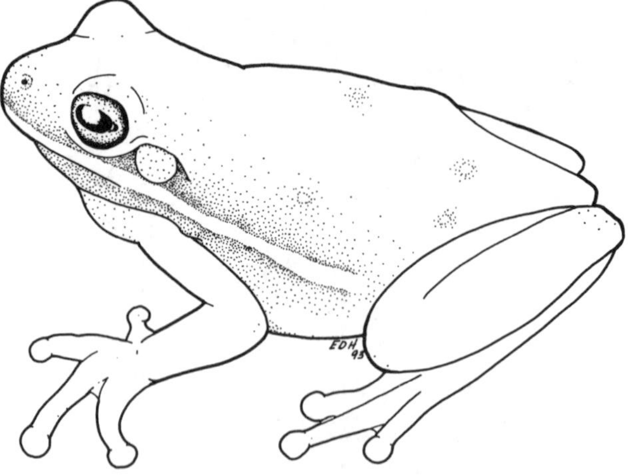

Figure 85. Dorsolateral view of *Hyla cinerea*, showing location of lateral stripe. Drawn from a 35mm color slide (KU).

9b. Webbing extends at least ½ length of the toes; from eastern U. S. .... ***H. andersonii***
CC 532, pl 47 **CAAR** 54 (1967)

10a. Light streak on upper lip sharply defined (Fig. 84); skin granular; dorsum frequently with numerous dark, rounded spots .... ***H. gratiosa***
CC 533, pl 47 **CAAR** 298 (1982)

10b. Light streak on upper lip absent or vaguely defined (Fig. 84); skin smooth; dorsum unicolored or with blotches 11

11a. Size moderately large (to more than 60 mm SVL); often with a light, distinctly bordered lateral stripe extending back below tympanum, usually to groin, and with green below; small light flecks usually present on back (Fig. 85); no dark spot between eyes .... ***H. cinerea***
CC 532, pl 47

11b. Size small (never more than 41 mm SVL); light lateral stripe, if present, indistinctly bordered below, blending with ventral coloration (no green below stripe), and usually not extending to groin; rarely with small light flecks on back; often with a dark spot between eyes .... ***H. squirella***
CC 535, pl 47 **CAAR** 168 (1975)

Note that specimens of *Hyla cinerea* and *H. squirella* are often difficult to distinguish when body sizes fall within the range of the latter. The calls may be used when calling males are encountered in the field (the call of *H. cinerea* is a hoarse quack, whereas that of *H. squirella* is a nasal trill or a grating sound likened by some to the scolding of a squirrel).

Ranidae

Rana

Key to Species of *Rana*

1a. Dorsolateral folds indistinct, absent, or not extending onto body (Fig. 86) 2

1b. With distinct dorsolateral folds extending at least ⅔ of the length of the body (Fig. 86) 10

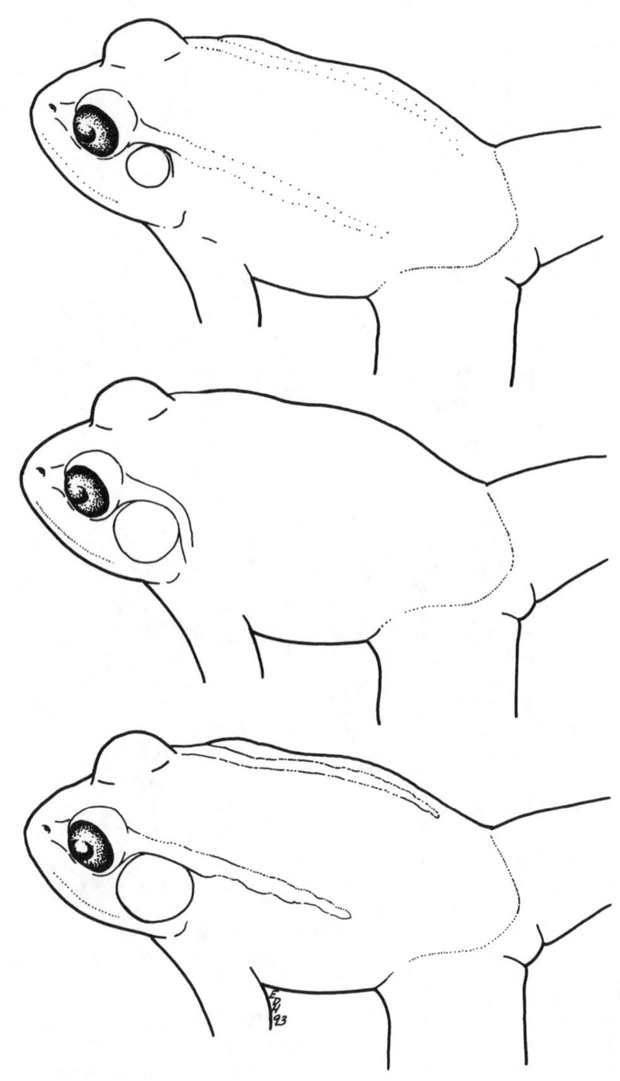

Figure 86. Dorsolateral views of three species of the genus *Rana*, showing absence, presence, and location of dorsolateral folds. Upper: *R. boylii*. Middle: *R. catesbeiana*. Lower: *R. clamitans*. Frog outline adapted from a photograph in Behler (1988).

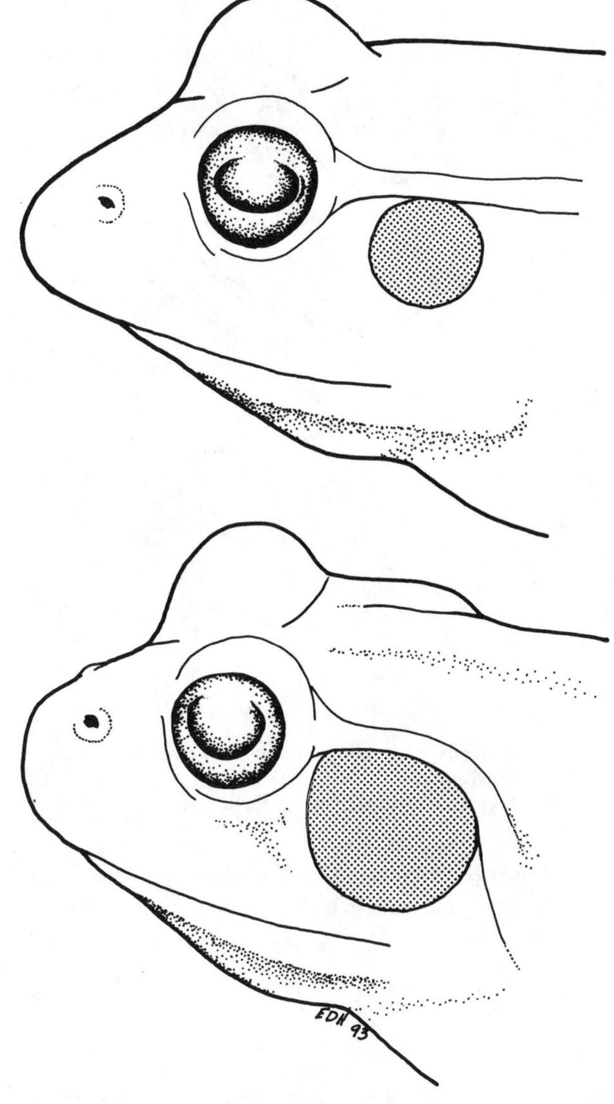

Figure 87. Lateral views of the heads of two species of the genus *Rana*, comparing tympanum and eye size. Upper: *R. tarahumarae*, showing tympanum smaller than eye. Lower: *R. catesbeiana*, showing tympanum larger than eye.

2a. Tympanum smaller than eye (Fig. 87) 3
2b. Tympanum at least as large as eye (Fig. 87) 5

3a. Tips of digits rounded with small discs; no outer metatarsal tubercle (Fig. 88); no light lines on upper lips; throat light (often suffused with light brown) .... ***R. tarahumarae***
S 53, pl 13 **CAAR** 66 (1968)
3b. Tips of digits pointed; an outer metatarsal tubercle present; at least indistinct light lines on the upper lips; throat usually spotted 4

4a. Usually with a pale spot on snout; toe tips not dark .... ***R. boylii***
S 49, pl 14 **CAAR** 71 (1968)
4b. Usually lacking a pale spot on snout; toe tips dark .... ***R. muscosa***
S 52, pl 13, 14 **CAAR** 65 (1968)

5a. With light dorsolateral stripes 6
5b. Lacking light dorsolateral stripes 7

6a. Toes fully webbed (Fig. 89) .... ***R. grylio*** (part)
CC 559, pl 48, p 557 **CAAR** 286 (1982)
6b. Two segments of longest toe free of web (Fig. 89) .... ***R. virgatipes***
CC 559, pl 48 **CAAR** 67 (1968)

7a. With light spots on lips; venter dark with light spots or short, wavy lines .... ***R. heckscheri***
CC 557, pl 48, p 558 **CAAR** 348 (1984)
7b. Lacking light spots on lips; venter light (may have variable amounts of dark markings) 8

8a. Hind toe webs indented on both sides of 4th toe; 4th toe extends beyond web by about one toe segment (Fig. 89) .... ***R. catesbeiana***
CC 555, pl 48, p 557, 558 **S** 47, pl 14

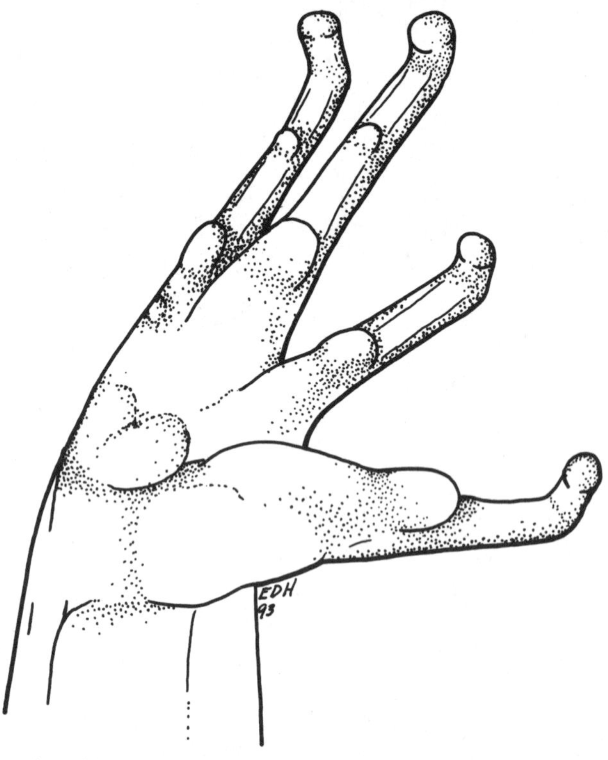

Figure 88. Ventral view of forefoot of *Rana tarahumarae*, showing tips of toes with small rounded discs and absence of terminal tubercle on each toe. Drawn from a preserved specimen (KU 200794).

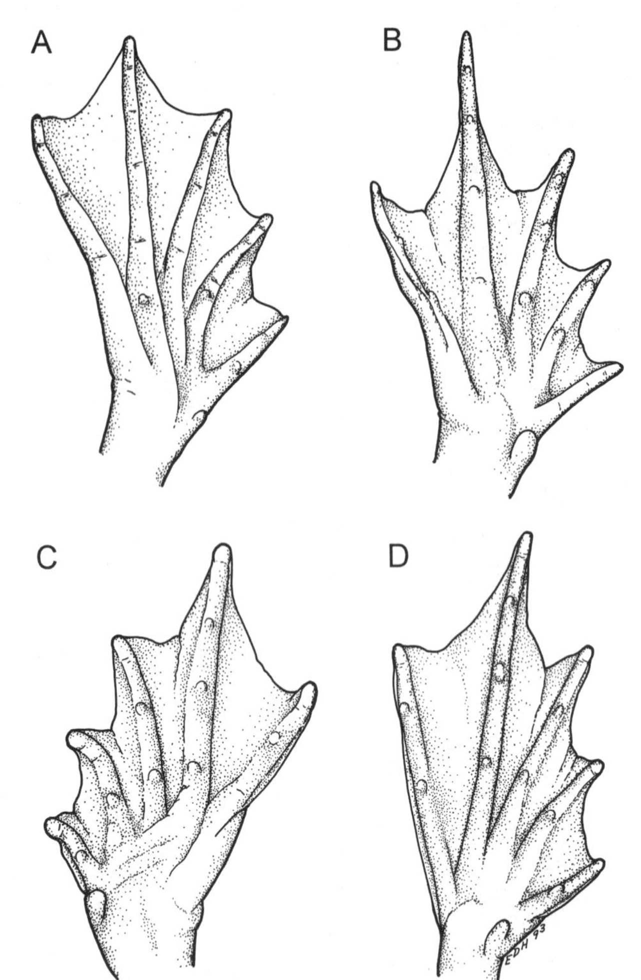

Figure 89. Ventral views of hind feet of four species of the genus *Rana*, showing extent and configuration of webbing. (A) *R. grylio* (KU 265921); (B) *R. virgatipes* (KU 153615); (C) *R. catesbeiana* (KU 214121); (D) *R. hecksheri* (KU 188312). All drawn from the preserved specimens indicated.

8b. Hind toe webs not indented (essentially straight between 3rd and 4th toes); toes webbed virtually to tips (Fig. 89) 9

9a. Small (to 76 mm); from the northern U.S. and Canada (largely north of the Great Lakes) .... ***R. septentrionalis*** (part)
CC 562, pl 48, p 562 **CAAR** 202 (1977)

9b. Large (to more than 160 mm); from the southeastern U.S. .... ***R. grylio*** (part)
CC 559, pl 48, p 557 **CAAR** 286 (1982)

10a. Rectangular spots in two rows between dorsolateral folds (Fig. 90)..................................... ***R. palustris***
CC 570, pl 48 **CAAR** 117 (1971)

10b. Spots absent or not rectangular and in two rows 11

11a. Dorsolateral folds extending only ⅔ the length of the body (Figs. 86, 91); from east of the 100th meridian (or if west, distinctly spotted) 12

11b. Dorsolateral folds nearly always extending to groin (although these may be interrupted) (Fig. 91); range variable 14

12a. Dorsal pattern of distinct spots on a light ground color .... ***R. onca*** (part)
S 55

12b. Dorsal pattern not as above, darker ground color and without distinct spots (dark flecks or small blotches may be present) 13

13a. Webbing on feet reduced, at least three segments of 4th toe free of web (Fig. 92) .... ***R. okaloosae***
CC 560, pl 48 **CAAR** 561 (1993)

13b. Webbing not reduced, no more than two segments of 4th toe free of web (Fig. 92) ....***R. clamitans***
CC 560, pl 48 **S** 49 **CAAR** 337 (1983)

14a. Lacking a light line on the upper lip (Fig. 93); stocky habitus; from east of the 100th meridian 15

14b. With a light line on the upper lip (may be obscure in some specimens) (Fig. 93); habitus and range variable 17

15a. Webbing extends to last joint of 4th toe and tip of 5th toe; from the northern U.S. and Canada (largely north of the Great Lakes) .... ***R. septentrionalis*** (part)
CC 562, pl 48, p 562 **CAAR** 202 (1977)

15b. Webbing less extensive; from the central and southern U.S. 16

Figure 90. Dorsolateral view of *Rana palustris*, showing location and configuration of dorsal spotting. Drawn from a photograph in Behler (1988).

16a. Venter largely unmarked; dorsal pattern of rounded spots with light borders .... ***R. areolata***
CC 572, pl 48, p 571 **CAAR** 324 (1983)

16b. At least chin and throat spotted, entire venter often heavily pigmented; dorsal pattern not of rounded spots with light borders .... ***R. capito***
CC 572, pl 48, p 571
CAAR 324 (1983) (as *R. areolata capito*)

17a. With a dark mask through eye and tympanum (mask sometimes obscure) (Fig. 94) 18

17b. With no trace of a mask 23

18a. Dorsal pattern of spots, usually distinct; mask distinct or not; from west of the 100th meridian 19

18b. Dorsum uniform or with faint stripes (if broken into spots, these are very indistinct); mask well defined; range variable 22

19a. Eyes turned upward (not fully covered by lids when viewed from above) (Fig. 95); no mottling on sides; heels of adpressed limbs do not reach nostrils; slightly granular skin 20

19b. Eyes not turned upward; with mottling on sides; heels of adpressed limbs reach or extend beyond nostrils; smooth skin 21

The following two species are difficult to distinguish using external morphology and are best differentiated by geography and measures of protein divergence.

20a. Plantar and palmar tubercles present (but may be obscure); venter red (in life); from extreme southwestern British Columbia south along the eastern shore of Puget Sound and the Willamette Valley, and from the Columbia River Gorge in southern Washington to the Cascades and the Klamath Valley in west-central Oregon .... ***R. pretiosa***
S 50, pl 13 **CAAR** 119 (1972) (part)

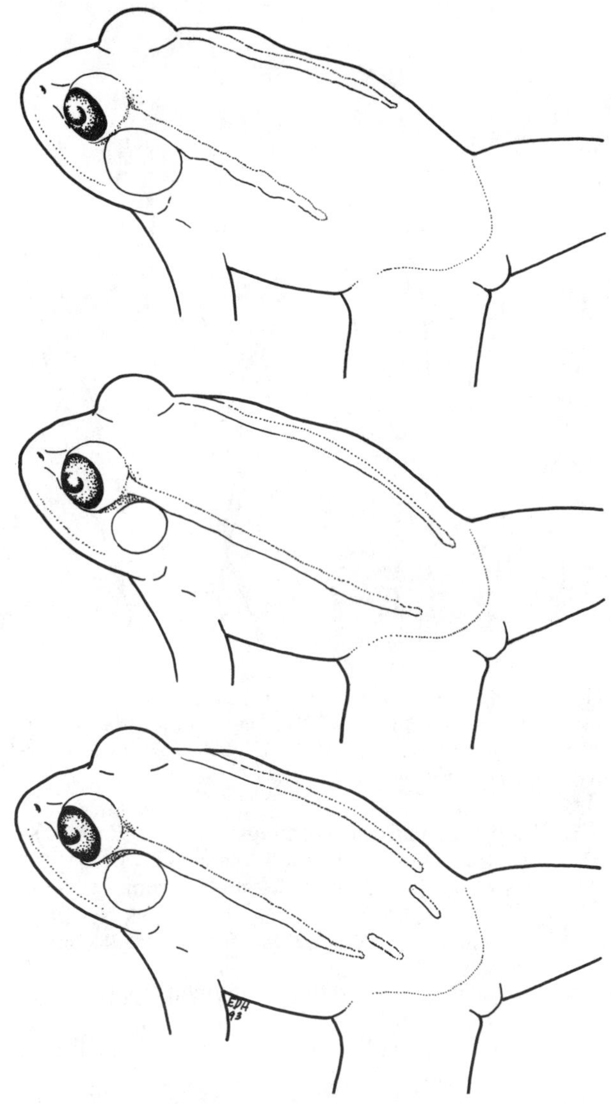

Figure 91. Dorsolateral views of the genus *Rana*, showing location and configuration of dorsolateral folds. Upper: folds extend only two-thirds length of body; Middle: folds extending unbroken nearly to groin; Lower: folds extending nearly to groin but broken and inset posteriorly. Frog outline adapted from a photograph in Behler (1988).

20b. Some large individuals from southern part of range lack plantar and palmar tubercles; venter orange to reddish (in life); from extreme southwestern Yukon through Alaska panhandle, most of British Columbia, and Washington east of Cascades to eastern Oregon, Idaho, western Montana, and northwestern Wyoming, with isolated populations in Nevada and Utah .... ***R. luteiventris***
S 50, pl 13 **CAAR** 119 (1972) (part)

21a. Dorsal spots intense, distinctly outlined; groin yellow or green (in life) .... ***R. cascadae***
S 51, pl 13 **CAAR** 105 (1971)

21b. Dorsal spots vague, indistinctly outlined; groin reddish (in life) .... ***R. aurora***
S 48, pl 13, 14 **CAAR** 160 (1972)

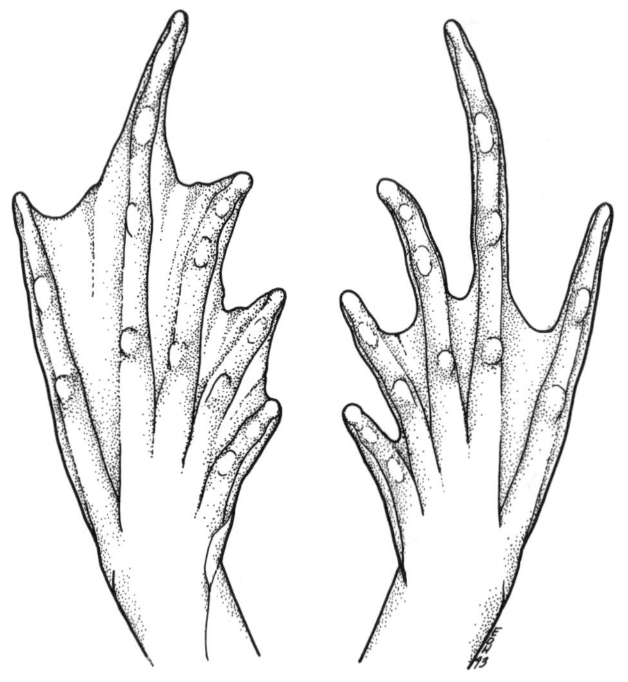

Figure 92. Ventral view of the hind feet of species of the genus *Rana*, showing the extent and location of webbing. Left: *R. clamitans* (KU 203611). Right: *R. okaloosae* (KU 204172). Both drawn from the preserved specimens indicated.

22a. From Colorado or Wyoming .... ***R. maslini***
CC 562 (as western subspecies), **S** 46, pl 14 **CAAR** 86 (1970)

22b. Range outside of Colorado and Wyoming .... ***R. sylvatica***
CC 562, pl 48 **S** 46, pl 14 **CAAR** 86 (1970)

23a. Dorsolateral fold continuous through groin (Fig. 91) 24

23b. Dorsolateral fold interrupted and inset medially near groin (Fig. 91) 25

24a. Snout pointed and head somewhat elongated; spots on sides scattered and few in number; usually a light spot in the center of the tympanum (Fig. 96); snout usually without a spot .... ***R. sphenocephala***
CC 567, pl 48, p 565 (as *Rana utricularia*)

24b. Snout rounded, head not elongated; numerous spots on sides; usually lacking a light spot in the center of the tympanum; snout usually with a spot .... ***R. pipiens***
CC 567, pl 48, p 565 **S** 54, pl 15

25a. Rear of thigh with light specks on a dark background or with a dark reticulum (Fig. 97); usually lacking a light spot on the tympanum; pattern of small numerous spots; stripe on upper lip usually fades in front of eye; from Arizona and adjacent areas south into México 26

25b. Rear of thigh not as above; usually with a light spot on the tympanum; pattern of larger and/or fewer spots; stripe on upper lip variable; range variable 28

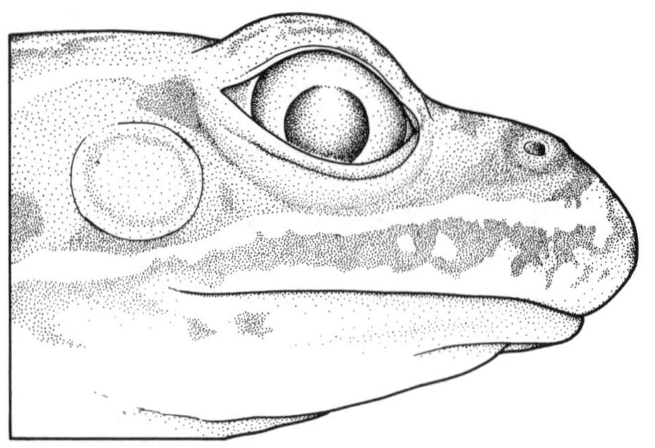

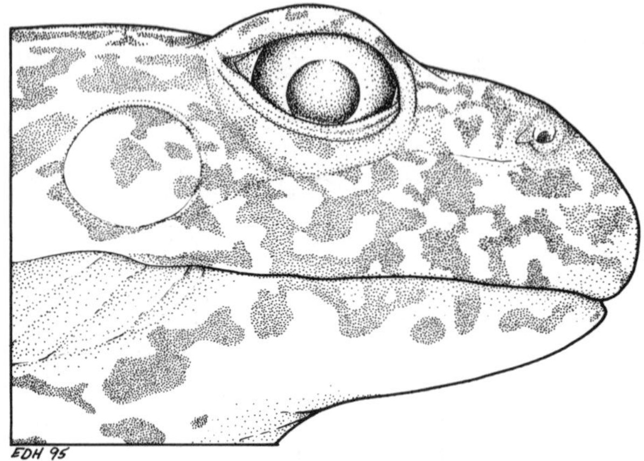

Figure 93. Lateral views of the heads of two species of the genus *Rana*. Upper: *R. pipiens*, showing presence of a light line on the upper lip. Drawn from a preserved specimen (KU 129627). Lower: *R. areolata*, showing lack of a light line on the upper lip. Drawn from a preserved specimen (KU 157902).

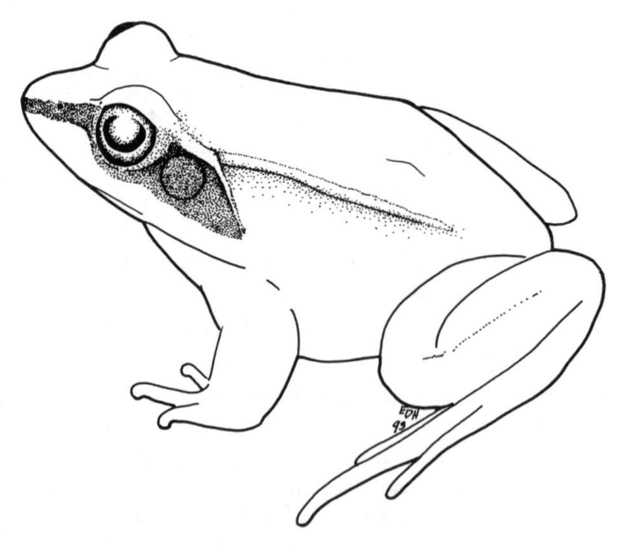

Figure 94. Dorsolateral view of *Rana sylvatica*, showing dark mask through eye. Drawn from a photograph in Johnson (1977).

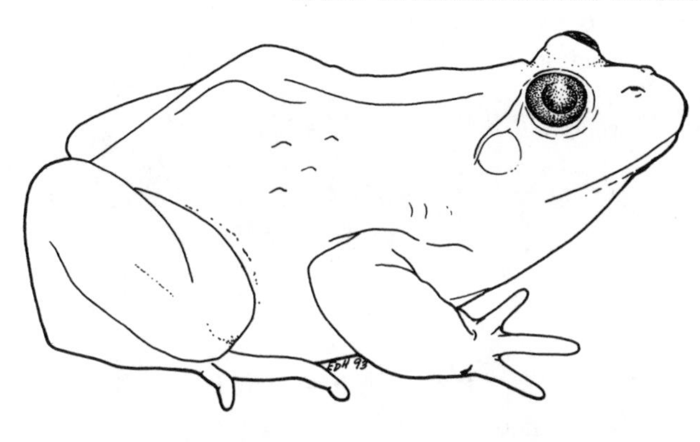

Figure 95. Lateral view, showing eyes turned upward, typical of *Rana pretiosa* and *R. luteiventris*. Drawn from a photograph in Baxter and Stone (1985).

26a. Rear of thigh with light specks on a dark background (Fig. 97); usually with a mottled venter 27

26b. Rear of thigh with a dark reticulum (Fig. 97); venter usually immaculate .... ***R. yavapaiensis***
S 54, pl 14, 15 **CAAR** 418 (1988)

27a. Toe tips expanded, knoblike in large adults; males call underwater; range restricted to the Huachuca Mountains, Arizona.................. ***R. subaquavocalis***

27b. Toe tips not expanded, or barely so in large adults; males call from surface; range includes portions of central and southwestern Arizona and southeastern New Mexico south to southern Durango, México .... ***R. chiricahuensis***
S 55, pl 15 **CAAR** 347 (1984)

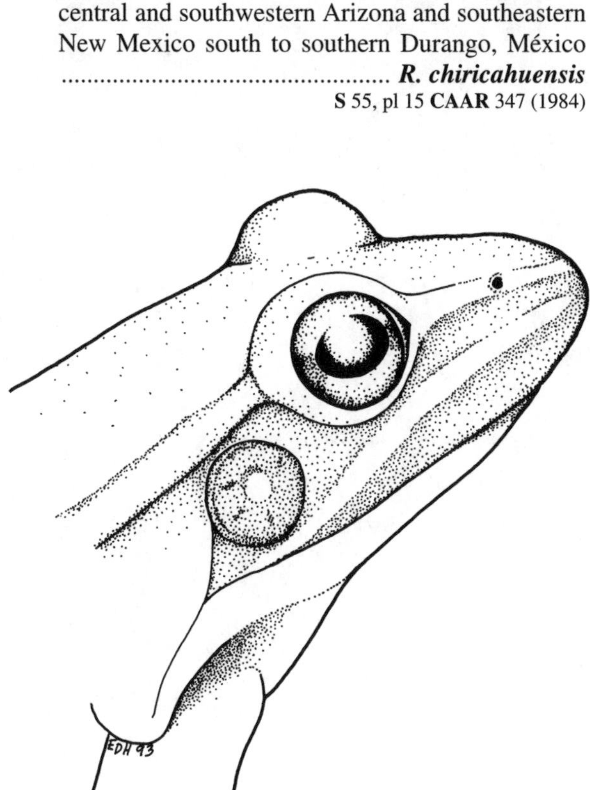

Figure 96. Lateral view of head of *Rana sphenocephala*, showing spot in center of tympanum. Drawn from a photograph in Johnson (1977).

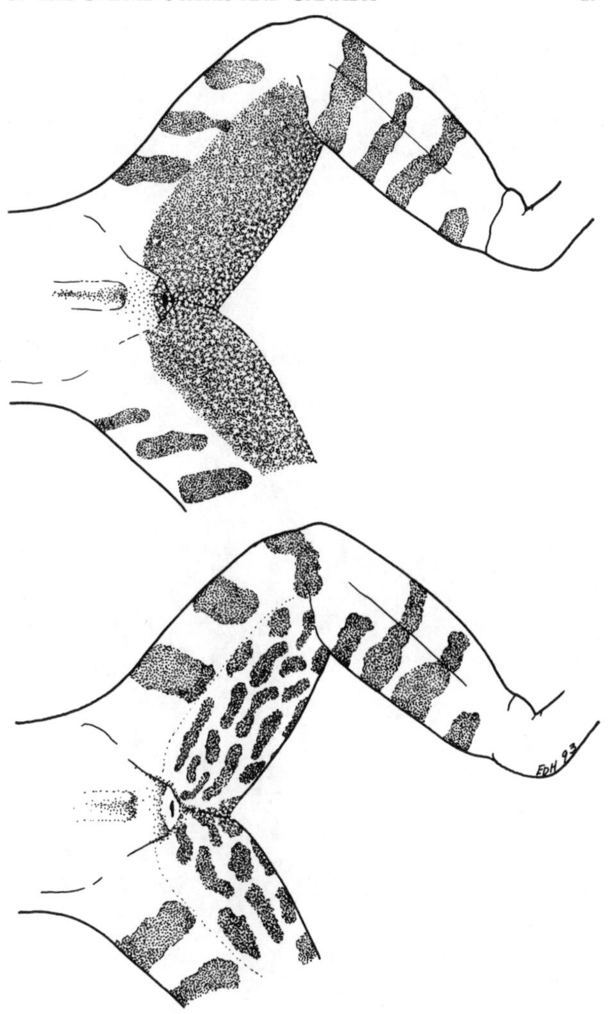

Figure 97. Dorsal views of posterior half of two species of the genus *Rana*, showing patterns on rear of thighs. Upper: *R. chiricahuensis* (KU 192872). Lower: *R. yavapaiensis* (KU 78181). Drawn from preserved specimens as indicated.

28a. Dorsolateral folds indistinct (may end short of groin); light stripe on upper lip usually fades in front of eye; from southern Nevada and adjacent northwestern Arizona and extreme southwestern Utah .... ***R. onca*** (part)
S 55

28b. Dorsolateral folds distinct; light stripe on upper lip variable; range not as above 29

29a. Ground color and pattern usually pale; stripe on upper lip usually fades in front of eye; from western and southern Texas south into México .... ***R. berlandieri***
CC 569, pl 48, p 565 **S** 54, pl 15 **CAAR** 508 (1991)

29b. Ground color often bright, pattern usually distinct; stripe on upper lip usually distinct in front of eye; native range is east of Arizona-New Mexico border (introduced populations are established in eastern Arizona) .... ***R. blairi***
CC 569, pl 48, p 565 **S** 54, pl 15 **CAAR** 536 (1992)

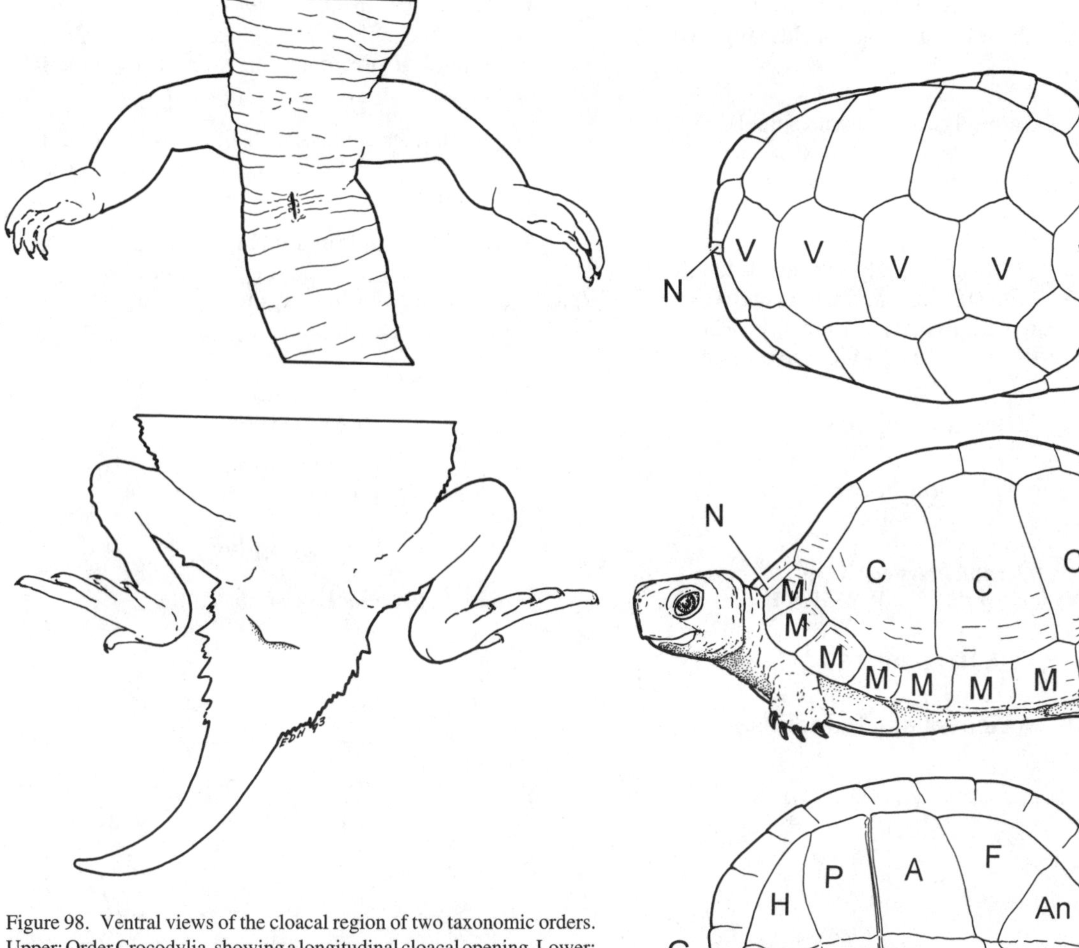

Figure 98. Ventral views of the cloacal region of two taxonomic orders. Upper: Order Crocodylia, showing a longitudinal cloacal opening. Lower: Order Squamata, showing a transverse cloacal opening.

REPTILIA

KEY TO ORDERS OF REPTILIA

1a. Bony or leathery shell present .... **Testudines** (p. 49)
1b. No bony or leathery shell 2

2a. Cloacal opening a longitudinal slit (Fig. 98) .... **Crocodylia** (p. 65)
2b. Cloacal opening a transverse slit (Fig. 98)............ .... **Squamata** (p. 67)

TESTUDINES
(Turtles)

KEY TO FAMILIES OF TESTUDINES

1a. Limbs oarlike (Fig. 100); digits elongated, flattened, and bound together 2
1b. Limbs not oarlike (Fig. 100); digits not bound together (but may be webbed) 3

Figure 99. Generalized drawings of a turtle (*Terrapene carolina*), showing three aspects of the shell and its components: Upper: Dorsal view of the upper shell (carapace). Middle: Lateral view of the entire turtle. Lower: Ventral view of lower shell (plastron). Scutes of the shell are: N = nuchal; V = vertebral; M = marginal; C = costal; G = gular; H = humeral; P = pectoral; A = abdominal; F = femoral; An = anal. Drawn from photographs by Suzanne L. Collins and Tom R. Johnson.

2a. Shell leathery; without claws .... **Dermochelyidae** (p. 50)
2b. Shell bony and with horny shields; with claws...... .... **Cheloniidae** (p. 50)

3a. No horny shields on shell, edges flexible; claws 3–3 (front-back) .... **Trionychidae** (p. 51)
3b. With horny shields on shell, edges rigid; claws 4–3 or 5–4 4

4a. Plastron with twelve shields (Fig. 101)................ 5
4b. Plastron with eleven or fewer shields (Fig. 101) 6

5a. Top of head entirely covered with shields (Fig.102) .... **TESTUDINIDAE** (p. 52)
5b. Top of head covered anteriorly with skin .... **EMYDIDAE** (p. 53)

6a. Rear margin of shell strongly serrate; bridge shorter than broad (Fig. 103)............. **CHELYDRIDAE** (p. 62)
6b. Rear margin of shell nearly smooth; bridge at least as long as broad (Fig. 103) ..... **KINOSTERNIDAE** (p. 63)

DERMOCHELYIDAE

Dermochelys

CAAR 238 (1980)

Dermochelys coriacea

CC pl 9 **S** pl 19 **CAAR** 238 (1980)

CHELONIIDAE

Key to Genera of Cheloniidae

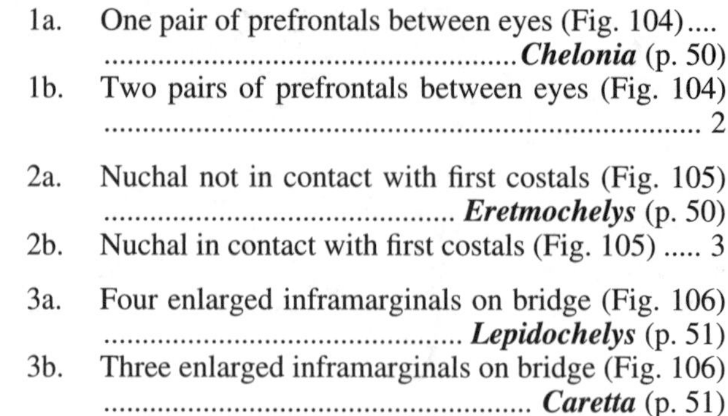

1a. One pair of prefrontals between eyes (Fig. 104).... .... ***Chelonia*** (p. 50)
1b. Two pairs of prefrontals between eyes (Fig. 104) 2

2a. Nuchal not in contact with first costals (Fig. 105) .... ***Eretmochelys*** (p. 50)
2b. Nuchal in contact with first costals (Fig. 105) 3

3a. Four enlarged inframarginals on bridge (Fig. 106) .... ***Lepidochelys*** (p. 51)
3b. Three enlarged inframarginals on bridge (Fig. 106) .... ***Caretta*** (p. 51)

Chelonia

CAAR 248 (1980)

Chelonia mydas

CC pl 9, p 191 **S** pl 19 **CAAR** 249 (1980)

Eretmochelys

Eretmochelys imbricata

CC pl 9, p 191 **S** p 107

Figure 100. Dorsolateral views of two species of turtles, showing different kinds of limbs. Upper: Oarlike limbs typical of families Dermochelyidae and Cheloniidae. Adapted from a photograph of *Caretta caretta* in Pope (1956). Lower: Limbs typical of other turtle families addressed in this key. Adapted from a photograph of *Clemmys marmorata* in Ernst and Barbour (1972).

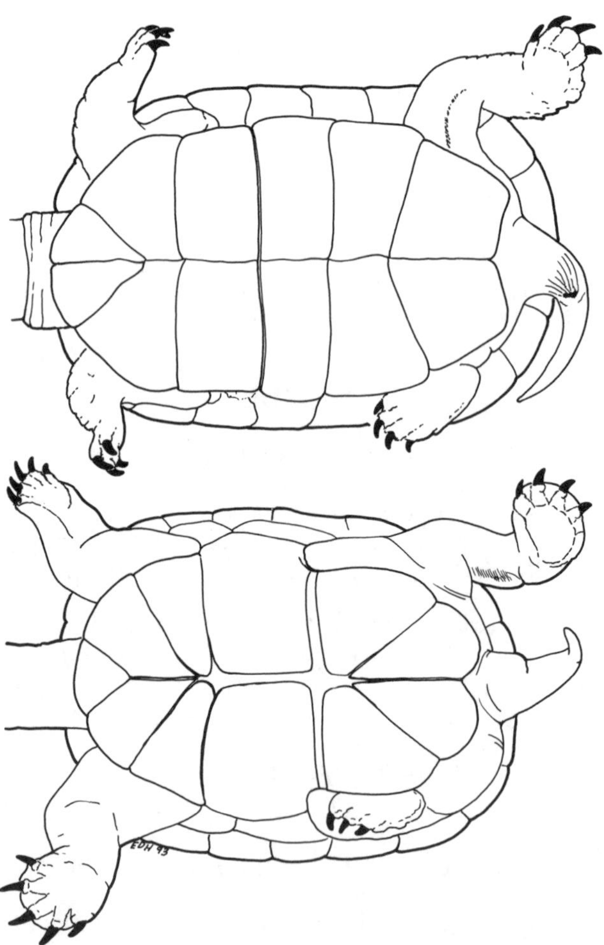

Figure 101. Ventral views of the plastron of two species of turtles, showing number of scutes. Upper: *Emydoidea blandingii*, showing 12 scutes and their location. Lower: *Kinosternon subrubrum*, showing 11 scutes and their location. Both drawn from plates in Babcock (1919).

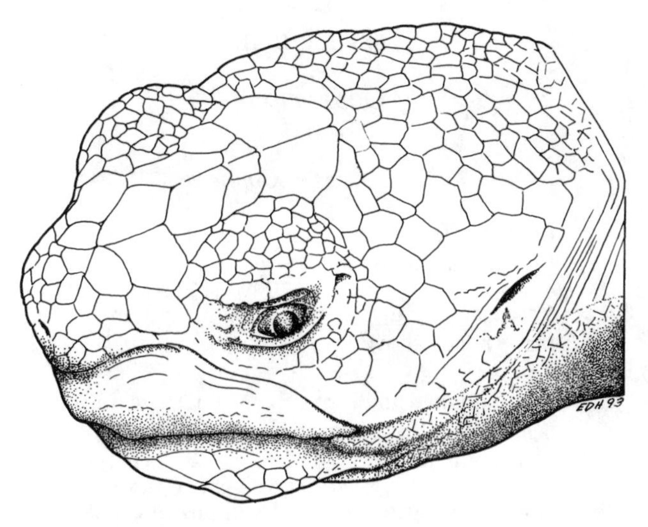

Figure 102. Dorsolateral view of the head of *Gopherus polyphemus*, showing head entirely covered by shields (scales). Drawn from a photograph by Suzanne L. Collins.

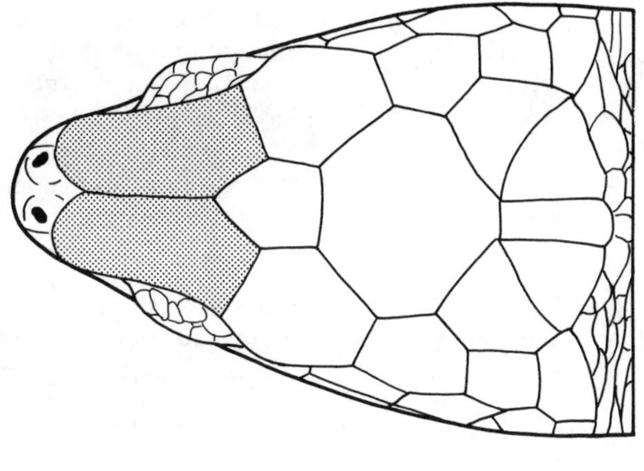

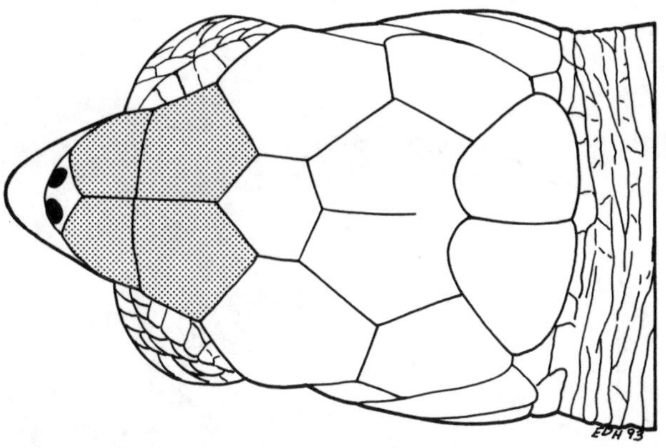

Figure 104. Dorsal views of the heads of two genera of sea turtles. Upper: *Chelonia mydas*, showing one pair of prefrontal scales. Lower: *Eretmochelys imbricata*, showing two pairs of prefrontal scales. Redrawn from artwork in Carr (1952).

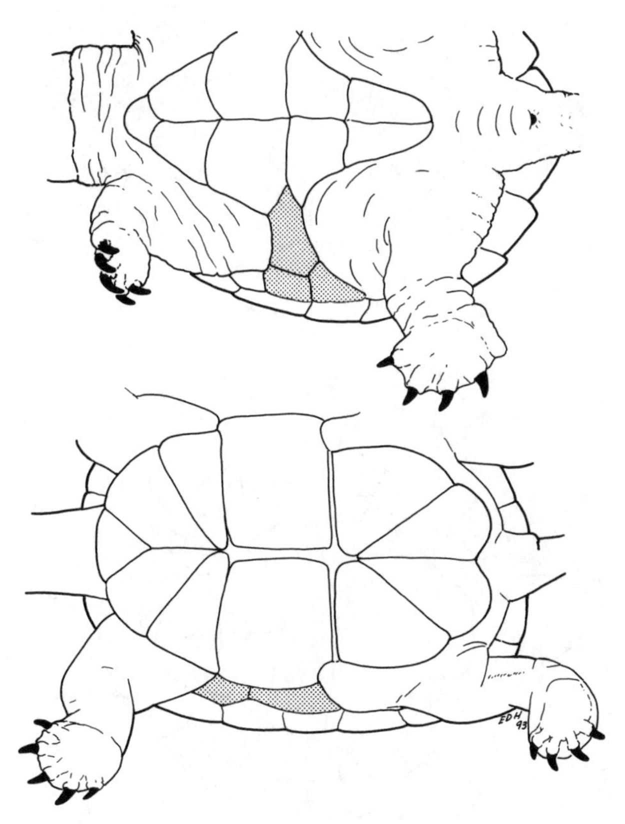

Figure 103. Ventral views of bridges (shaded) of two families of turtles. Upper: Family Chelydridae, showing bridge shorter than broad as in *Chelydra serpentina*. Lower: Family Kinosternidae, showing bridge at least as long as broad as in *Kinosternon subrubrum*. Both drawn from plates in Babcock (1919).

Lepidochelys

CAAR 587 (1994)

Key to Species of ***Lepidochelys***

1a. Five costals (Fig. 107); from the Atlantic Ocean ... . ***L. kempii***
CC pl 9 **CAAR** 509 (1991)

1b. Six to eight costals (Fig. 107); from the Pacific Ocean .... ***L. olivacea***
S pl 19 **CAAR** 653 (1998)

Caretta

CAAR 482 (1990)

Caretta caretta

CC pl 9 **S** pl 19 **CAAR** 483 (1990)

TRIONYCHIDAE

Apalone

CAAR 487 (1990) (as *Trionyx*)

Key to Species of ***Apalone***

1a. Nasal septum with lateral ridges projecting into nostrils (Fig. 108); anterior margin of shell tuberculate (Fig. 109) 2

1b. Nasal septum without lateral ridges projecting into nostrils (Fig. 108); anterior margin of shell not tuberculate .... ***A. mutica***
CC 195, pl 4, 10 **S** 67 **CAAR** 139 (1973) (as *Trionyx muticus*)

2a. Flattened tubercles in more than one row along anterior edge of carapace (Fig. 109); marginal ridge present .... ***A. ferox***
CC 199, pl 4, 10 **CAAR** 138 (1973) (*as Trionyx ferox*)

2b. Spines or conical projections in single row along anterior edge of carapace (Fig. 109); no marginal ridge .... ***A. spinifera***
CC 196, pl 4, 10, p 196
S 66, pl 18 **CAAR** 140 (1973) (as *Trionyx spiniferus*)

TESTUDINIDAE

Gopherus

CAAR 211 (1978)

Key to Species of ***Gopherus***

1a. Carapace length less than twice maximum height; from Texas south into México .... ***G. berlandieri***
CC 190, pl 2 **CAAR** 213 (1978)

1b. Carapace length at least twice maximum height; range not as above 2

2a. Forelimbs and hindlimbs about equal in width; plastron not bent upward in front; from the southwestern U.S. south into México .... ***G. agassizii***
S 58, pl 18 **CAAR** 212 (1978)

2b. Hindlimbs narrower than forelimbs; plastron bent upward in front; from the southeastern U.S. .... ***G. polyphemus***
CC 189, pl 2 **CAAR** 215 (1978)

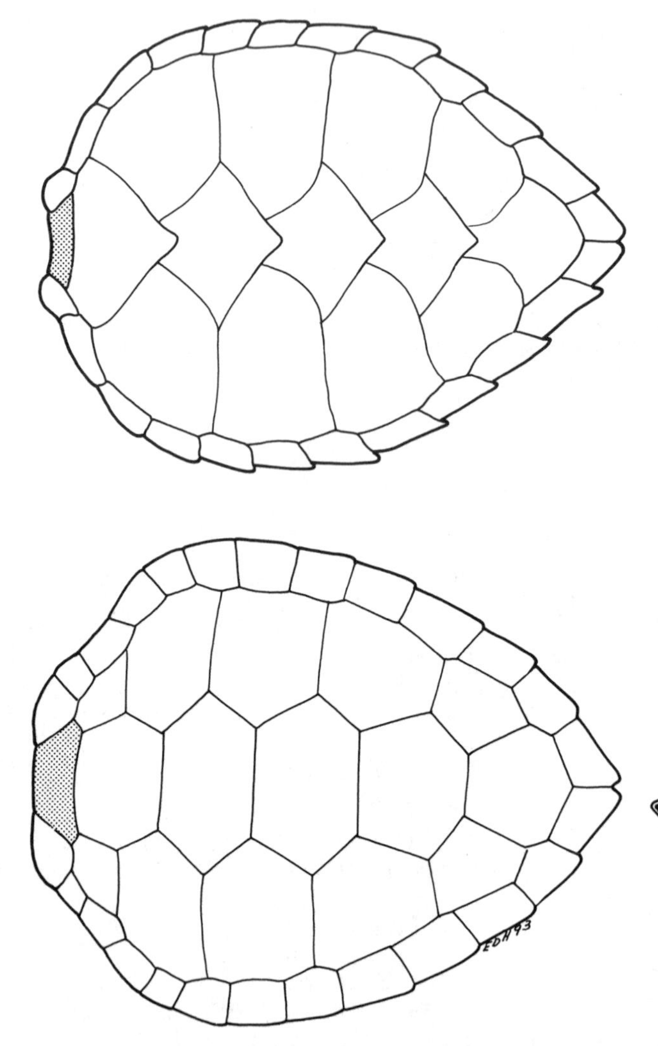

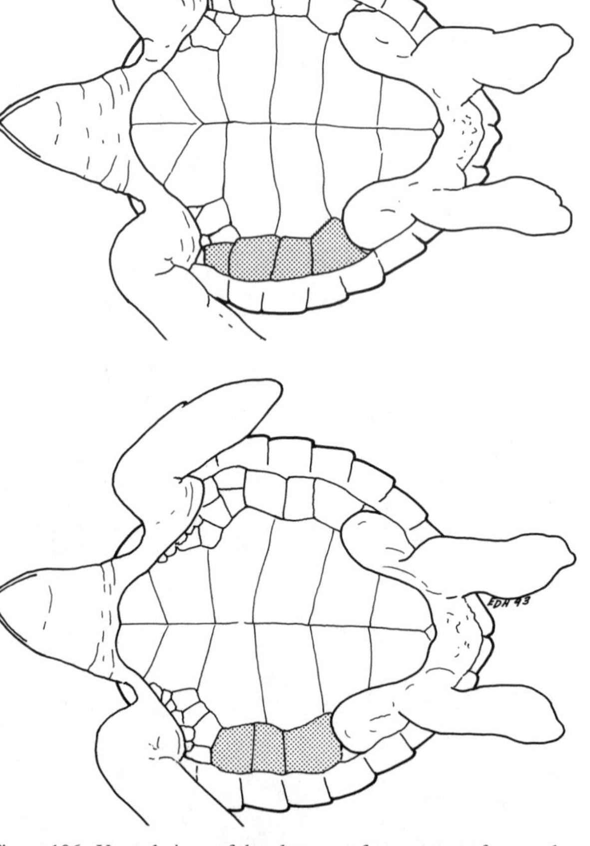

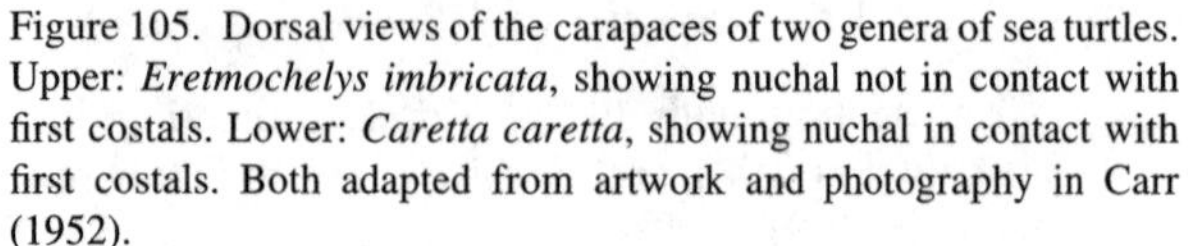

Figure 105. Dorsal views of the carapaces of two genera of sea turtles. Upper: *Eretmochelys imbricata*, showing nuchal not in contact with first costals. Lower: *Caretta caretta*, showing nuchal in contact with first costals. Both adapted from artwork and photography in Carr (1952).

Figure 106. Ventral views of the plastrons of two genera of sea turtles. Upper: *Lepidochelys kempii,* showing the presence of four inframarginals on bridge. Lower: *Caretta caretta,* showing the presence of three inframarginals on bridge. Both adapted from artwork and photography in Carr (1952).

Figure 107. Dorsolateral views of two species of the genus *Lepidochelys*. Upper: *L. kempii*, showing the presence of five costals. Drawn from a 35 mm color slide (KU 542). Lower: *L. olivacea*, showing the presence of more than five costals. Drawn from a photograph in Behler and King (1979).

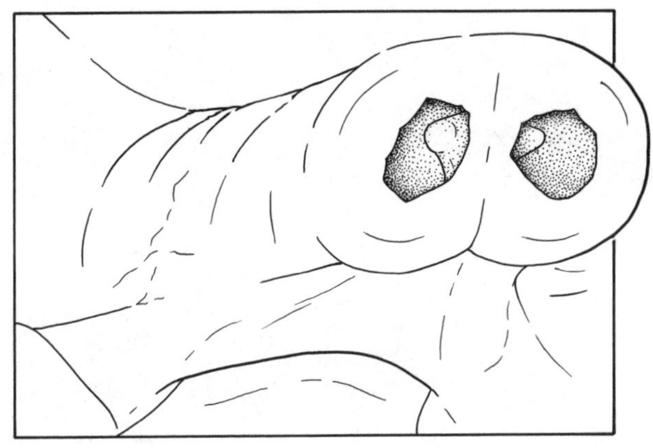

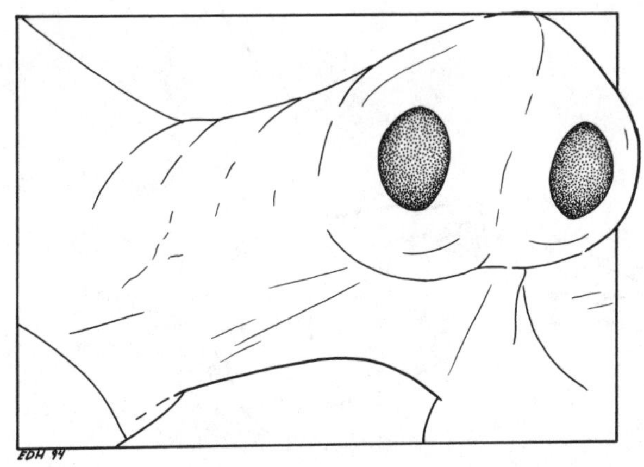

Figure 108. Frontal views of the snouts of two species of the genus *Apalone*. Upper: *A. spinifera*, showing presence of nasal septa projecting laterally from the midline of the snout. Drawn from a preserved specimen (KU 88854). Lower: *A. mutica*, showing lack of nasal septa. Drawn from a preserved specimen (KU 1450).

Emydidae

Key to Genera of Emydidae

1a. Plastron with one transverse hinge (Figs. 101, 110) 2

1b. Plastron with two transverse hinges or none (Fig. 110) 3

2a. Front of upper jaw with hooked beak .... ***Terrapene*** (p. 54)

2b. Front of upper jaw without hooked beak (curved upward when viewed from side) (Fig. 111) .... ***Emydoidea*** (p. 57)

3a. With a pattern of vertical bands on the rear surface of thighs (Fig. 112) and a broad light stripe along the anterior margin of each forelimb (Fig. 113); neck very long (about equal to plastron length) .... ***Deirochelys*** (p. 56)

3b. Without such a pattern (if vertical bands present on the rear surface of thighs, light stripes on anterior margins of forelimbs are narrow; neck short (less than plastron length) 4

4a. Axillaries and inguinals absent or rudimentary (Fig. 114) .... ***Clemmys*** (p. 56)

4b. Axillaries and inguinals well developed (Fig. 114) 5

5a. Entire head and limbs spotted; all large carapacial scutes with well-defined concentric rings (Fig. 115) .... ***Malaclemys*** (p. 57)

5b. Head and limbs with at least some stripes or lines; carapacial scutes smooth or with furrows (if in the form of concentric rings, these are very indistinct or restricted to only some scutes) 6

6a. Crushing surface of upper jaw is broad (Fig. 116); nearly always with some light vertical lines on sides of head; posterior marginal scutes may be notched .... ***Graptemys*** (p. 57)

6b. Crushing surface of upper jaw is narrow (Fig. 116); light lines on sides of head longitudinal or diagonal, not vertical; posterior marginal scutes usually not notched 7

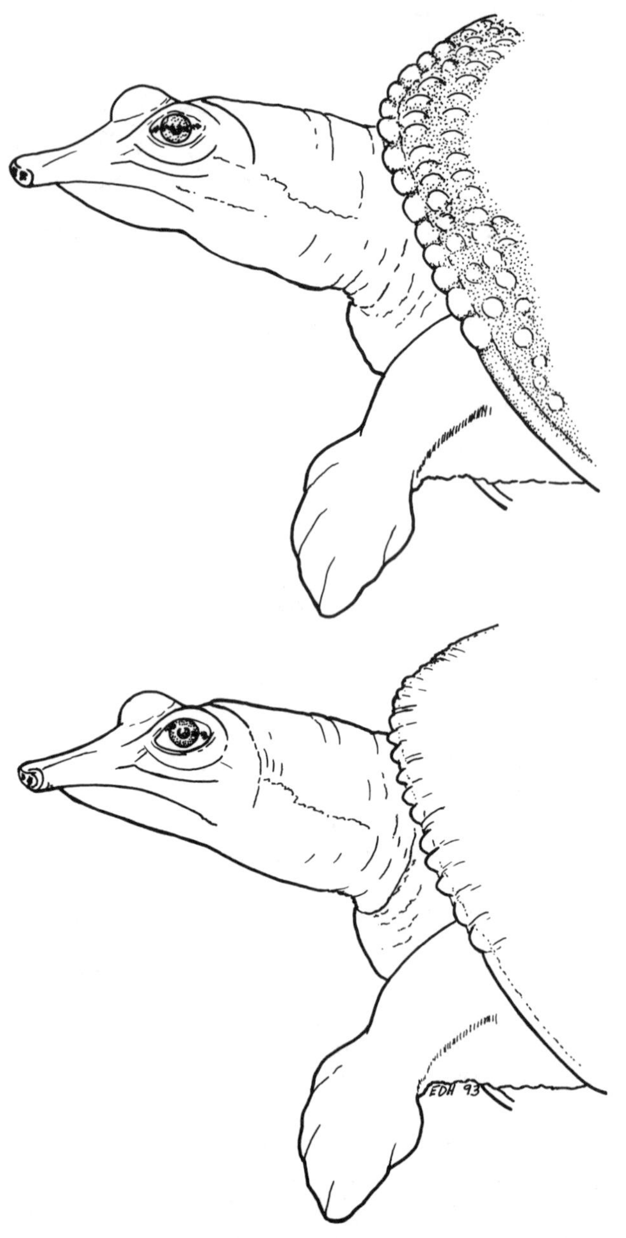

Figure 109. Dorsolateral views of the anterior portions of two species of the genus *Apalone*. Upper: *A. ferox*, showing the presence of rows of tubercles along the anterior edge of the carapace. Lower: *A. spinifera*, showing the presence of a single row of spines or conical projections along the anterior edge of the carapace. Both adapted from photographs in Ashton and Ashton (1985).

7a. First marginal scute usually not extending beyond suture between first vertebral scute and first costal scute (Fig. 117); posterior margin of carapace not serrated; carapacial scutes smooth .... ***Chrysemys*** (p. 60)

7b. First marginal scute extends beyond suture between first vertebral scute and first costal scute (Fig. 117); posterior margin of carapace serrated; carapacial scutes with furrows 8

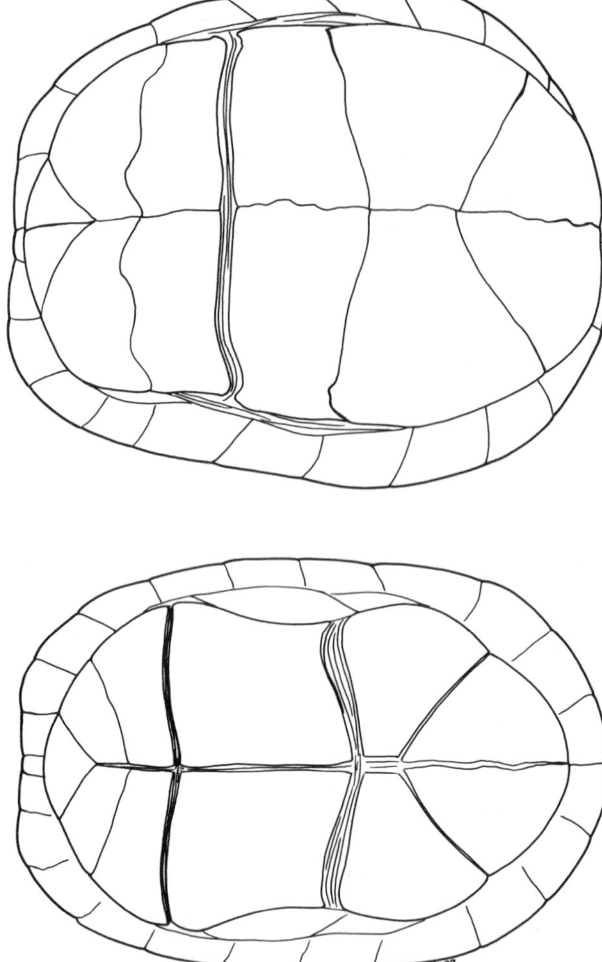

Figure 110. Ventral views of the plastrons of two families of turtles, showing the number and location of hinges. Upper: Emydidae (genus *Terrapene*), with a single hinge. Lower: Kinosternidae (genus *Kinosternon*), with two hinges. Both adapted from photographs in Johnson (1987) and Ernst and Barbour (1972).

8a. Underside of jaw rounded (Fig. 118); postorbital spot or line at least ½ diameter of orbit (usually a prominent patch of red or yellow on each side of head in life) .... ***Trachemys*** (p. 61)

8b. Underside of jaw flattened (Fig. 118); postorbital spot or line nearly always less than ½ diameter of orbit .... ***Pseudemys*** (p. 60)

Terrapene

CAAR 511 (1991)

Key to Species of *Terrapene*

1a. Hinge opposite 6th marginal or between 5th and 6th; no trace of a middorsal keel; interfemoral sulcus greater than ½ as long as interabdominal (Fig. 119) .... ***T. ornata***

CC 164, pl 5, 163 **S** 60, pl 17 **CAAR** 217 (1978)

Figure 111. Lateral view of the head of *Emydoidea blandingii*, showing the upward curved jaw. Adapted from a photograph in Ernst and Barbour (1972).

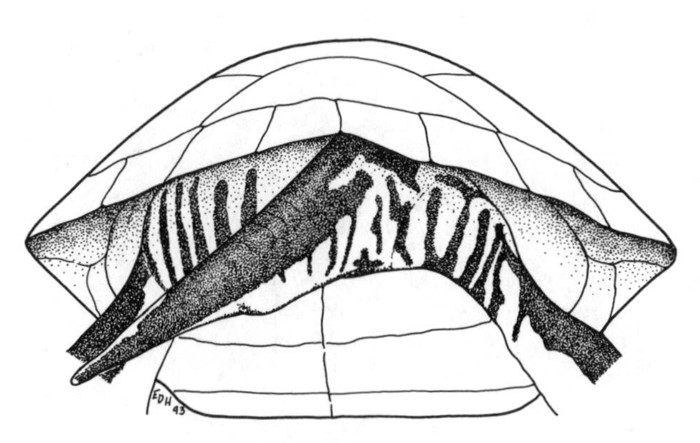
Figure 112. Posterior view of the thighs of *Deirochelys reticularia*, showing the pattern of vertical dark streaks typical of this genus. Adapted from a photograph in Ernst and Barbour (1972).

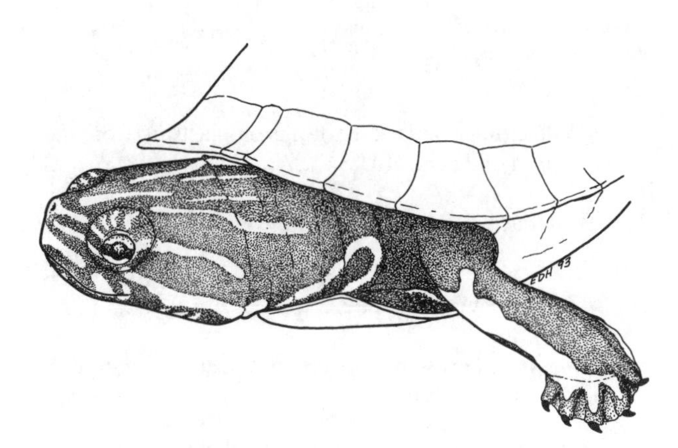
Figure 113. Frontolateral view of *Deirochelys reticularia*, showing presence of a broad light stripe along the anterior margin of the forelimb. Adapted from a photograph in Ashton and Ashton (1985).

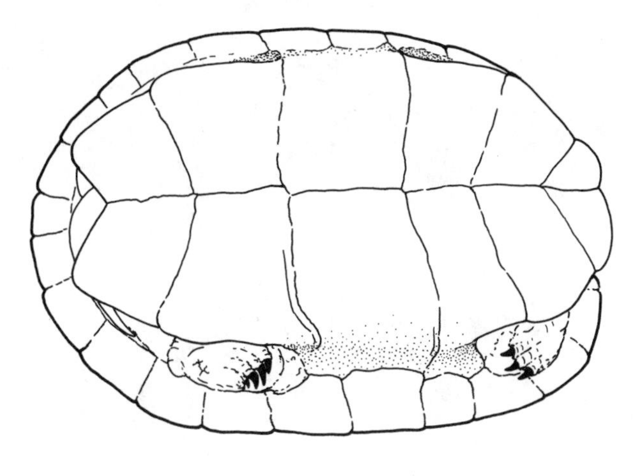
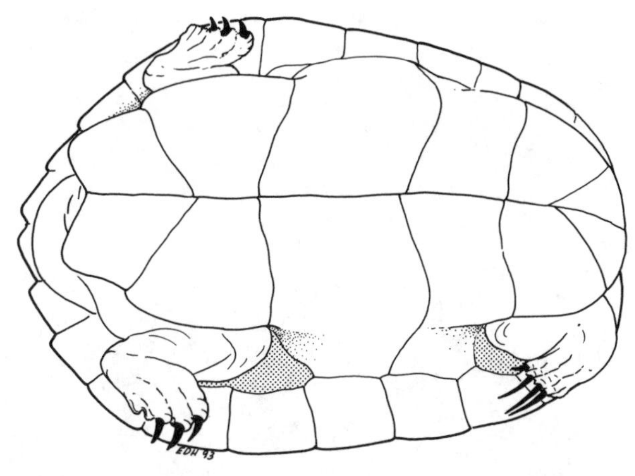
Figure 114. Ventral views of the plastrons of two genera of the family Emydidae. Upper: *Clemmys*, showing the presence of rudimentary axillary and inguinal patches. Lower: *Pseudemys*, showing the well-developed axillary and inguinal scutes (shaded). Adapted from Smith (1956).

1b. Hinge opposite 5th marginal; with at least a trace of a middorsal keel; interfemoral sulcus less than ½ as long as interabdominal (Fig. 119) .... ***T. carolina***
CC 161, pl 3, 5 **CAAR** 512 (1991)

Some authorities have suggested that *Terrapene carolina* constitutes a species complex.

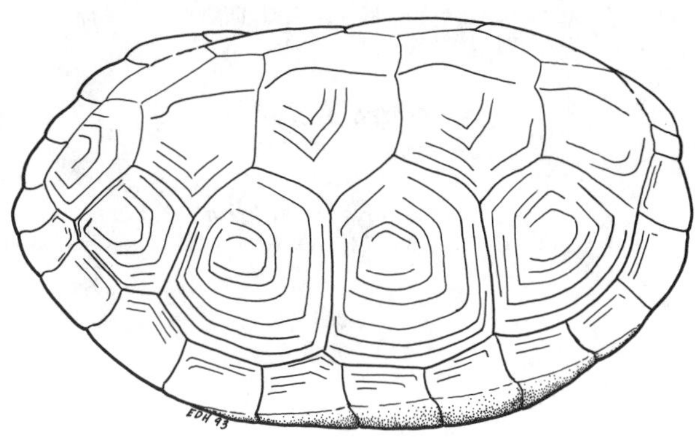
Figure 115. Dorsolateral view of the carapace of *Malaclemys terrapin*, showing concentric rings on the scutes. Drawn from a 35 mm color slide (KU 427).

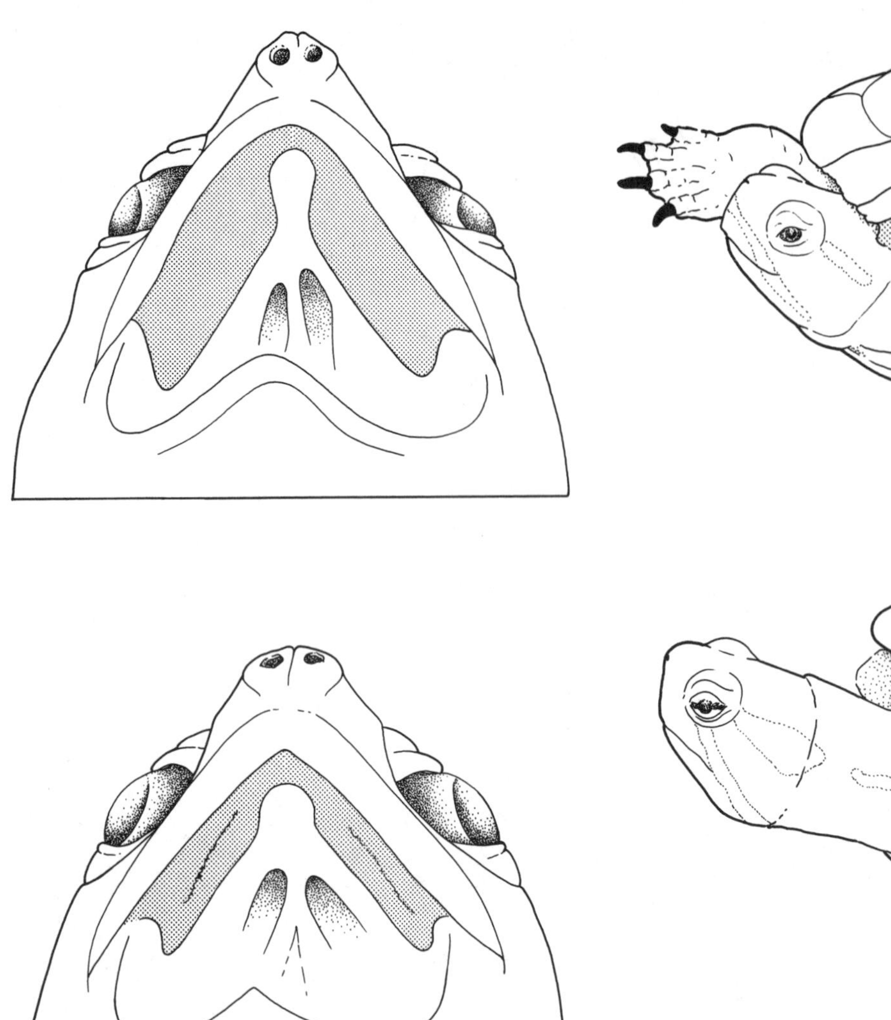

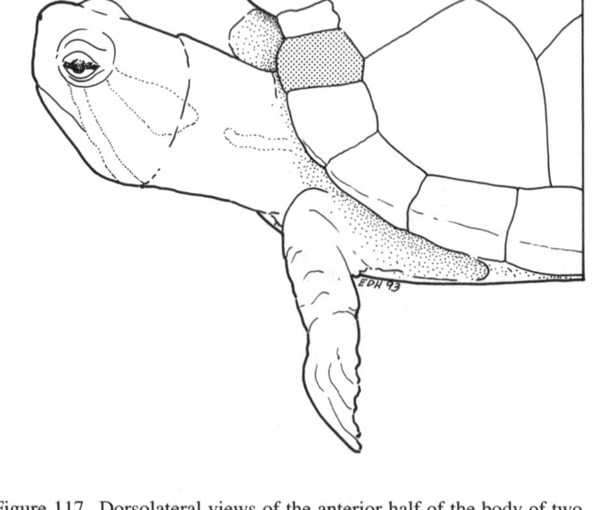

Figure 116. Views of the roofs of the mouth of two genera of the family Emydidae. Upper: *Graptemys pseudogeographica*, showing the broad crushing surface of the upper jaw (shaded). Drawn from a preserved specimen (KU 187862). Lower: *Chrysemys picta*, showing the narrow crushing surface of the upper jaw (shaded). Drawn from a preserved specimen (KU 177127).

Figure 117. Dorsolateral views of the anterior half of the body of two genera of the family Emydidae. Upper: *Chrysemys picta*, showing first marginal (shaded) not extending beyond suture between first vertebral and first costal. Lower: First marginal (shaded) extends beyond suture between first vertebral and first costal, typical of species of the genera *Pseudemys* and *Trachemys*.

Deirochelys

CAAR 107 (1971)

Deirochelys reticularia

CC 187, pl 4, 7 **CAAR** 107 (1971)

Clemmys

CAAR 203 (1977)

Key to Species of *Clemmys*

1a. With few dark markings on plastron; from the western U.S. .... ***C. marmorata***
S 59, pl 17 **CAAR** 100 (1970)

1b. With extensive dark markings on plastron; from the eastern and central U.S. 2

2a. Carapace dark with scattered yellow (in life) spots .... ***C. guttata***
CC 158, pl 3, 5 **CAAR** 124 (1972)

2b. Carapace without spots 3

3a. Prominent growth rings on dorsal scutes pyramidal; no blotch on temple .... ***C. insculpta***
CC 159, pl 3, 5 **CAAR** 125 (1972)

3b. Growth rings on dorsal scutes neither prominent nor pyramidal; with a large blotch on temple (Fig. 120) .... ***C. muhlenbergii***
CC 159, pl 5 **CAAR** 204 (1977)

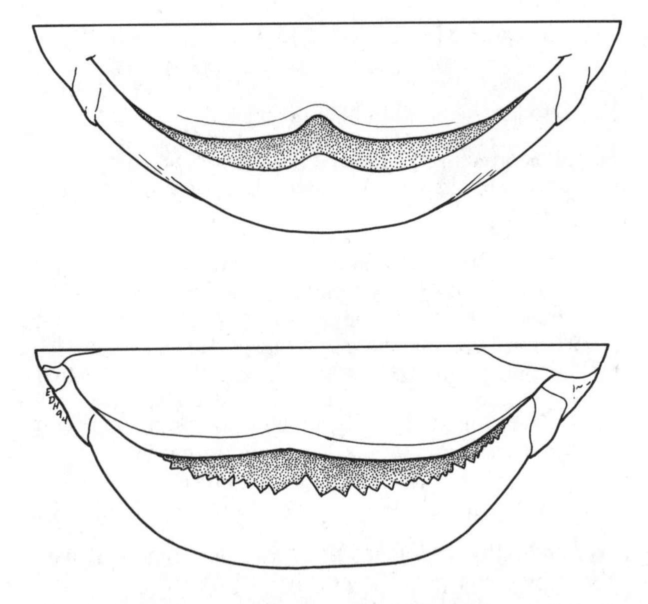

Figure 118. Frontal views of the lower jaws of two genera of the family Emydidae. Upper: *Trachemys*, showing the rounded underside of the lower jaw. Drawn from a preserved specimen (*T. scripta*, KU). Lower: *Pseudemys*, showing the flattened underside of the lower jaw. Drawn from a preserved specimen (*P. concinna*, KU 199735).

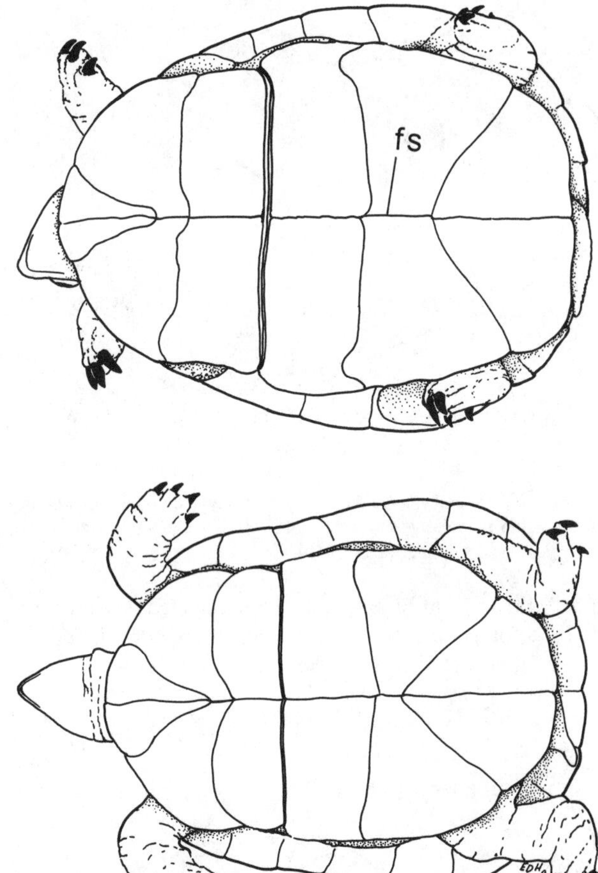

Figure 119. Ventral views of the plastron of two species of the genus *Terrapene*. Upper: *T. ornata*, showing interfemoral sulcus greater than half the length of the interabdominal sulcus anterior to it. Lower: *T. carolina*, showing interfemoral sulcus less than half the length of the interabdominal sulcus. Drawn from photographs in Smith (1956).

Emydoidea
CAAR 136 (1973)

Emydoidea blandingii
CC 188, pl 3, 5 **CAAR** 136 (1973)

Malaclemys
CAAR 299 (1982)

Malaclemys terrapin
CC 166, pl 4, 5 **CAAR** 299 (1982)

Graptemys
CAAR 584 (1994)

Key to Species of ***Graptemys***

1a. Vertebral projections not prominent, height of 2nd vertebral projection less than ⅙ length of scute (Fig. 121) 2

1b. Vertebral projections prominent, height of 2nd vertebral projection greater than ⅙ length of scute (Fig. 121) 7

2a. Small spot behind eye isolated and separated from eye by 2–3 diagonal lines (Fig. 122A) .... ***G. geographica***
CC 168, pl 3, 6 **CAAR** 484 (1990)

2b. No small isolated spot behind eye 3

3a. With distinctly convex costals; postorbital line or spot terminating above eye and extending back as a neck stripe (Fig. 122B); from the Colorado River system, Texas .... ***G. versa***
CC 172, pl 6, p 169 **CAAR** 280 (1981)

3b. Costal not distinctly convex; postorbital line not as above; range variable 4

4a. Postorbital spot large and C-shaped (Fig. 122C); with a dark-edged light bar or crescent across chin; from the San Antonio–Guadalupe River system, Texas .... ***G. caglei***
CC 172, p 169 **CAAR** 184 (1976)

4b. Without either a large C-shaped postorbital spot or a transverse bar or crescent across chin; range variable 5

5a. Chin with transverse light and dark bars .... ***G. ouachitensis*** (part)
CC 171, pl 6, p 171 (as *G. pseudogeographica sabinensis*) **CAAR** 603 (1995) (part)

5b. Chin without transverse light and dark bars 6

6a. Postorbital blotch narrow or forming a narrow crescent (Fig. 122D); up to six lines entering the orbit; no large spots on lower jaw ... ***G. pseudogeographica***
CC 171, pl 3, 6 (as *G. kohnii,* in part)

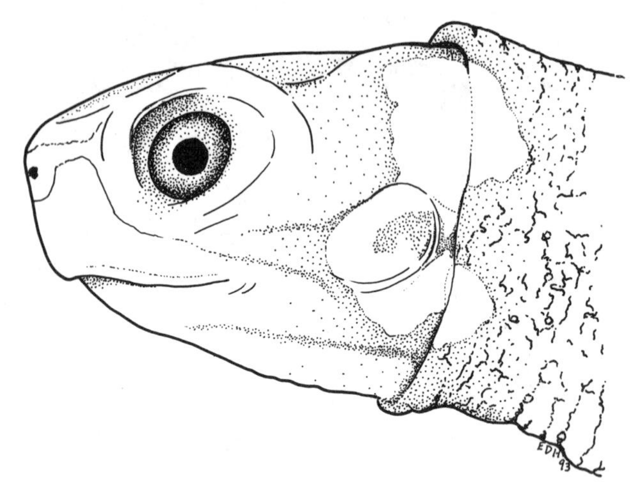

Figure 120. Lateral view of the head of *Clemmys muhlenbergii*, showing the location and relative size of the light-colored temporal blotch. Drawn from a photograph in Zappalorti (1976).

6b. Postorbital blotch wide, or connected to subocular spot to form a wide crescent, or no more than three lines entering the orbit; with four large spots on lower jaw (Fig. 122E) .... ***G. ouachitensis*** (part)
CC 171, pl 6, p 171 (as *G. pseudogeographica ouachitensis*)
CAAR 603 (1995) (part)

7a. With a large central light area on each costal; from the Pascagoula River system, Mississippi (Fig. 122F) .... ***G. flavimaculata***
CC 173, pl 6

Figure 121. Dorsolateral views of two species of the genus *Graptemys*, showing the location of the 2nd vertebral scute (shaded). Upper: *G. geographica*, with a low 2nd vertebral projection that is less than one-sixth the length of the scute. Drawn from a photograph by Suzanne L. Collins. Lower: *G. barbouri*, with a high 2nd vertebral scute that is greater than one-sixth the length of the scute. Adapted from a photograph in Ernst and Barbour (1972).

7b. No large central light area on each costal 8

8a. Postorbital spot large; no neck stripes reach eye 9

8b. Postorbital spot smaller; at least some neck stripes reach eye 12

9a. With a transverse light bar (often curved) on chin; light marks on marginals less than ½ width of marginals; from the Apalachicola River system (Florida panhandle and adjacent Georgia and Alabama) (Fig. 122G) .... ***G. barbouri***
CC 168, pl 3, 6, p 169

9b. With a longitudinal light bar on chin; light marks on marginals greater than ½ width of marginals; from extreme western Florida to extreme eastern Louisiana north into Mississippi and Alabama 10

10a. Interorbital blotches not connected or rarely, narrowly connected to postorbital blotches; nasal trident well developed; supraoccipital spots or expansions of dorsal paramedian neck stripes present (Fig. 122H) .... ***G. ernsti***
CAAR 585 (1994)

10b. Interorbital blotches connected to postorbital blotches; nasal trident present or not; supraoccipital spots or expansions of dorsal paramedian neck stripes rarely present 11

11a. Nasal trident usually present; single wide, yellow bar (greater than 15% scute width) on dorsal surfaces of marginal scutes (Fig. 122I) .... ***G. gibbonsi***
CC pl 6 (as *G. pulchra*) **CAAR** 585 (1994)

11b. Nasal trident usually absent; narrow concentric ocelli (less than 10% scute width) on dorsal surfaces of marginal scutes (Fig. 122J) .... ***G. pulchra***
CC 170, p 169
CAAR 360 (1985) (includes *G. ernsti* & *G. gibbonsi*)

12a. Vertebral spines broad, knob-like; postorbital line continuous with neck stripe (Fig. 122K); from the Alabama–Tombigbee–Black Warrior River system, Alabama and Mississippi .... ***G. nigrinoda***
CC 173, pl 6 **CAAR** 396 (1986)

12b. Vertebral spines not broad, knob-like; postorbital line not continuous with neck stripe (Fig. 122L); from the Pearl River system, Mississippi and Louisiana .... ***G. oculifera***
CC 173, pl 6

Figure 122 (facing page). Lateral and dorsolateral views of the head patterns of twelve species of the turtle genus *Graptemys*. (A) *G. geographica*. Drawn from a photograph by Suzanne L. Collins. (B) *G. versa*. Drawn from a photograph by Suzanne L. Collins. (C) *G. caglei*. Drawn from a photograph in Dixon (1987). (D) *G. pseudogeographica*. Drawn from a photograph by Suzanne L. Collins. (E) *G. ouachitensis*. Drawn from a photograph by Suzanne L. Collins. (F) *G. flavimaculata*. Drawn from a photograph by Suzanne L. Collins. (G) *G. barbouri*. Drawn from a photograph in Ernst and Barbour (1972). (H) *G. ernsti*. Drawn from a photograph by Suzanne L. Collins. (I) *G. gibbonsi*. Drawn from a photograph in Ashton and Ashton (1985). (J) *G. pulchra*. Drawn from a photograph in Behler and King (1979). (K) *G. nigrinoda*. Drawn from a photograph by Suzanne L. Collins. (L) *G. oculifera*. Drawn from a photograph by Suzanne L. Collins.

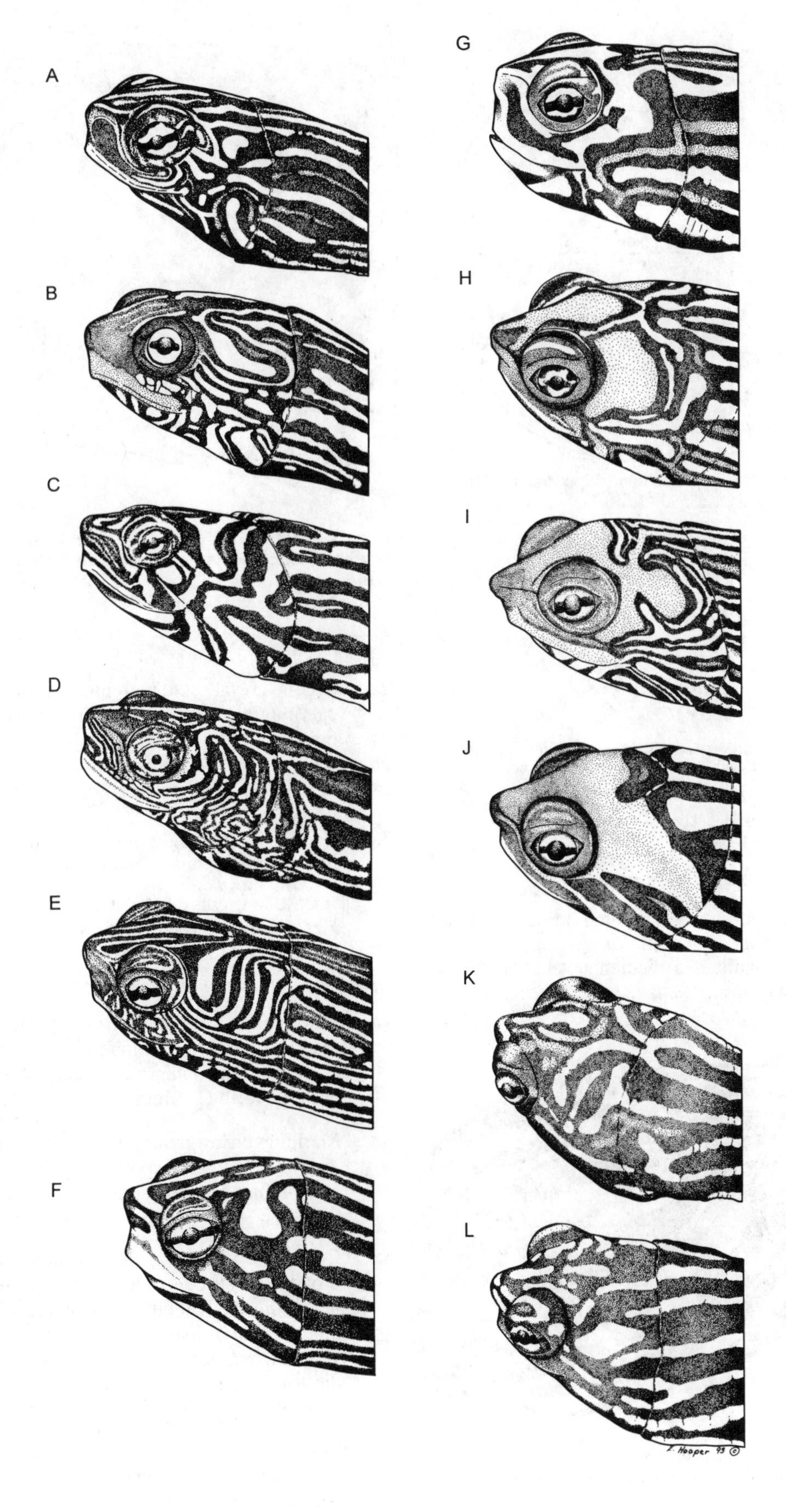
A
B
C
D
E
F
G
H
I
J
K
L
J. Hooper 93 ©

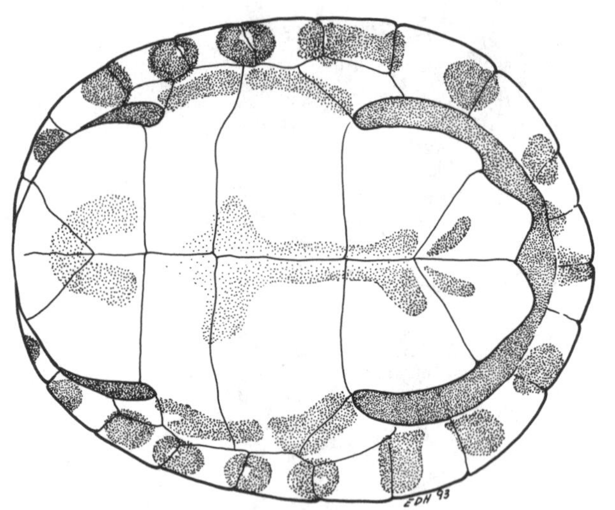

Figure 123. Ventral view of the plastron of *Pseudemys rubriventris*, showing distinctive ocelli on marginal scutes.

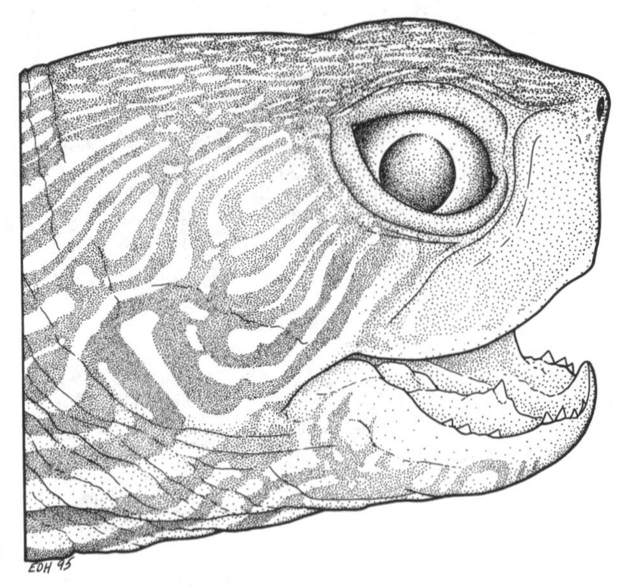

Figure 125. Lateral view of the head of *Pseudemys texana*, showing the broad light stripe extending posteroventrally from below the eye with a vertical branch extending dorsally, and the small postorbital blotch. Drawn from a preserved specimen (MES 1912).

Chrysemys

CAAR 438 (1988)

Chrysemys picta

CC 185, pl 4, 7 **S** 63, pl 18 **CAAR** 106 (1971)

Pseudemys

CAAR 625 (1996)

The genus *Pseudemys* contains species whose relationships are quite complex and whose traits often are exceedingly variable; therefore, these taxa are difficult to distinguish consistently across the full geographic range of the genus. Particular emphasis should be placed on geography in identifying these turtles.

Key to Species of *Pseudemys*

1a. Upper jaw with a median notch bordered laterally with prominent cusps 2
1b. Upper jaw without a median notch bordered by prominent cusps 6

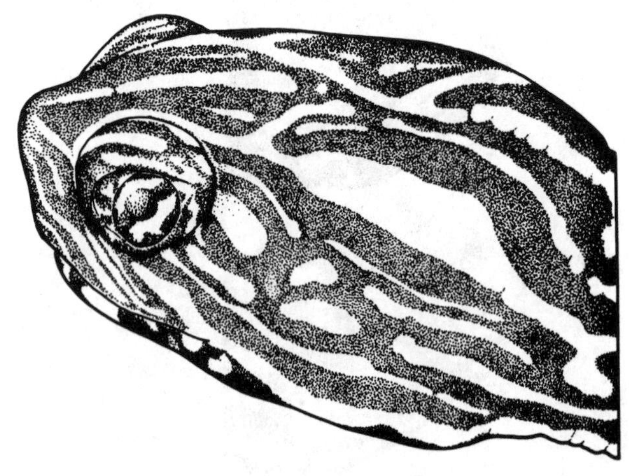

Figure 124. Dorsolateral view of the head pattern of *Pseudemys gorzugi*, showing the wide midsagittal stripe and distinct dark-bordered postorbital blotch.

2a. Second costal scute with blotches or a broad, vertical, light bar; plastral ground color (in life) usually orange or reddish 3
2b. Second costal scute with distinct, concentric light whorls; plastral ground color (in life) yellow 5

3a. Head with three light interorbital lines; paramedian lines ending behind eyes; from Florida .... ***P. nelsoni***
CC 183 pl 8, p 183 **CAAR** 210 (1978) (as *Chrysemys nelsoni*)
3b. Head with five light interorbital lines; paramedian lines extending between eyes; not from Florida 4

4a. Markings on lower marginals form ocelli (Fig. 123); markings on bridge oblong or concentric with light centers; from southern New Jersey south to northeastern North Carolina .... ***P. rubriventris***
CC 183, pl 7, p 183 **CAAR** 510 (1991)
4b. Markings on lower marginals solid; bridge unmarked or with a few large spots or bars; from the Mobile Bay drainage in Alabama .... ***P. alabamensis***
CC 183 **CAAR** 371 (1985)

5a. Centers of whorls on 2nd costal scute are light; a broad light stripe extending posteroventrally from below the eye without a vertical branch; postorbital blotch often a distinct dark-bordered ocellus (Fig. 124); the temporal stripe broadens anteriorly; from the Rio Grande and Pecos river systems in Texas, New Mexico, and adjacent México ...... ***P. gorzugi***
CC 179 (as *P. concinna gorzugi*) **CAAR** 461 (1990)
5b. Centers of whorls on 2nd costal scute are dark; a broad light stripe extending posteroventrally from below the eye, often with a vertical branch (Fig.

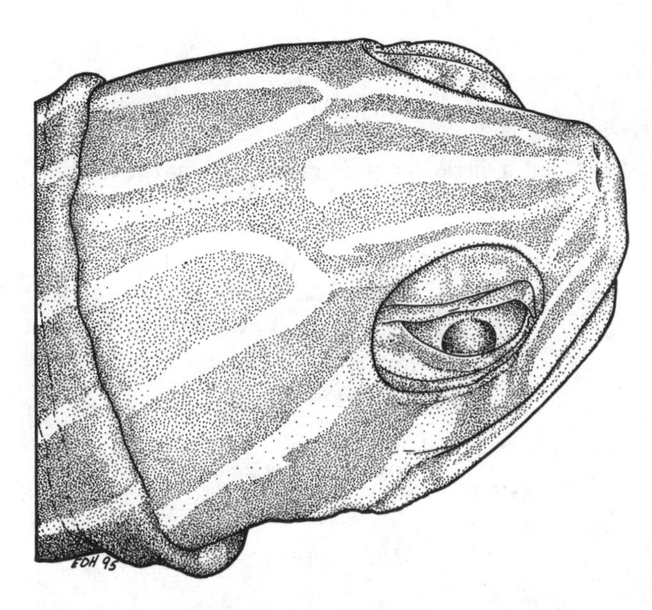

Figure 126. Dorsolateral view of the head of *Pseudemys peninsularis*, showing the "hairpin" pattern formed by the convergence of the supratemporal and paramedian light lines. Drawn from a preserved specimen (KU 69075).

125); the postorbital spot is small; from the Colorado, Brazos, Guadalupe, and San Antonio River systems in Texas .... ***P. texana***
CC 181, pl 8 **CAAR** 485 (1990)

6a. Supratemporal and paramedian lines on nape converge to form a *hairpin* pattern on the back of the head (lines on one or both sides may be incomplete) (Fig. 126); from peninsular Florida .... ***P. peninsularis***
CC 181, pl 4, 8 (as *P. floridana peninsularis*)

6b. Without a *hairpin* pattern on the back of the head; from peninsular Florida or not 7

7a. Plastron unmarked; second costal scute with vertical light lines; range excludes peninsular Florida .... ***P. floridana***
CC 181, pl 8 **CAAR** 626 (as *P. concinna floridana*)

7b. Plastron with dark markings; second costal scute with a pale C-shaped mark; range may include peninsular Florida 8

Some authorities believe that *P. floridana* and *P. concinna* constitute a single polymorphic species.

8a. Range excludes peninsular Florida; light C-shaped mark on second costal scute indistinct .... ***P. concinna***
CC 179, pl 8 **CAAR** 626 (as *P. concinna concinna*)

8b. From peninsular Florida (Suwannee River region of the upper Florida Gulf Coast south to the vicinity of Tampa Bay); light C-shaped mark on second costal scute distinct .... ***P. suwanniensis***
CC 179, pl 8 **CAAR** 626 (as *P. concinna suwanniensis*)

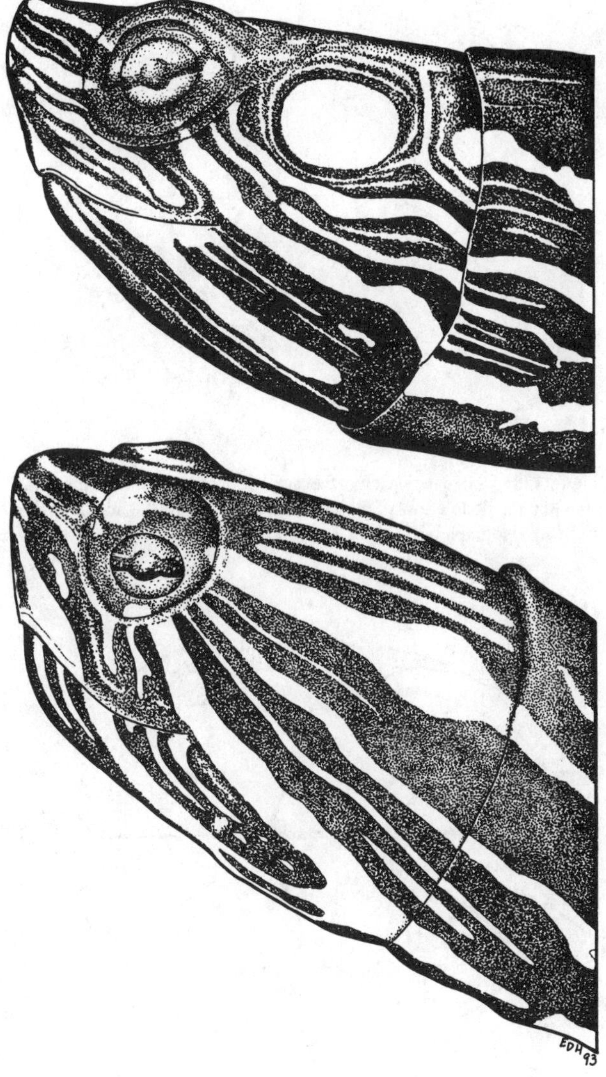

Figure 127. Lateral views of the head patterns of two species of the genus *Trachemys*. Upper: *T. gaigeae*. Drawn from a photograph in Pritchard (1979). Lower: *T. scripta*. Drawn from a photograph in Zappalorti (1976).

Trachemys

Key to Species of *Trachemys*

1a. With a small postorbital spot followed by a large dark-bordered spot (orange in life) (Fig. 127), usually evident even in melanistic individuals; from the Rio Grande system in New Mexico and the Big Bend region of Texas and from the Río Conchos system in Chihuahua, México .... ***T. gaigeae***
CC 177, p 171 **CAAR** 538 (1992)
S 64, pl 18 (*as Pseudemys scripta gaigeae*)

1b. Usually with a broad postorbital line or bar (Fig. 127) (often obscure in melanistic specimens); range excludes the upper Rio Grande system in New Mexico, western Texas, and adjacent México .... ***T. scripta***
CC 177, pl 4, 7 **S** 64, pl 18 (as *Pseudemys scripta*)

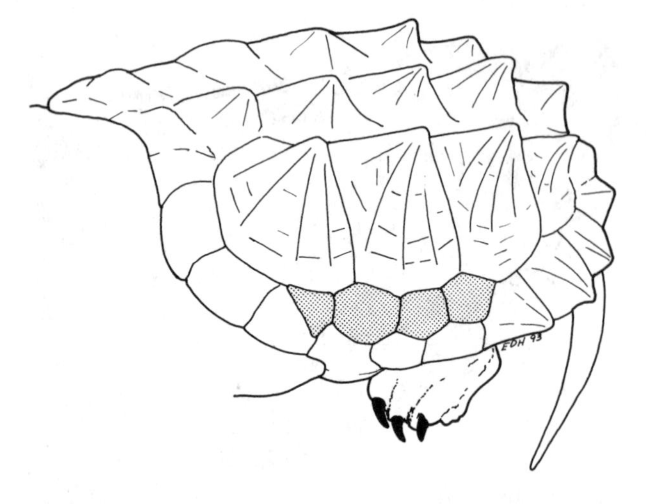

Figure 128. Dorsolateral view of the carapace of *Macroclemys temminckii*, showing the double row of marginal scutes (upper row shaded). Drawn from a photograph by Suzanne L. Collins.

CHELYDRIDAE

Key to Genera of Chelydridae

1a. With a single row of marginals; tail with two rows of large scales underneath and one series of tubercles above .... ***Chelydra*** (p. 62)

1b. Fifth to eighth marginals doubled; tail with many small scales underneath and three series of tubercles above .... ***Macrochelys*** (p. 62)

Chelydra

CAAR 419 (1988)

Chelydra serpentina

CC 147, pl 3, 9 **S** 57, pl 17 **CAAR** 420 (1988)

Macrochelys

Macrochelys temminckii

CC 148, pl 3, 9 (as *Macroclemys temminckii*)

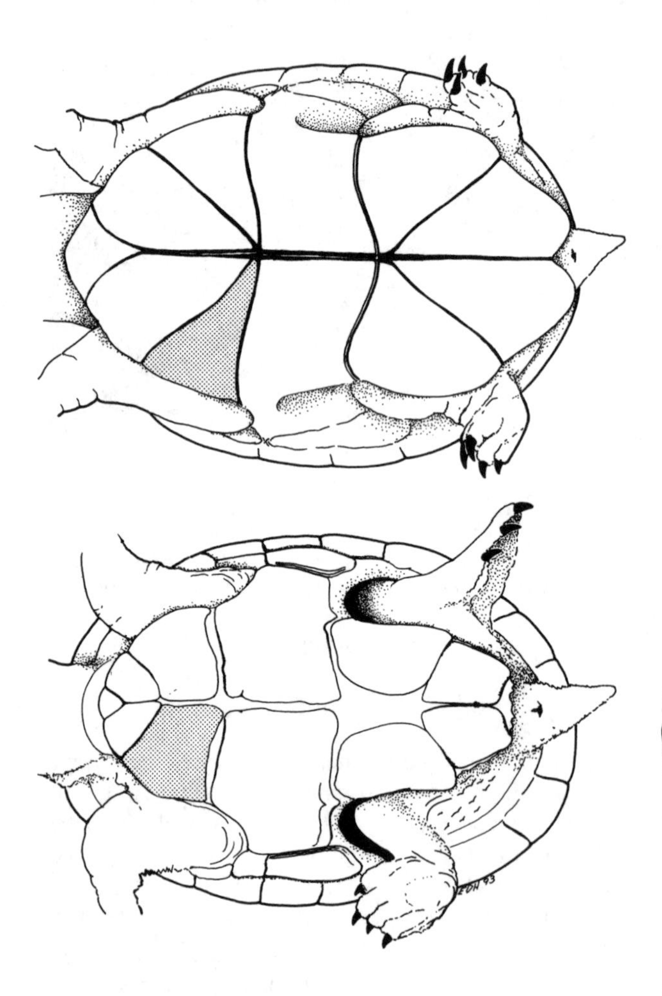

Figure 129. Ventral views of the plastron of two genera of the family Kinosternidae. Upper: *Kinosternon subrubrum*, showing the triangular pectoral scute (shaded). Lower: *Sternotherus odoratus*, showing the four-sided pectoral scute (shaded).

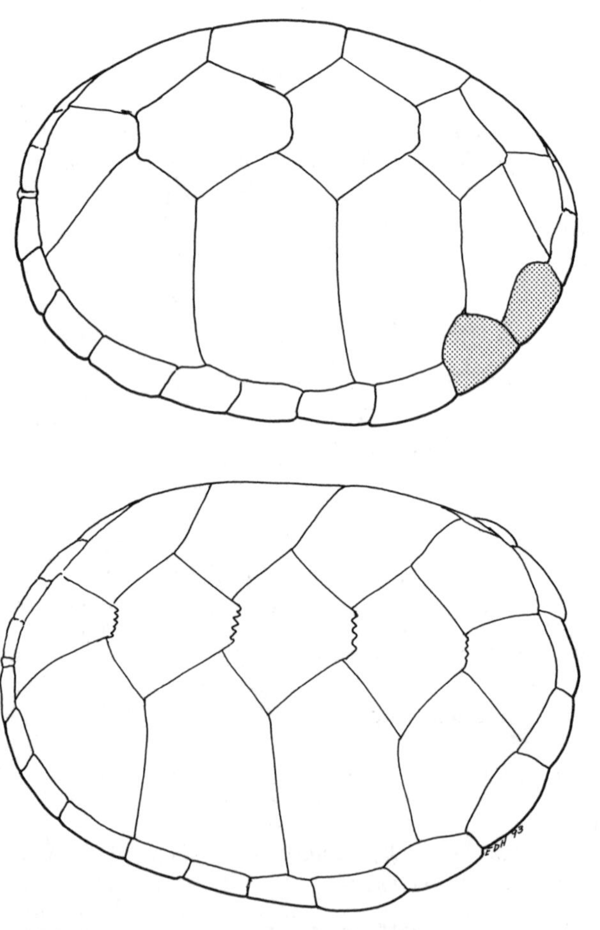

Figure 130. Dorsolateral views of the carapace of two species of the genus *Kinosternon*. Upper: *K. flavescens*, showing the upward extension of the 9th and 10th marginals (shaded). Drawn from a 35 mm color slide (KU 13). Lower: *K. subrubrum*, showing the lack of an upward extension of the 9th and 10th marginals (shaded), typical of other members of the genus. Drawn from a 35 mm color slide (KU 157).

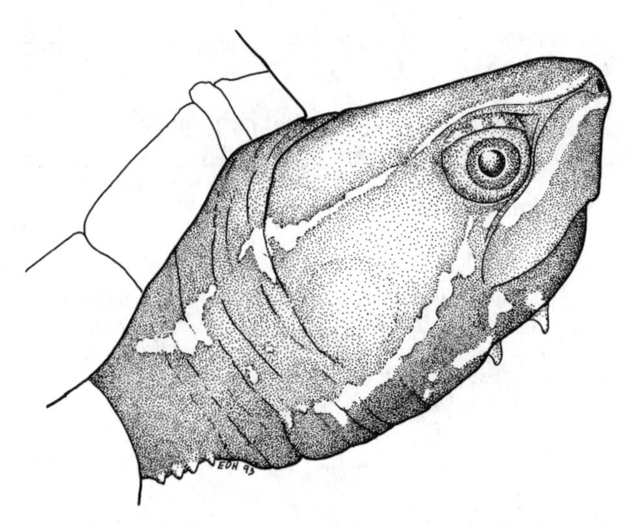

Figure 131. Lateral view of the head of *Sternotherus odoratus*, showing the two light lines on the side of the head and the presence of barbels on the chin and throat. Drawn from a photograph in Zappalorti (1976).

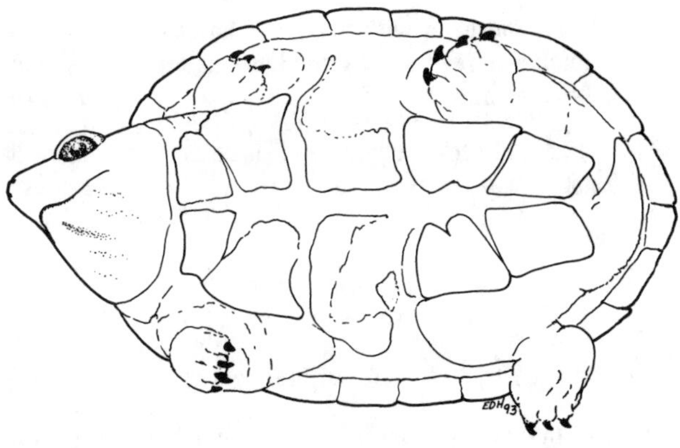

Figure 132. Ventral view of the plastron of *Sternotherus carinatus*, showing lack of a gular scute. Drawn from a photograph in Ernst and Barbour (1972).

KINOSTERNIDAE

Key to Genera of Kinosternidae

1a. Length of interfemoral suture less than length of interhumeral suture; pectoral triangular; with little uncornified skin along median longitudinal suture (Fig. 129)............................. ***Kinosternon*** (p. 63)

1b. Length of interfemoral suture at least equal to length of interhumeral; pectoral four-sided; usually with considerable uncornified skin along median longitudinal suture (Fig. 129).......... ***Sternotherus*** (p. 64)

Kinosternon

Key to Species of *Kinosternon*

A recent study has suggested that the genus *Sternotherus* should be in the synonymy of *Kinosternon*. Those authorities accepting this treatment would recognize only one genus north of México, and all species listed below under *Sternotherus* would be relegated to *Kinosternon*.

1a. Often with three longitudinal light lines on carapace; two longitudinal light lines present on each side of head; from peninsular Florida and the Coastal Plain of Georgia and extreme southern South Carolina .. .. ***K. baurii***
CC 155, pl 2 **CAAR** 161 (1974)

1b. Without longitudinal light lines on carapace or head; if light lines present on head (but not on carapace), not from peninsular Florida or the Coastal Plain of Georgia and extreme southern South Carolina 2

2a. Ninth and tenth marginals extend upward (Fig. 130) .... ***K. flavescens***
CC 157, pl 2, p 154, 156 **S** 62, pl 17 **CAAR** 216 (1978)

Some authorities have suggested that *Kinosternon flavescens* consists of three species, *K. flavescens*, *K. arizonense*, and *K. spooneri*.

2b. Ninth marginal does not extend upward (Fig. 130) 3

3a. First vertebral not in contact with second marginals; 10th marginal not much higher than 9th (Fig. 130) .... ***K. subrubrum***
CC 155, pl 2, 3, p 154 **CAAR** 193 (1977)

3b. First vertebral in contact with second marginals; 10th marginal distinctly higher than 9th (Fig. 130) 4

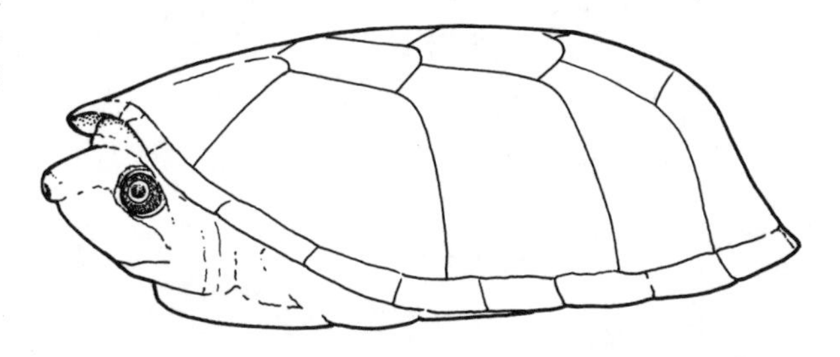

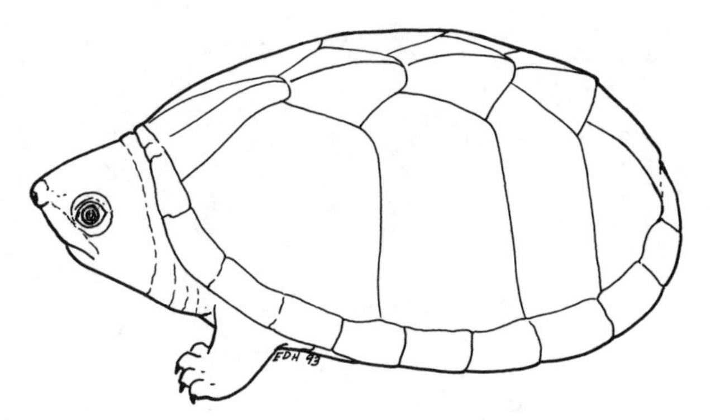

Figure 133. Dorsolateral views of two species of the genus *Sternotherus*. Upper: *S. depressus*, showing the flattened carapace without keels. Drawn from a photograph in Ernst and Barbour (1972). Lower: *S. minor*, showing the raised carapace with a middorsal keel. Drawn from a photograph in Niering (1985).

4a. Head and neck with fine reticulations; head shield notched at rear; shell arched, not flattened (less than twice as high as wide) .... ***K. hirtipes***
CC 157, pl 2, p 154 **CAAR** 361 (1985)

4b. Head and neck mottled; head shield not notched; shell flattened (twice as wide as high) .... ***K. sonoriense***
S 61, pl 17 **CAAR** 176 (1976)

Sternotherus

CAAR 397 (1986)

Key to Species of ***Sternotherus***

1a. With two light lines on sides of head; barbels on chin and throat (Fig. 131) .... ***S. odoratus***
CC 151, pl 2, 3 **CAAR** 287 (1982)

1b. Without light lines on sides of head; barbels on chin only 2

2a. Gular absent or nearly so (Fig. 132); with a prominent middorsal keel .... ***S. carinatus***
CC 151, pl 2, p 152, 156 **CAAR** 226 (1979)

2b. Gular present and obvious (Fig. 129); keel variable (rarely prominent) 3

3a. Carapace low and flat (more than twice as wide as high); no middorsal keel (Fig. 133) ... ***S. depressus***
CC 151, pl 2, p 152 **CAAR** 194 (1977)

3b. Carapace rounded, not low and flat; with a sharp middorsal keel or three keels (Fig. 133) .... ***S. minor***
CC 153, pl 2, 3, p 152 **CAAR** 195 (1977)

CROCODYLIA
(Alligators and Crocodiles)

KEY TO FAMILIES OF CROCODYLIA

1a. Snout broad (Fig. 134); 4th tooth not visible when the mouth is closed .... ALLIGATORIDAE (p. 65)
1b. Snout narrower and tapered (Fig. 134); 4th tooth visible when the mouth is closed .... CROCODYLIDAE (p. 65)

Some authorities recognize only one family and categorize the above as subfamilies.

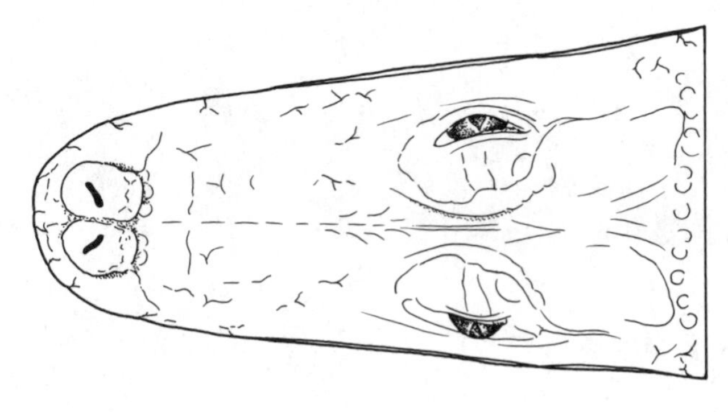

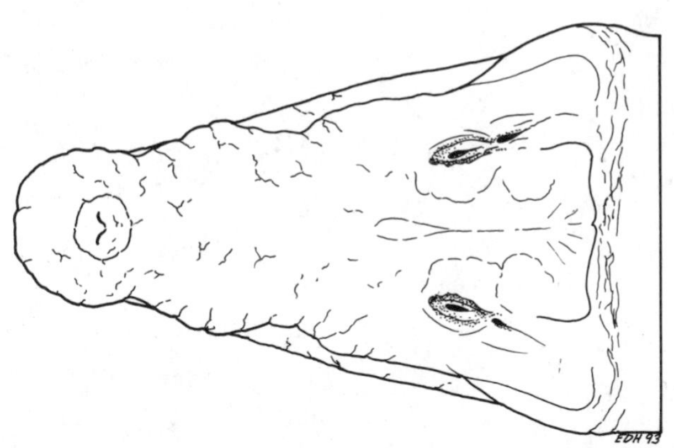

Figure 134. Dorsal views of the snouts of two families of crocodilians. Upper: Family Alligatoridae, showing the broad snout of *Alligator mississippiensis*. Redrawn from a photograph in Zappalorti (1976). Lower: Family Crocodylidae, showing the narrow snout of *Crocodylus acutus*.

ALLIGATORIDAE

Key to Genera of Alligatoridae

1a. With a curved ridge between the anterior edges of orbits (Fig. 135) .... ***Caiman*** (p. 65)
1b. No curved ridge between anterior edges of orbits .... ***Alligator*** (p. 65)

Caiman

***Caiman crocodilus* (A)**
CC 145, pl 1

Alligator

Alligator mississippiensis
CC 143, pl 1 **CAAR** 600 (1994)

CROCODYLIDAE

Crocodylus

Crocodylus acutus
CC 143, pl 1

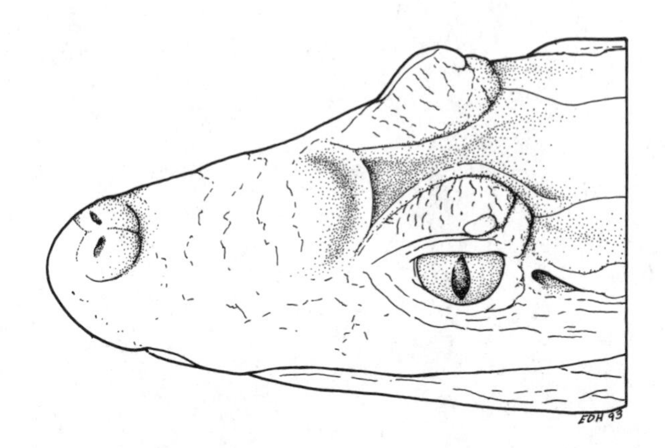

Figure 135. Dorsolateral view of the head of *Caiman crocodilus*, showing the presence of a curved ridge between the anterior edges of the orbits. Drawn from a photograph in Zappalorti (1976).

SQUAMATA
(Amphisbaenians, Lizards, and Snakes)

KEY TO SUBORDERS OF SQUAMATA

Because recent systematic studies have shown conclusively that "snakes" are nested within various groups historically referred to as "lizards," the traditional classification of suborders is not an accurate reflection of their relationship. We retain them solely for the sake of convenience.

1a. Rings of scales encircling body; eyes covered by scales; without limbs or external ears (Fig. 136) ... . **Amphisbaenia** (p. 67)
1b. Scales not forming rings which encircle the body; eyes not covered by scales; limbs and external ears present or absent 2

2a. With at least one of the following: limbs, moveable eyelids, external ears .... **Sauria** (p. 67)
2b. Lacking limbs, moveable eyelids, and external ears .... **Serpentes** (p. 99)

AMPHISBAENIA
(Amphisbaenians)

Amphisbaenidae

Rhineura
CAAR 42 (1967)

Rhineura floridana
CC 281, pl 17 **CAAR** 43 (1967)

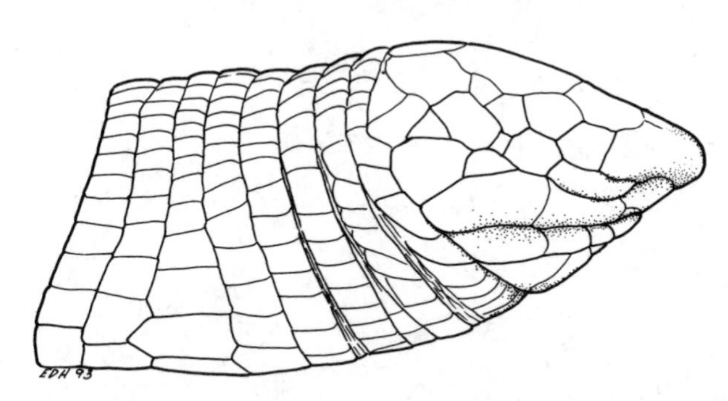

Figure 136. Lateral view of the head and neck of *Rhineura floridana*, showing rings of scales encircling the body, scales covering the eye, and the lack of an external ear. Drawn from a preserved specimen (KU 61842).

SAURIA
(Lizards)

KEY TO FAMILIES OF SAURIA

1a. Limbs absent .... **Anguidae** (part) (p. 69)
1b. Two pairs of limbs present 2

2a. Digits fused into sets of two and three, forming opposable, mittenlike hands and feet (Fig. 138) .... **Chamaeleonidae** (p. 70)
2b. Digits not fused as above 3

3a. Scales around body relatively large, about equal in size and smooth (shiny) (Fig. 139) .... **Scincidae** (p. 71)
3b. Some scales small or not smooth (Fig. 139), or scales around body not equal in size 4

4a. With a lateral skin fold with very small scales (distinctly smaller than either dorsal or ventral scales) (Fig. 140) .... **Anguidae** (part) (p. 69)
4b. Without a lateral skin fold with very small scales 5

5a. Eyelids absent 6
5b. Moveable eyelids present.................................... 7

6a. Ventral scales rectangular (Fig. 141); 1–2 scales between eyes .... **Xantusiidae** (p. 77)
6b. Ventral scales rounded (Fig. 141); more than ten scales between eyes............... **Gekkonidae** (p. 78)

7a. Head covered with small scales about equal in size 8
7b. Head scales not equal in size 9

8a. Dorsal scales beadlike (Fig. 142); ventral scales squarish and arranged in transverse rows .... **Helodermatidae** (p. 81)
8b. Dorsal scales not beadlike; ventral scales not squarish or arranged in transverse rows .... **Eublepharidae** (p. 77)

9a. Ventral scales quadrangular, arranged in ≤ twelve longitudinal rows (Fig. 143) 10
9b. Ventral scales not quadrangular, arranged in more than twelve rows across the venter (Fig. 143) ... 11

10a. With one large preanal scale (Fig. 144) .... **Lacertidae** (p. 81)
10b. With several enlarged preanal scales (Fig. 144) .... **Teiidae** (p. 82)

The following couplets diagnose families that are grouped into the single family Iguanidae by some authorities.

11a. Toepads expanded (Fig. 145)................................ .... **Polychrotidae** (p. 88)

Note that some species of the family Gekkonidae may also have expanded toepads. These have been previously identified on the basis of lacking eyelids.

11b. Toepads not expanded...................................... 12

12a. Femoral pores present (Fig. 146) 13
12b. Femoral pores absent 15

13a. Middorsal scale row enlarged or, if not, rostral scale divided (Fig. 147) .... **Iguanidae** (p. 90)
13b. Middorsal scale row not enlarged (several rows of enlarged middorsal scales may be present); rostral scale not divided 14

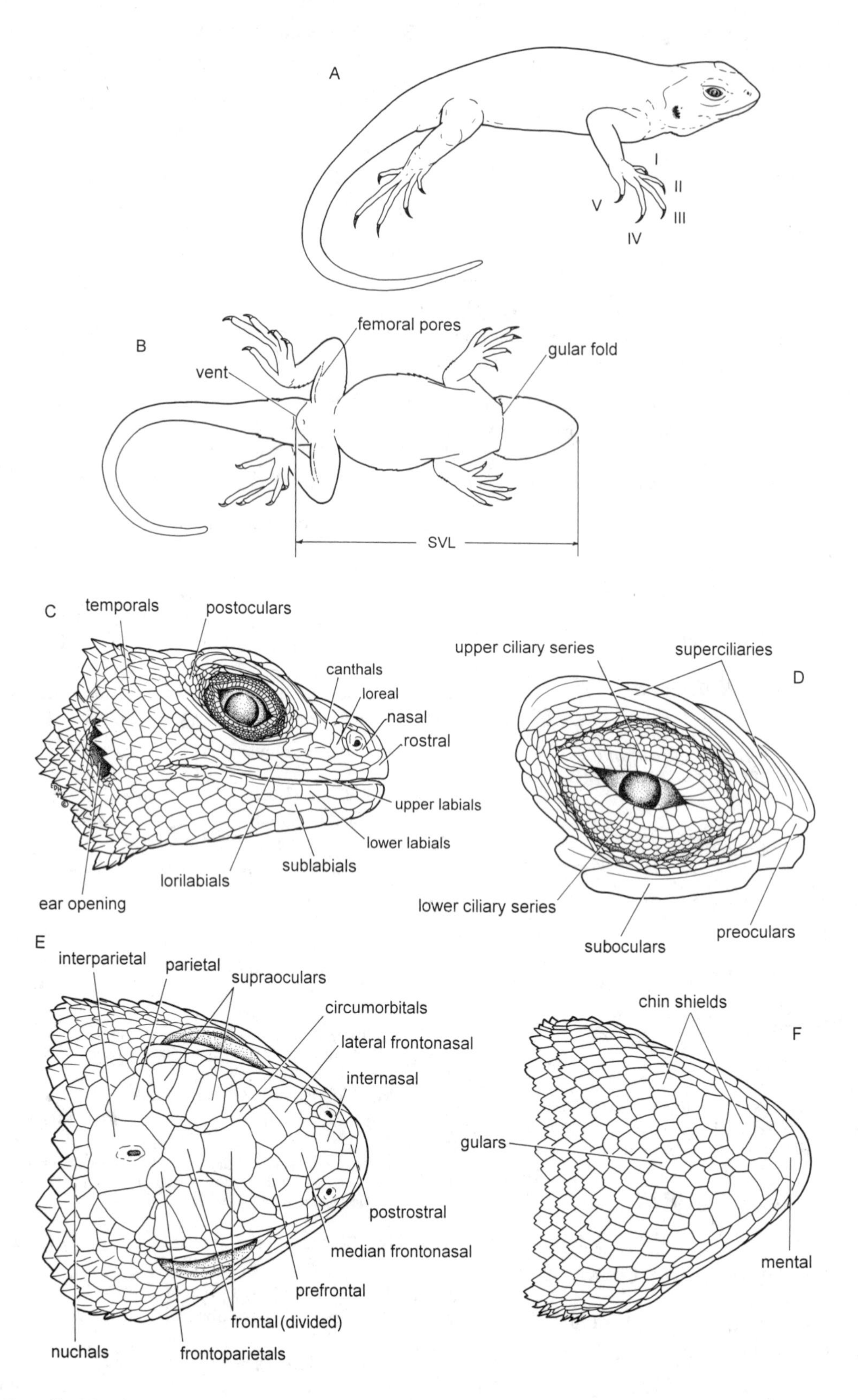

Figure 137. Generalized drawings of the shape of a lizard (upper) and detailed drawings of the head of a lizard (lower), identifying features and scales used in this key. Drawn from a preserved specimen of *Sceloporus undulatus* (KU).

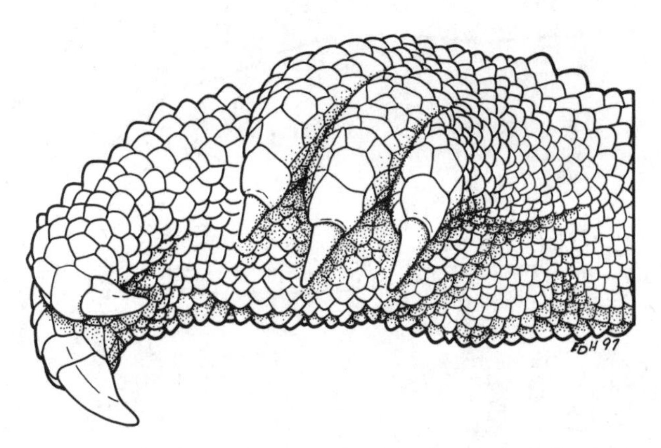

Figure 138. Ventrolateral view of the right front foot of *Chamaeleo jacksonii*, showing digits fused into sets of two and three. Drawn from preserved specimen (BWMC 1248).

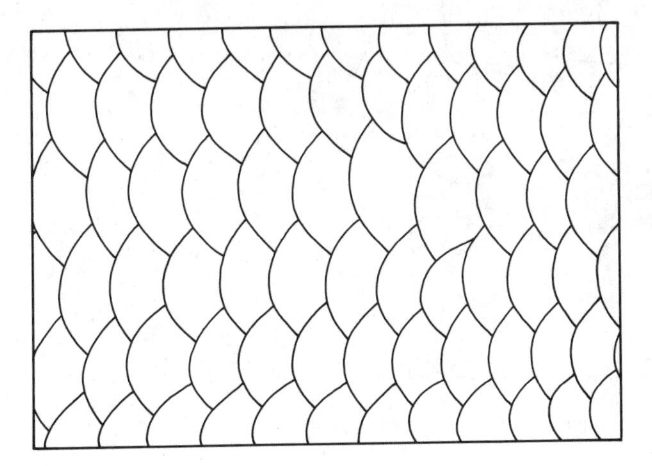

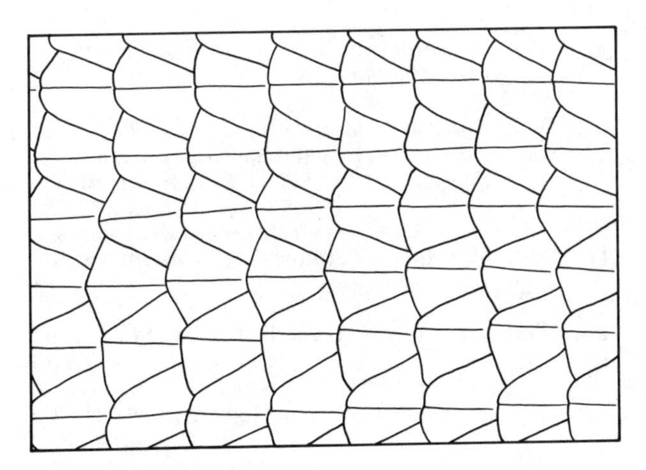

Figure 139. Drawings of dorsal body scales of lizards. Upper: Smooth scales. Drawn from a preserved specimen of *Eumeces gilberti* (KU 211266). Lower: Keeled scales. Drawn from a preserved specimen of *Elgaria multicarinata* (KU 23106).

14a. Interparietal scale large (nearly as large as or larger than ear opening or no ear opening present) or, if interparietal scale smaller than ear opening, with bony spines or a projecting ridge on the head or with scales projecting over the ear openings and with scales forming a distinct fringe on sides of toes (Fig. 148) .... **PHRYNOSOMATIDAE** (p. 91)
14b. Interparietal scale small (distinctly smaller than ear opening); head without bony spines or a projecting ridge; no scales projecting over the ears; no scales forming a prominent fringe on sides of toes .... **CROTAPHYTIDAE** (p. 96)

15a. Gular fold complete (Fig. 149) .... **CORYTOPHANIDAE** (p. 97)
15b. Gular fold incomplete medially (Fig. 149) .... **TROPIDURIDAE** (p. 97)

ANGUIDAE

KEY TO GENERA OF ANGUIDAE

1a. Limbs absent 2
1b. Two pairs of limbs present.................................. 3

2a. External ear openings present .... ***Ophisaurus*** (p. 69)
2b. External ear openings absent .... ***Anniella*** (p. 70)

Some authorities recognize a separate monotypic family Anniellidae.

3a. Median postrostral scale present; nasal scales separated from the rostral scale (Fig. 150)................... ....***Gerrhonotus*** (p. 70)
3b. Median postrostral scale absent; nasal scales in contact with the rostral scale (Fig. 150)...................... .... ***Elgaria*** (p. 70)

Ophisaurus

CAAR 110 (1971)

Key to Species of *Ophisaurus*

1a. One or two upper labial scales in contact with the orbit (or separated from it by very small scales) (Fig. 151) 2
1b. Upper labial scales separated from the orbit by a row of enlarged lorilabial scales (Fig. 151) 3

2a. Several longitudinal dark stripes present above the lateral fold and on much of the tail ..... ***O. mimicus***
CC 278, p 276 **CAAR** 543 (1992)
2b. Usually only one longitudinal dark stripe present above the lateral fold and on the tail .... ***O. compressus***
CC 278, p 276 **CAAR** 113 (1971)

3a. Middorsal stripe present; dark markings present below the lateral fold on the neck........ ***O. attenuatus***
CC 277, pl 17, p 276 **CAAR** 111 (1971)

Some authorities have suggested that this taxon consists of two species, *O. attenuatus* and *O. longicaudus*.

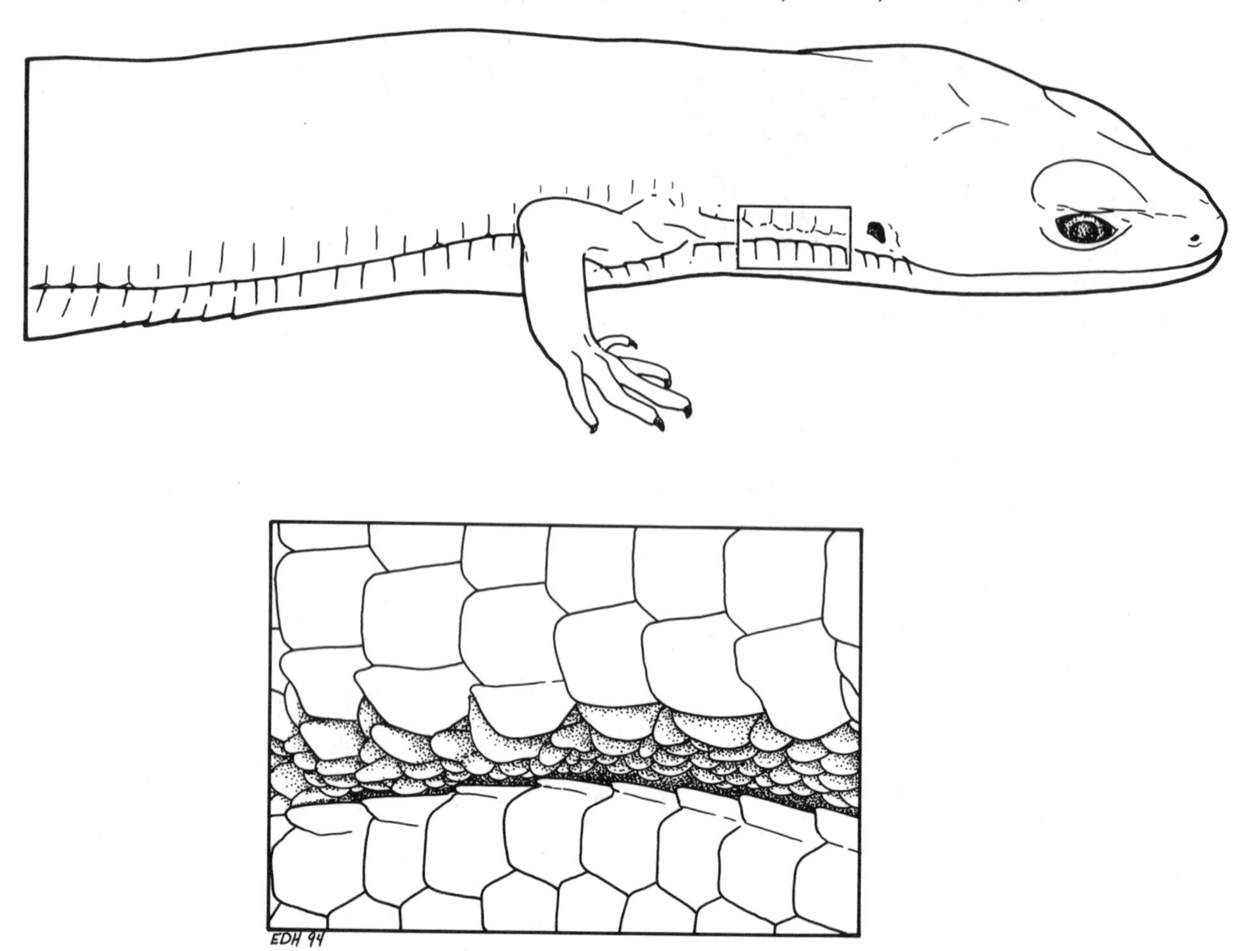

Figure 140. Lateral views of the head, body, and forelimb of a member of the family Anguidae, with an inset, showing tiny scales within the lateral skin fold. Drawn from a preserved specimen of *Elgaria kingii* (KU 6972).

3b. Middorsal stripe absent; no dark markings below the lateral fold on the neck .... ***O. ventralis***
CC 277, pl 17, p 276 **CAAR** 115 (1971)

Anniella

Anniella pulchra
S 122, pl 34

Gerrhonotus

Gerrhonotus infernalis
CC 279, pl 17 (as *G. liocephalus*)

Elgaria

Key to Species of ***Elgaria***

1a. With 16–17 dorsal scale rows; crossbands absent, indistinct, or broken; ventral scales of tail keeled .... ***E. coerulea***
S 121, pl 29 **CAAR** 178 (1976) (as *Gerrhonotus coeruleus*)

1b. With fourteen dorsal scale rows; crossbands distinct; ventral scales of tail not keeled 2

2a. Dark crossbands narrow (less than 2 scale rows wide); young with longitudinal stripes .... ***E. multicarinata***
S 119, pl 29 **CAAR** 187 (1976) (as *Gerrhonotus multicarinatus*)

2b. Dark crossbands wide (more than 2 scale rows wide); young with crossbands 3

3a. With contrasting dark and light spots on upper lip .... ***E. kingii***
S 119, pl 29 **CAAR** 97 (1970) (as *Gerrhonotus kingii*)

3b. Without contrasting dark and light spots on upper lip .... ***E. panamintina***
S 119, pl 29 **CAAR** 629 (1996) (as *Gerrhonotus panamintinus*)

CHAMAELEONIDAE

Chamaeleo

Chamaeleo jacksonii (A)

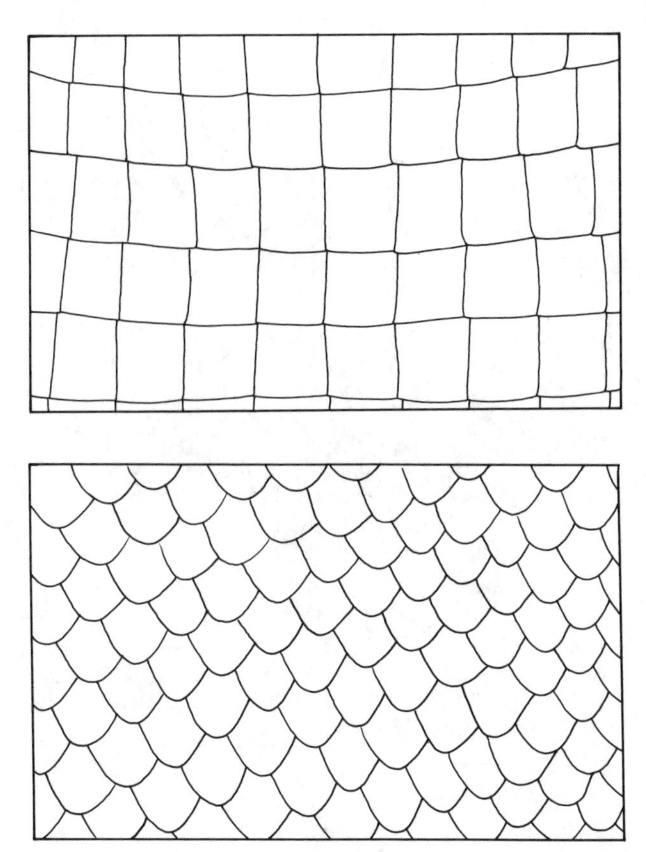

Figure 141. Drawings of ventral scales of lizards. Upper: Scales rectangular. Drawn from a preserved specimen of *Xantusia vigilis* (KU 154129). Lower: Scales rounded. Drawn from a preserved specimen of *Coleonyx brevis* (KU 145773).

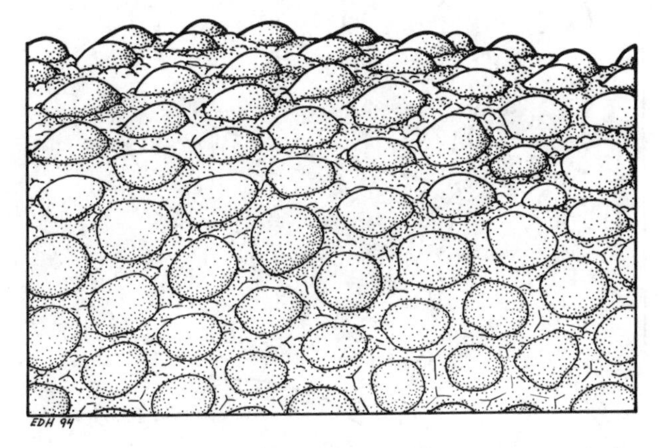

Figure 142. Lateral view, showing the beadlike dorsal body scales of a member of the family Helodermatidae. Drawn from a preserved specimen of *Heloderma suspectum* (KU 152635).

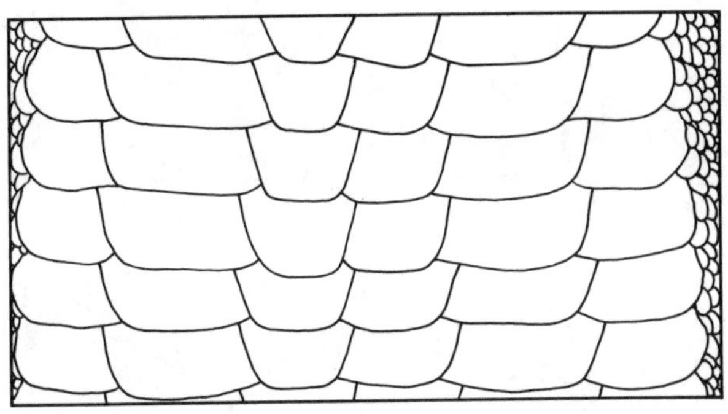

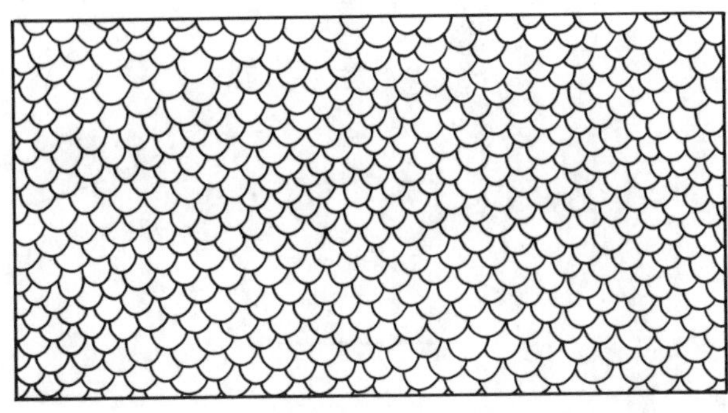

Figure 143. Ventral view of bellies of lizards. Upper: Scales quadrangular and arranged in rows along the long axis of the body, typical of the families Lacertidae and Teiidae. Drawn from a preserved specimen of *Podarcis sicula* (KU 190469). Lower: Scales neither quadrangular nor arranged in longitudinal rows, typical of other families of iguanian lizards. Drawn from a preserved specimen of *Anolis carolinensis* (KU 153684).

Scincidae

Key to Genera of Scincidae

1a. Limbs very short, with no more than two digits .... ***Neoseps*** (p. 71)
1b. Limbs long or short, with five digits 2

2a. Supranasal scales absent (Fig. 152); lower eyelid with a translucent window (Fig. 153) .... ***Scincella*** (p. 71)
2b. Supranasal scales present (Fig. 152); lower eyelid scaled (Fig. 153) .... ***Eumeces*** (p. 74)

Neoseps

CAAR 80 (1969)

Neoseps reynoldsi
CC 274, pl 19 **CAAR** 80 (1969)

Scincella

Scincella lateralis
CC 263, pl 19 **CAAR** 169 (1975)

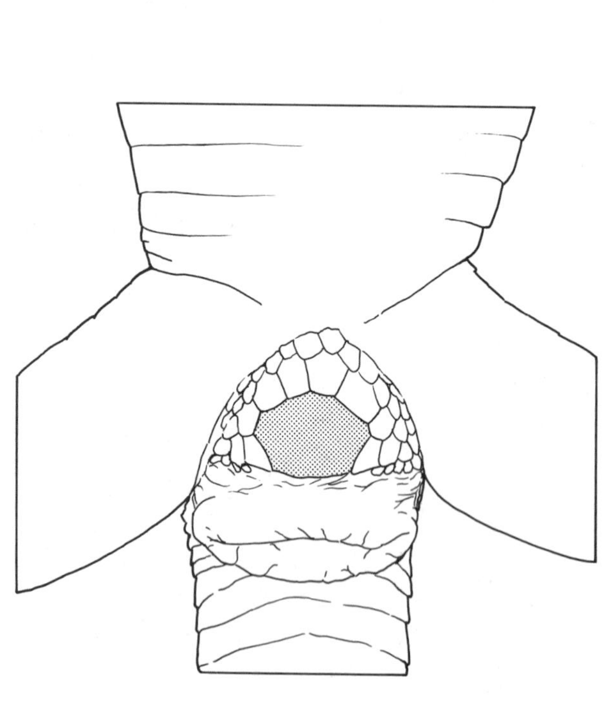

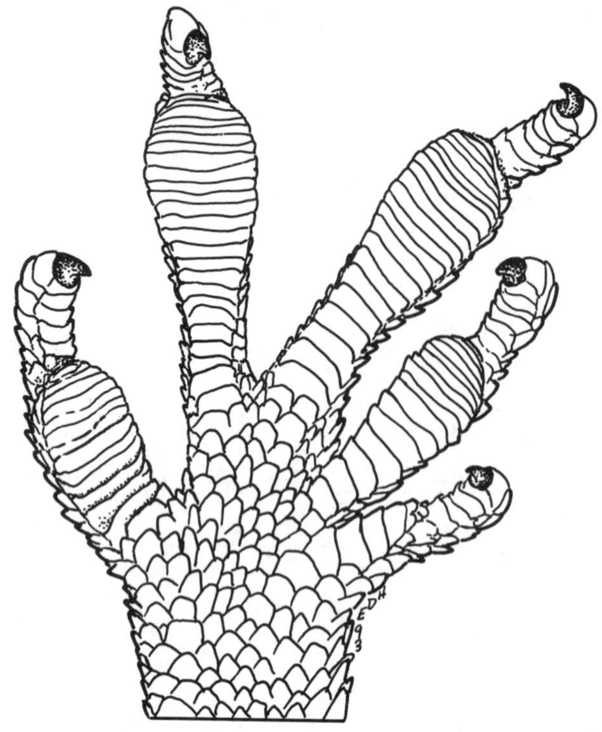

Figure 145. Ventral view of the foot of *Anolis carolinensis,* showing the expanded toe pads typical of the family Polychrotidae. Drawn from a preserved specimen (KU 153684).

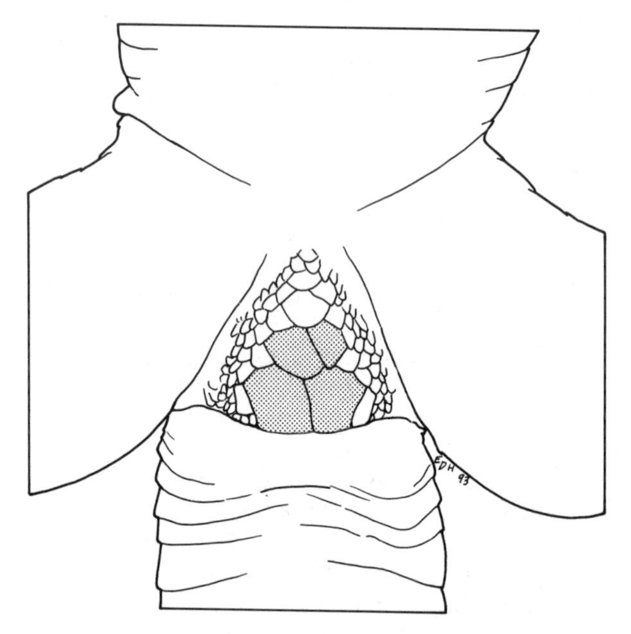

Figure 144. Ventral views of the cloacal regions of two families of lizards. Upper: A single enlarged preanal scale (shaded), typical of the family Lacertidae. Drawn from a preserved specimen of *Podarcis sicula* (KU 220087). Lower: Several enlarged preanal scales (shaded), typical of the family Teiidae. Drawn from a preserved specimen of *Cnemidophorus dixoni* (DAK 1127).

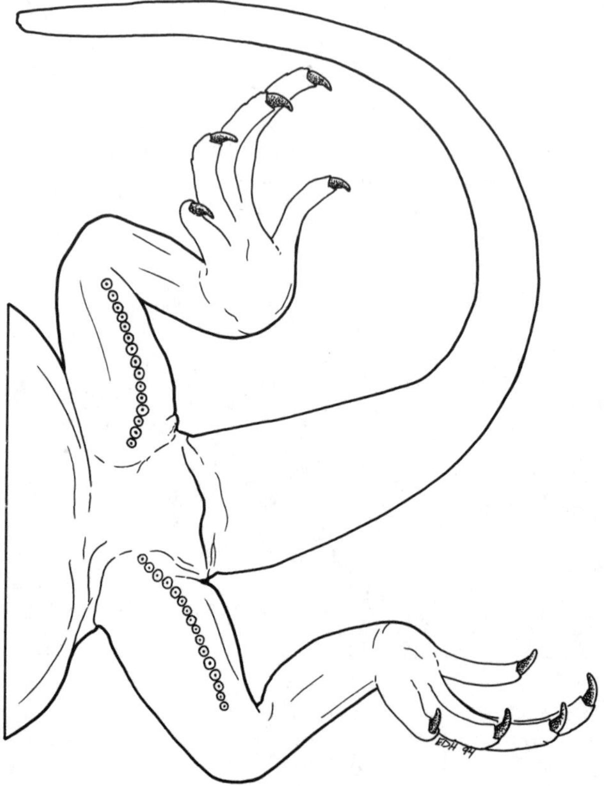

Figure 146. Ventral view of the hindquarters of *Sauromalus obesus*, showing the presence of a row of femoral pores on the underside of each thigh. Drawn from a preserved specimen (KU 88492).

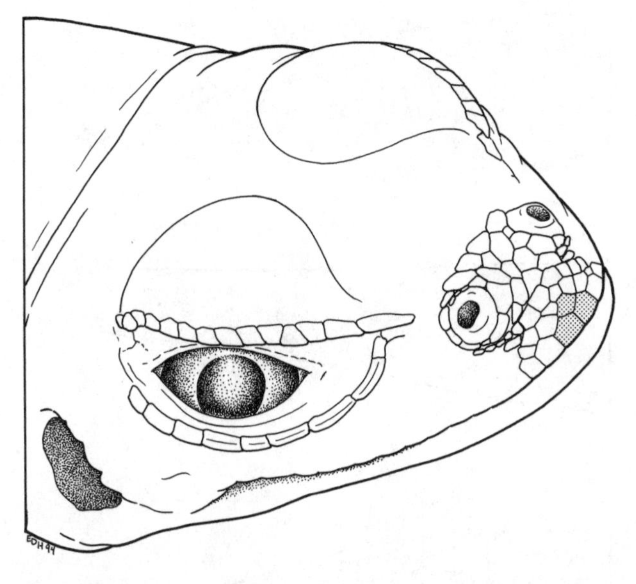

Figure 147. Dorsolateral view of the head of *Sauromalus obesus,* showing the divided rostral scale on the snout (shaded). Drawn from a preserved specimen (KU 61506).

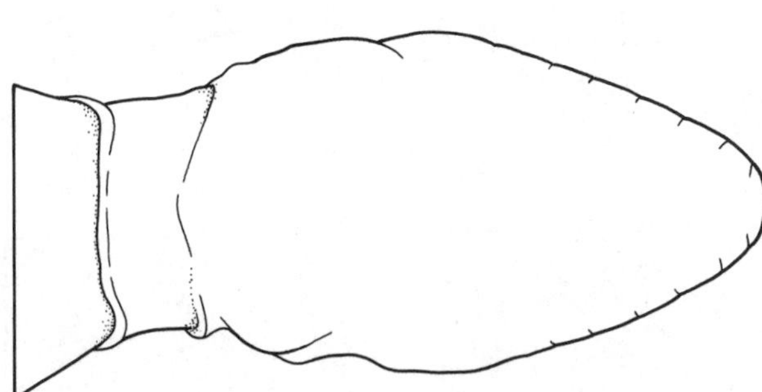

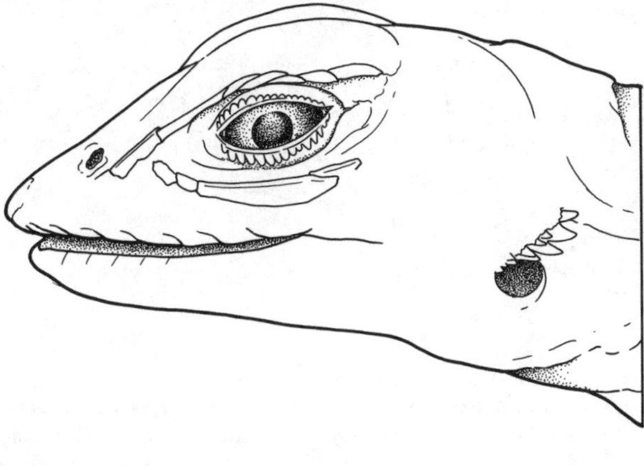

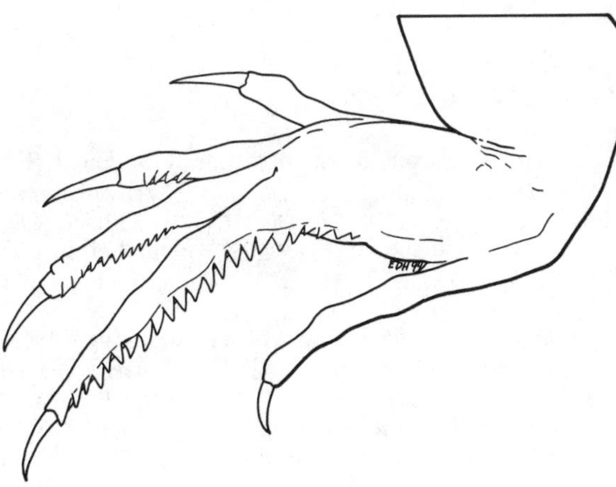

Figure 148. Two views of characteristics of a lizard of the family Phrynosomatidae. Upper: Lateral view of the head, showing scales projecting over the ear opening. Lower: Dorsolateral view of a hind foot, showing the presence of prominent fringes of scales on some of the toes. Both drawn from a preserved specimen of *Uma notata* (KU 61519).

Figure 149. Ventral views of the heads and necks of two families of lizards. Upper: Family Corytophanidae, showing a complete gular fold. Drawn from a preserved specimen of *Basiliscus vittatus* (KU 67195). Lower: Family Tropiduridae, showing an incomplete gular fold. Drawn from a preserved specimen of *Leiocephalus schreibersii* (KU 93354).

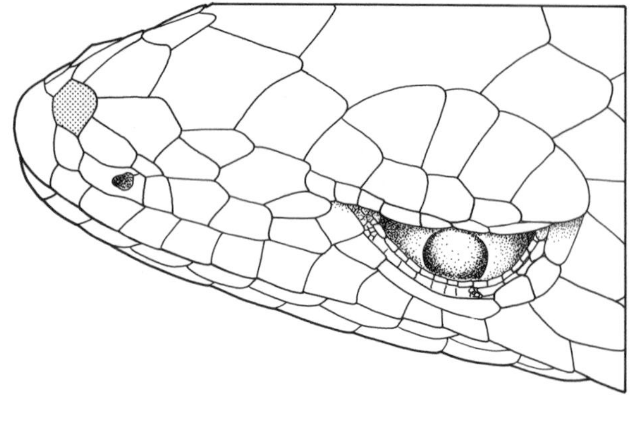

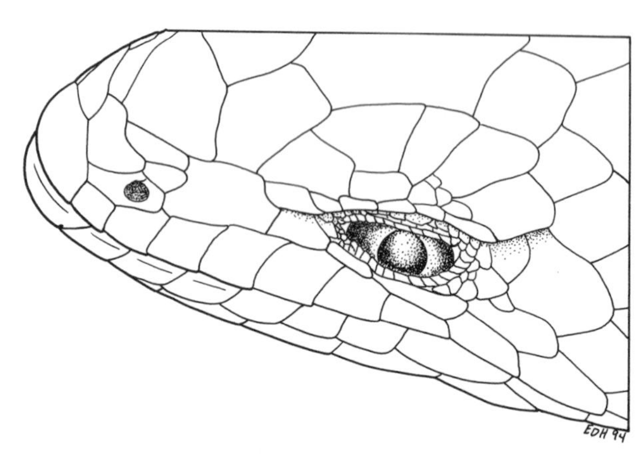

Figure 150. Dorsolateral views of the heads of two genera of lizards of the family Anguidae. Upper: *Gerrhonotus liocephalus,* showing the presence of a single median postrostral scale (shaded) not in contact with the nasal scale. Drawn from a preserved specimen (KU 9047). Lower: *Elgaria kingii,* showing the presence of a dorsolateral postrostral scale in contact with the nasal scale. Drawn from a preserved specimen (KU 6976).

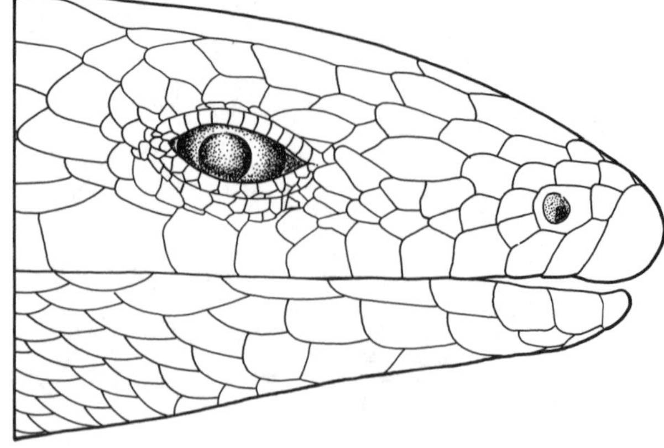

Figure 151. Lateral views of the heads of two species of the genus *Ophisaurus*. Upper: *O. compressus,* showing the upper labials contacting the small scales beneath the orbit. Drawn from a preserved specimen (NCSM 24609). Lower: *O. attenuatus,* showing the presence of a row of enlarged lorilabial scales (shaded) separating the upper labials from the orbit. Drawn from a preserved specimen (KU 33514).

Eumeces

Key to Species of *Eumeces*

1a. Lateral scales in oblique rows, not parallel to dorsal scale rows (Fig. 154) .... ***E. obsoletus***
CC 267, pl 19, p 266, 268 **S** 104, pl 28 **CAAR** 186 (1976)

1b. Lateral scale rows parallel to dorsal rows 2

2a. With one scale between last labial scale and parietal scale (Fig. 155) .... ***E. egregius***
CC 273, pl 19 **CAAR** 73 (1968)

Some authorities have suggested that this taxon consists of two species, *E. egregius* and *E. insularis*.

2b. With three (very rarely 2) scales between last labial scale and parietal scale (Fig. 155) 3

3a. Postmental scale not divided (Fig. 156) 4

3b. Postmental scale divided transversely (Fig. 156) ... 8

4a. Median subcaudals at most barely widened (Fig. 157) .... ***E. tetragrammus***
CC 267, pl 19 **CAAR** 492 (1990)

4b. Median subcaudals distinctly widened (Fig. 157) 5

5a. Dorsolateral light lines absent or restricted to 3rd scale row from middorsum .... ***E. multivirgatus*** (part)
CC 271, pl 19 **S** 106, pl 28 **CAAR** 241 (1980)

5b. Dorsolateral light lines usually present, not restricted to 3rd scale row from middorsum 6

6a. Dorsolateral light line present on 3rd and 4th scale rows from middorsum .... ***E. anthracinus***
CC 269, pl 19, p 268 **CAAR** 658 (1998)

6b. Dorsolateral light line present on 4th or on 4th and 5th scale rows from middorsum 7

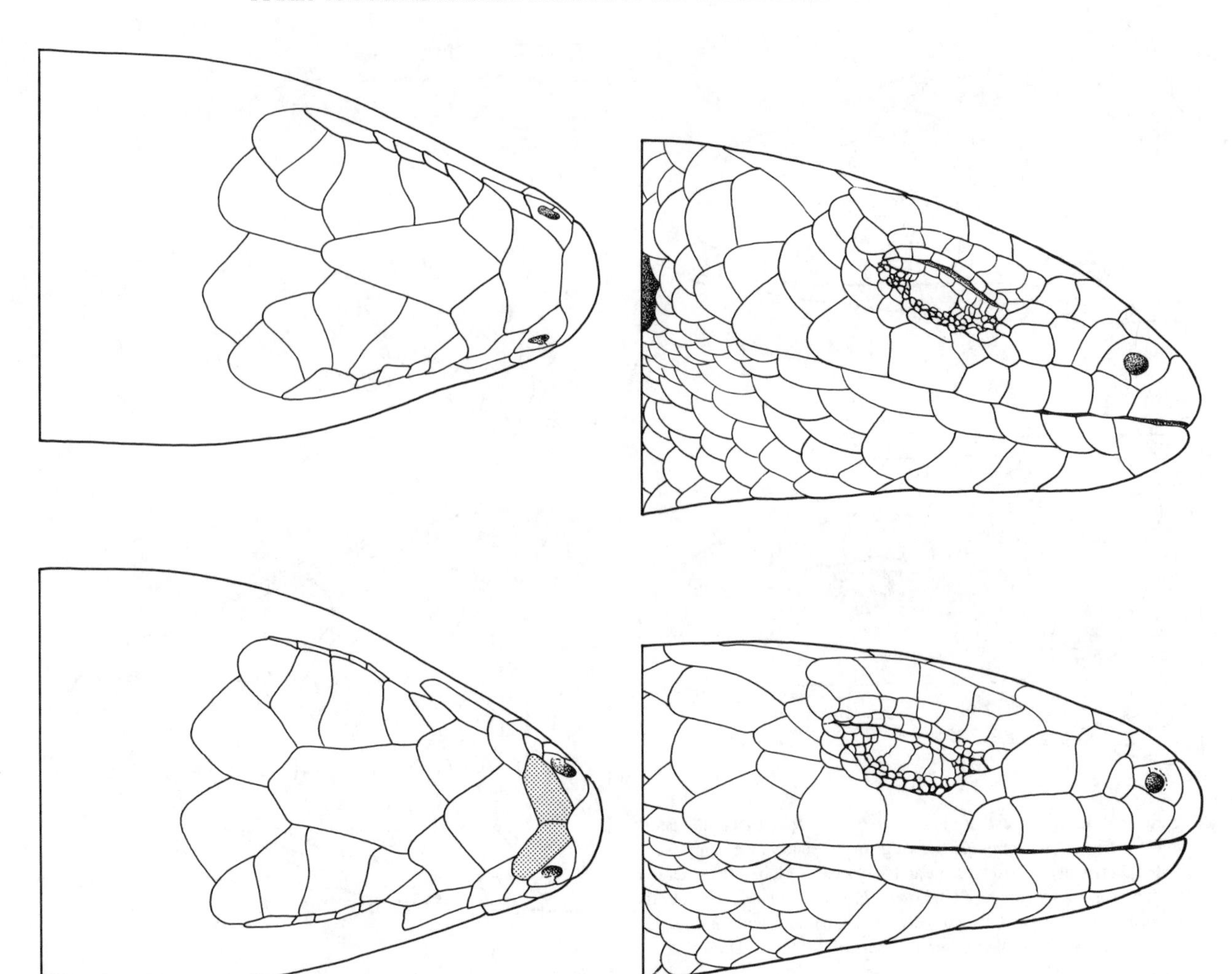

Figure 152. Dorsal view of the heads of two genera of the family Scincidae. Upper: *Scincella lateralis*, showing absence of supranasal scales. Drawn from a preserved specimen (KU 28797). Lower: *Eumeces fasciatus*, showing the presence of supranasal scales (shaded). Drawn from a preserved specimen (KU 185886).

Figure 153. Lateral view of the heads of two genera of the family Scincidae. Upper: *Scincella lateralis*, showing presence of a translucent window on the lower eyelid. Drawn from a preserved specimen (KU 28797). Lower: *Eumeces fasciatus*, showing the presence of scales on the lower eyelid. Drawn from a preserved specimen (KU 28883).

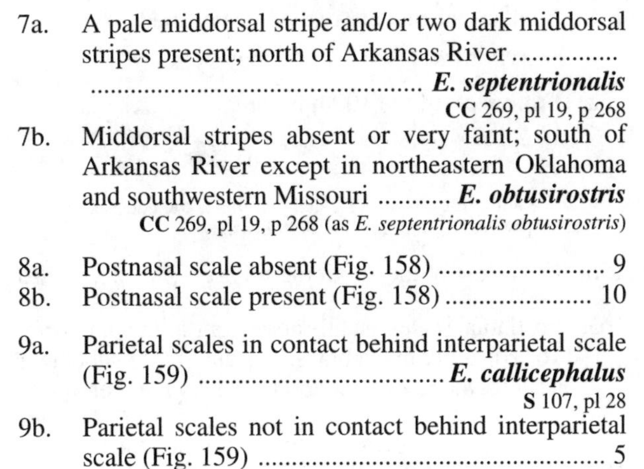

7a. A pale middorsal stripe and/or two dark middorsal stripes present; north of Arkansas River .... ***E. septentrionalis***
CC 269, pl 19, p 268

7b. Middorsal stripes absent or very faint; south of Arkansas River except in northeastern Oklahoma and southwestern Missouri .... ***E. obtusirostris***
CC 269, pl 19, p 268 (as *E. septentrionalis obtusirostris*)

8a. Postnasal scale absent (Fig. 158) 9

8b. Postnasal scale present (Fig. 158) 10

9a. Parietal scales in contact behind interparietal scale (Fig. 159) .... ***E. callicephalus***
S 107, pl 28

9b. Parietal scales not in contact behind interparietal scale (Fig. 159) 5

10a. Dorsolateral light lines present on head and body 11

10b. No dorsolateral light lines on head and body 14

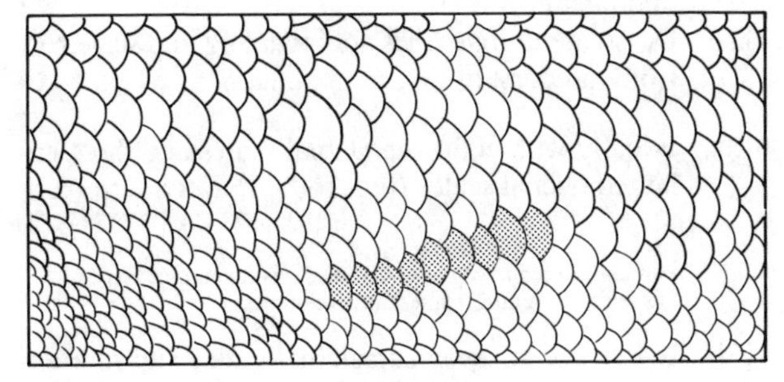

Figure 154. Lateral view of the body of *Eumeces obsoletus*, showing the oblique aspect of the scale rows (one row shaded). Drawn from a photograph by Suzanne L. Collins.

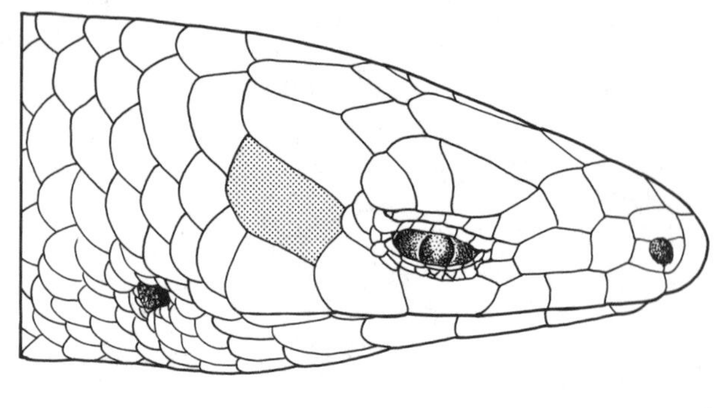

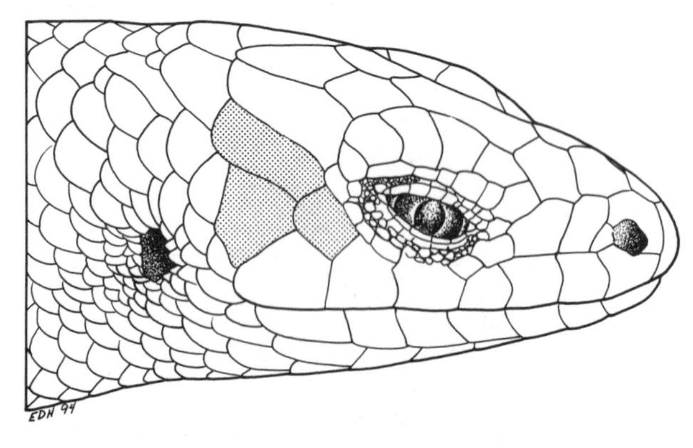

Figure 155. Lateral view of the heads of two species of the genus *Eumeces*. Upper*:* *E. egregius*, showing the presence of a single scale (shaded) between the parietal scale and the last upper labial scale. Drawn from a preserved specimen (KU). Lower: *E. tetragrammus*, showing the presence of three scales (shaded) between the parietal scale and the last upper labial scale, typical of other members of the genus. Drawn from a preserved specimen (KU 145823).

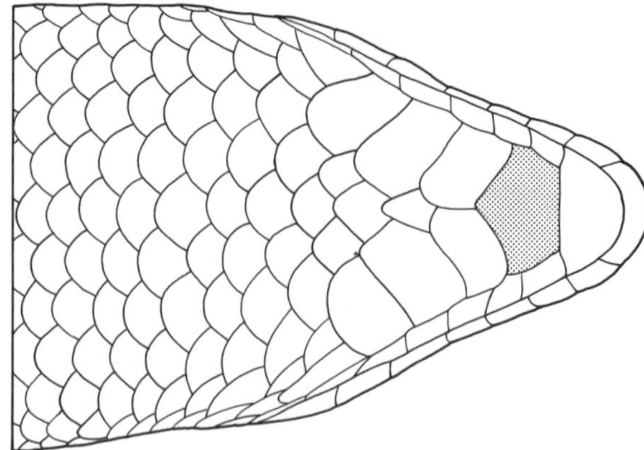

Figure 156. Ventral view of the heads of two species of the genus *Eumeces*. Upper: *E. tetragrammus*, showing the undivided postmental scale (shaded). Drawn from a preserved specimen (KU 145806). Lower: *E. callicephalus*, showing the divided postmental scale (shaded). Drawn from a preserved specimen (KU 6473).

11a. Dorsolateral light lines on 3rd scale row from middorsum only .... ***E. multivirgatus*** (part)
CC 271, pl 19 **S** 106, pl 28 **CAAR** 241 (1980)

11b. Dorsolateral light lines not restricted to 3rd scale row from middorsum 12

12a. Dorsolateral light lines including 2nd scale row from middorsum 13

12b. Dorsolateral light lines not including 2nd scale row from middorsum 14

13a. Usually with eight supralabial scales and three enlarged nuchal scales (Fig. 160) ..... ***E. gilberti*** (part)
S 103, pl 28 **CAAR** 372 (1985)

Some authorities have suggested that this taxon consists of two species, *E. gilberti* and *E. arizonensis*.

13b. Usually with seven supralabial scales and four enlarged nuchal scales (Fig. 160) ..... ***E. skiltonianus***
S 105, pl 28 **CAAR** 447 (1988)

14a. Median subcaudals barely widened (Fig. 157) .... ***E. inexpectatus***
CC 265, pl 19, p 264 **CAAR** 385 (1986)

14b. Median subcaudals distinctly widened (Fig. 157) 15

15a. From west of the 100th meridian .... ***E. gilberti*** (part)
S 103, pl 28 **CAAR** 372 (1985)

Some authorities have suggested that this taxon consists of two species, *E. gilberti* and *E. arizonensis*.

15b. From east of the 100th meridian 16

16a. Postlabial scales usually absent or rudimentary (Fig. 161); usually five labial scales anterior to subocular scale; scale rows at midbody usually number 30–32 .... ***E. laticeps***
CC 265, pl 19, p 264, 265 **CAAR** 445 (1988)

16b. With two postlabial scales (Fig. 161); usually four labial scales anterior to subocular scale; scale rows at midbody usually number 26–30.......... ***E. fasciatus***
CC 263, pl 19, p 264, 265

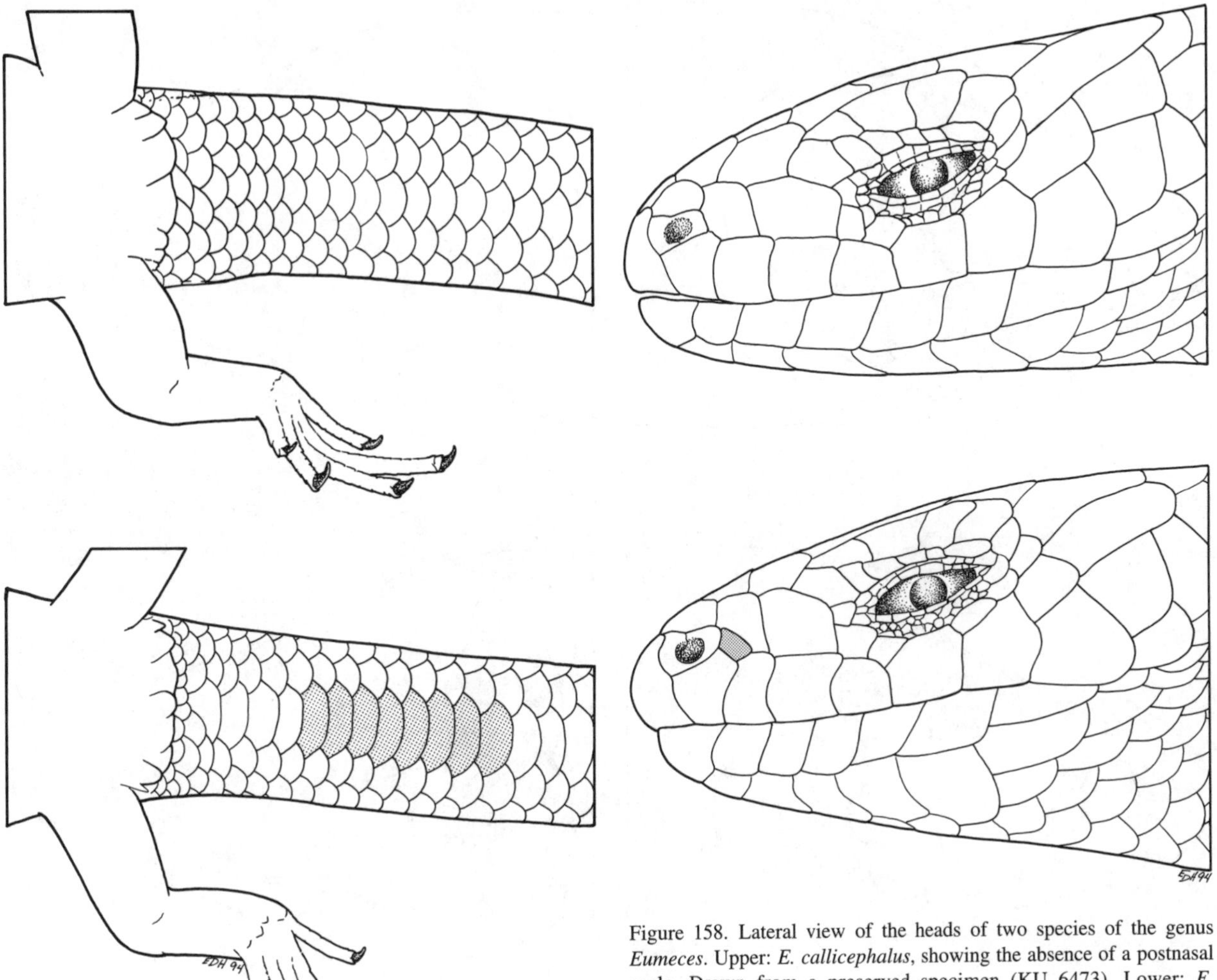

Figure 157. Ventral view of the pelvic areas of two species of the genus *Eumeces*. Upper: *E. tetragrammus*, showing the uniform size of the median subcaudal scales on the tail. Drawn from a preserved specimen (KU 188506). Lower: *E. multivirgatus*, showing enlarged subcaudal scales (shaded) on the tail. Drawn from a preserved specimen (KU 16344).

Figure 158. Lateral view of the heads of two species of the genus *Eumeces*. Upper: *E. callicephalus*, showing the absence of a postnasal scale. Drawn from a preserved specimen (KU 6473). Lower: *E. multivirgatus*, showing the presence of a postnasal scale (shaded). Drawn from a preserved specimen (KU 16343).

XANTUSIIDAE

Xantusia

Key to Species of ***Xantusia***

1a. Ventrals in sixteen longitudinal rows .... ***X. riversiana***
S 71, pl 30 **CAAR** 518 (1991)

1b. Ventrals in 12–14 longitudinal rows 2

2a. Ventrals in twelve longitudinal rows; dorsum with small flecks or spots .... ***X. vigilis***
S 71, pl 30 **CAAR** 302 (1982)

Some authorities have suggested that this taxon consists of two species, *X. vigilis* and *X. utahensis*.

2b. Ventrals in fourteen longitudinal rows; dorsum often with large dark blotches .... ***X. henshawi***
S 72, pl 30 **CAAR** 189 (1976)

EUBLEPHARIDAE

Coleonyx

CAAR 95 (1970)

Key to Species of ***Coleonyx***

1a. Dorsum with scattered enlarged tubercles (Fig. 162) 2

1b. Dorsum with uniform granular scales 3

2a. From California south into Baja California; often with distinct black and white bands on tail .... ***C. switaki***
S 70, pl 35 **CAAR** 464 (1990)

2b. From Texas; tail bands usually indistinct .... ***C. reticulatus***
CC 207, p 208 **CAAR** 89 (1970)

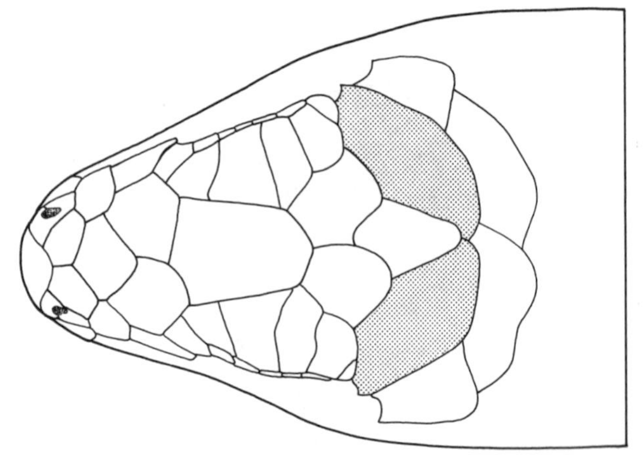

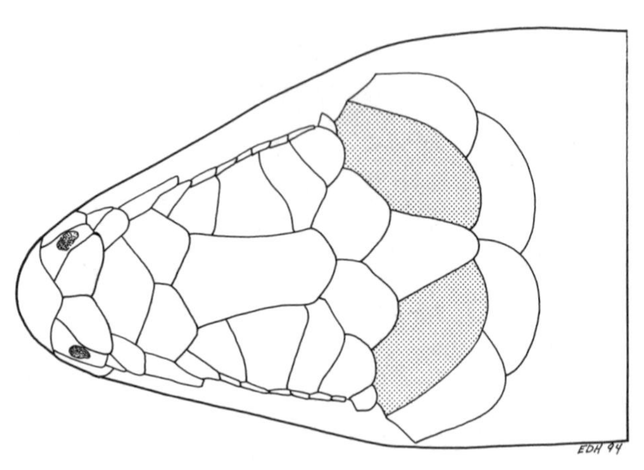

Figure 159. Dorsal view of the heads of two species of the genus *Eumeces*. Upper: *E. callicephalus*, showing the parietal scales (shaded) in contact behind the interparietal scale. Drawn from a preserved specimen (KU 6473). Lower: *E. multivirgatus*, showing the parietal scales (shaded) not in contact. Drawn from a preserved specimen (KU 16343).

3a. Scales bearing preanal pores separated by one or more midventral scales; usually four preanal pores (Fig. 163); from southeastern New Mexico south and east into Texas and México .... ***C. brevis***
CC 207, pl 11, p 208 **S** 69 **CAAR** 88 (1970)

3b. Scales bearing preanal pores continuous across midventral line; usually more than four preanal pores (Fig. 163); from west of extreme southwestern New Mexico .... ***C. variegatus***
S 69, pl 35 **CAAR** 96 (1970)

GEKKONIDAE

Key to Genera of Gekkonidae

1a. Tail strongly flattened dorsoventrally, almost leaflike; digital pads well developed and connected, almost weblike; range restricted to southern Florida ....
.... ***Cosymbotes*** (p. 79)

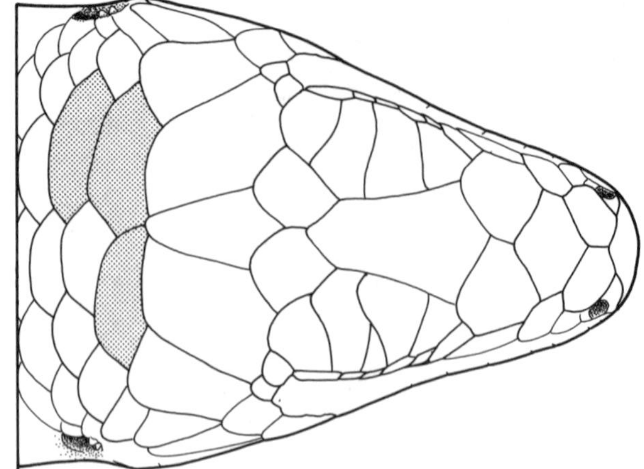

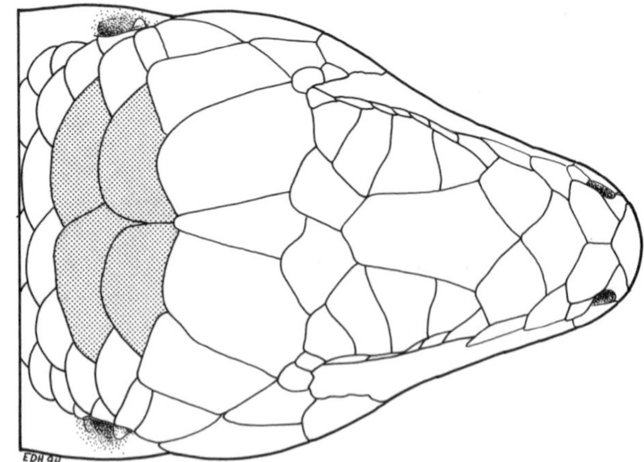

Figure 160. Dorsal view of the heads of two species of the genus *Eumeces*. Upper: *E. gilberti*, showing the presence of three enlarged nuchal scales (shaded). Drawn from preserved specimens (KU 211268–271). Lower: *E. skiltonianus*, showing the presence of four enlarged nuchal scales (shaded). Drawn from preserved specimens (KU 69597 and 69598).

1b. Tail round or oval in cross-section; digital pads variable, but never connected 2

2a. Digits without enlarged pads (Fig. 164); pupils round
.... ***Gonatodes*** (p. 79)

2b. Digits with at least terminal pads (Fig. 164); pupils elliptical 3

3a. With single round scales at tips of digits (Fig. 164); dorsal scales uniform in size ....
.... ***Sphaerodactylus*** (p. 79)

3b. No single round scales at tips of digits; dorsal scales uniform in size or not 4

4a. With two pads at tips of digits, claw between the pads (Fig. 164) .... ***Phyllodactylus*** (p. 80)

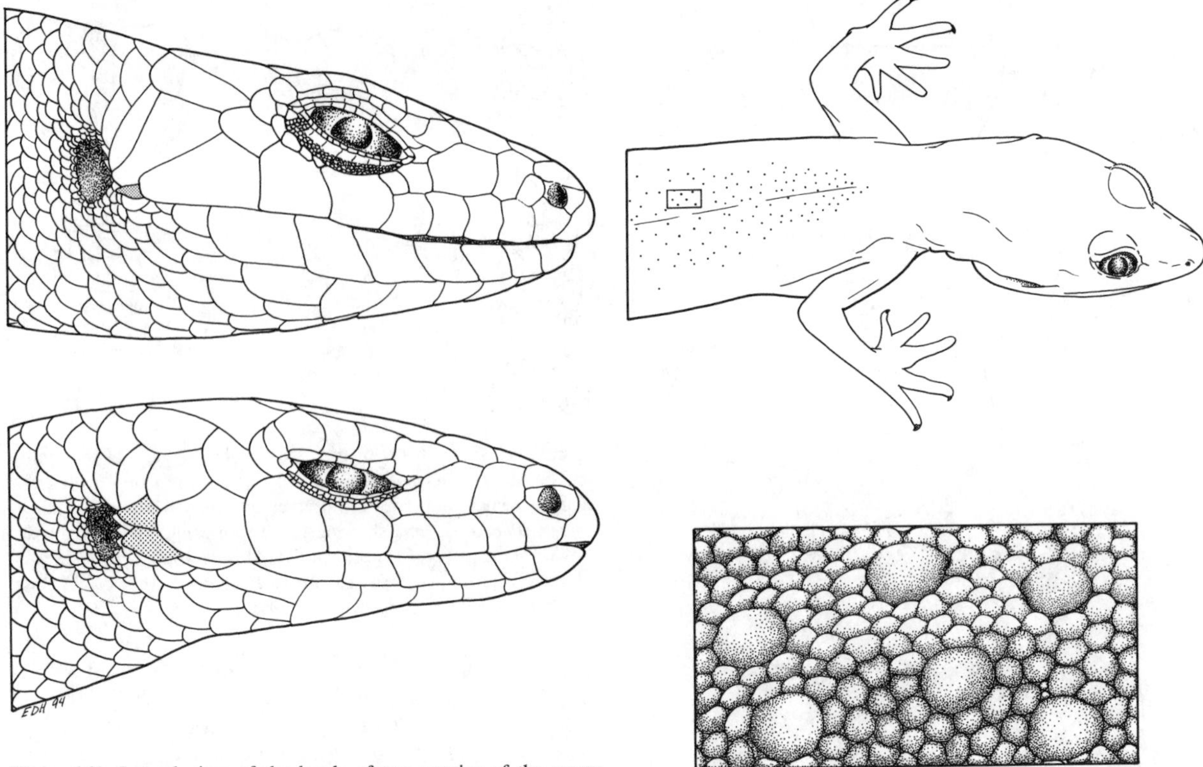

Figure 161. Lateral view of the heads of two species of the genus *Eumeces*. Upper: *E. laticeps*, showing the presence of a single reduced postlabial scale (shaded). Drawn from a preserved specimen (KU). Lower: *E. fasciatus*, showing the presence of two well-developed postlabial scales (shaded). Drawn from a preserved specimen (KU 176600).

Figure 162. Upper: Dorsolateral view of the anterior half of *Coleonyx reticulatus*. Lower: Close-up of the scattered, enlarged tubercles on the skin of this species. Drawn from a preserved specimen (UTAVC R25915).

4b. With most of the digits forming pads (Fig. 165) 5

5a. Tail with prominently enlarged, keeled scales .... ***Cyrtopodion*** (p. 80)

5b. Tail without prominently enlarged, keeled scales (some enlarged tubercles may be present) 6

6a. Dorsum and venter with small, irregular, reddish (in life) spots; with more than fifteen lamellae on 4th toe (Fig. 165); large (SVL to 150 mm) .... ***Gekko*** (p. 80)

6b. Dorsal pattern variable, venter usually without markings; with less than fifteen lamellae on 4th toe; smaller (SVL always less than 70 mm) 7

7a. Distal toe lamellae are divided medially into two parallel rows; all toes with distinctly visible claws .... ***Hemidactylus*** (p. 80)

7b. All toe lamellae extend undivided across the entire toe pad; only the third and fourth toes with distinctly visible claws (females have small, retractible claws on other toes) .... ***Tarentola*** (p. 81)

Cosymbotes

***Cosymbotes platyurus* (A)**

Gonatodes

***Gonatodes albogularis* (A)**
CC 206, pl 11

Sphaerodactylus

CAAR 142 (1973)

Key to Species of *Sphaerodactylus*

1a. Dorsal scales granular .... ***S. elegans* (A)**
CC 205, pl 11

1b. Dorsal scales overlapping, keeled 2

2a. With more than fifty scale rows at midbody; with many white spots on neck .... ***S. argus* (A)**
CC 206, pl 11

2b. With less than fifty scale rows at midbody; with two white spots on neck or none .... ***S. notatus***
CC 205, pl 11 **CAAR** 90 (1970)

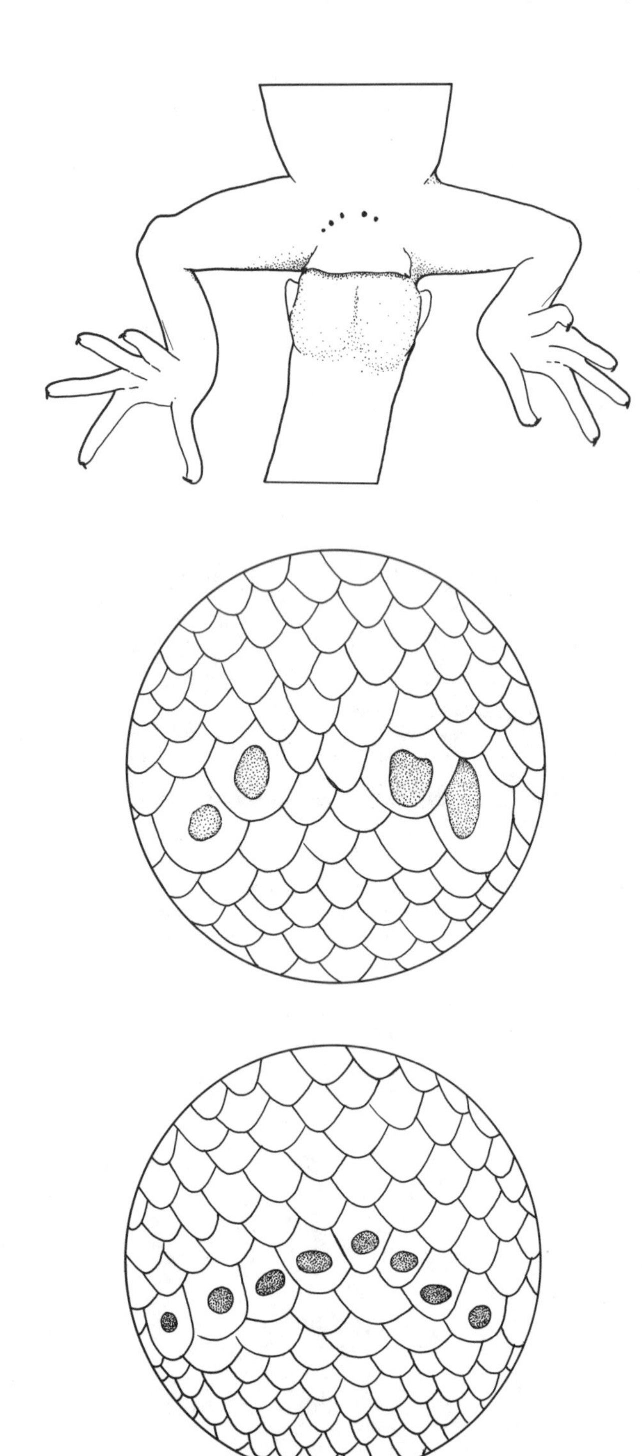

Figure 163. Upper: Ventral view of the pelvic areas of two species of the genus *Coleonyx*. Middle: *C. brevis*, showing the presence of four preanal pores (shaded) separated by two scale rows. Drawn from a preserved specimen (KU 9065). Lower: *C. variegatus*, showing the presence of more than four preanal pores (shaded), continuous across the midventral line. Drawn from a preserved specimen (KU 10738).

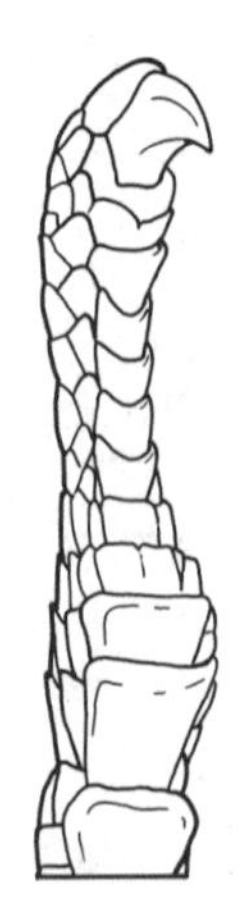

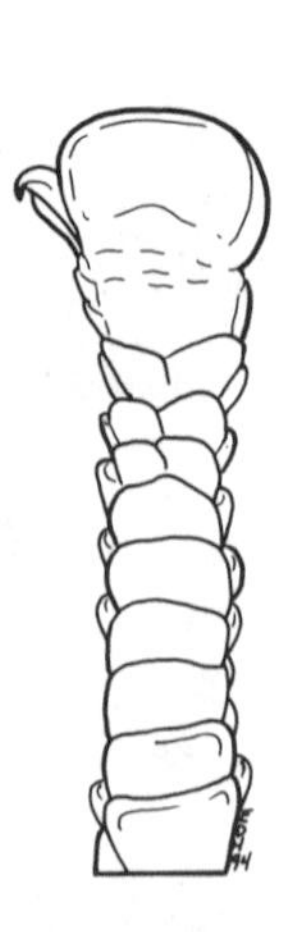

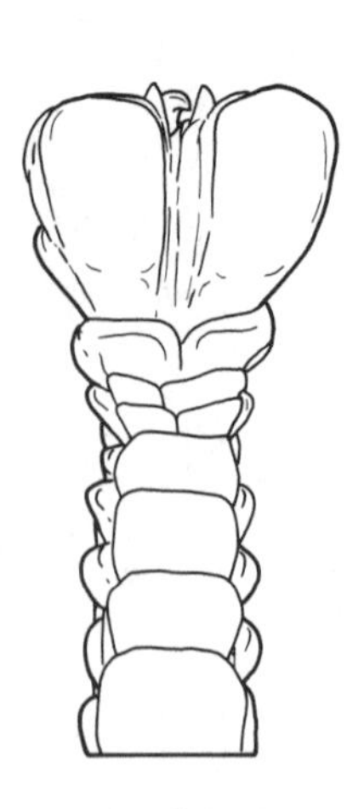

Figure 164. Ventral view of the digits of three genera of the family Gekkonidae. Left: *Gonatodes albogularis*, showing a digit without an enlarged toepad. Drawn from a preserved specimen (KU 68966). Middle: *Sphaerodactylus elegans*, showing a digit with a single enlarged toepad. Drawn from a preserved specimen (KU 88344). Right: *Phyllodactylus xanti*, showing a digit with two enlarged toepads divided by a claw. Drawn from a preserved specimen (KU 152619).

Phyllodactylus

CAAR 141 (1973)

Phyllodactylus nocticolus

S 68, pl 35 **CAAR** 79 (1969) (as *P. xanti*)

Cyrtopodion

Cyrtopodion scabrum (A)

CC 202, pl 11

Gekko

Gekko gecko (A)

CC 202, pl 11

Hemidactylus

Key to Species of *Hemidactylus*

1a. Subdigital lamellae of the 4th toe fail to reach the origin of the digit (Fig. 166) .... ***H. mabouia*** (A)

This recently introduced species is currently restricted to southern Florida.

1b. Subdigital lamellae of the 4th toe reach the origin of the digit (Fig. 166) 2

2a. Small dorsal tubercles restricted to dorsolateral rows .... ***H. garnotii*** (A)
CC 203, pl 11

2b. Dorsal tubercles larger and scattered (not restricted to dorsolateral rows) 3

3a. Femoral pores present; often uniformly colored, or with dark lateral stripes and a light line through eyes; known range currently restricted to southern Florida, and parts of Dallas, Texas .... ***H. frenatus*** (A)

3b. Femoral pores absent; pattern of irregular dark flecks and spots; widely distributed through the southern U.S. .... ***H. turcicus*** (A)
CC 203, pl 11 **S** 70, pl 35 **CAAR** 87 (1970)

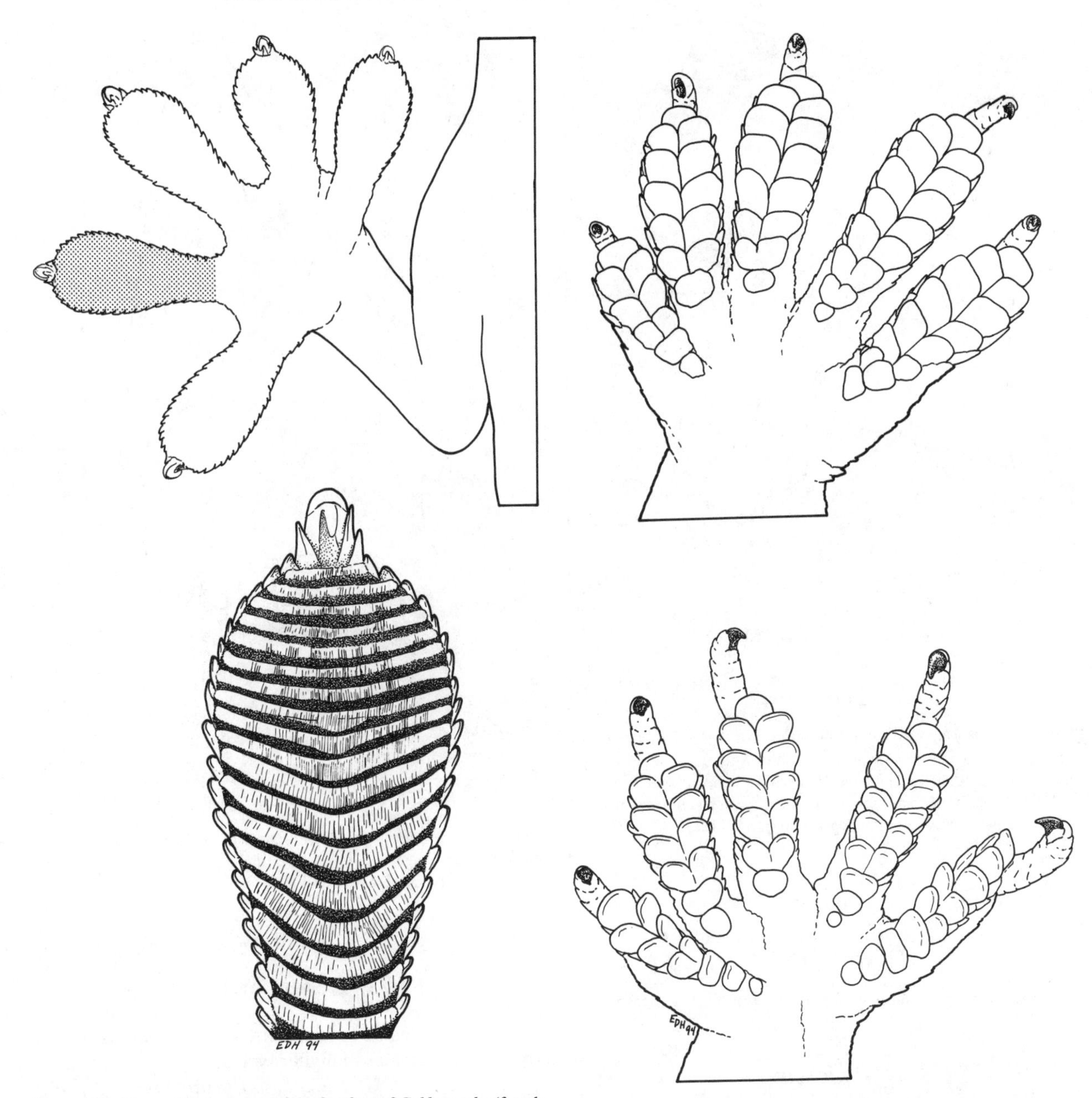

Figure 165. Upper: Ventral view of the forefoot of *Gekko gecko* (fourth toe shaded). Lower: Ventral view of the fourth toe of *Gekko gecko* enlarged, showing the presence of more than fifteen lamellae. Adapted from a photograph in Carmichael and Williams (1991).

Figure 166. Ventral view of feet of two species of the genus *Hemidactylus*, showing subdigital lamellae on the toes. Upper: *H. mabouia*, showing the lamellae on the fourth toe (counting from left to right) not reaching the origin of the digit. Drawn from a preserved specimen (KU 229543). Lower: *H. turcicus*, showing the lamellae on the fourth toe reaching the origin of the digit. Drawn from a preserved specimen (KU 206689).

Tarentola

Tarentola mauritanica (A)

Helodermatidae

Heloderma

Heloderma suspectum
S 120, pl 20

Lacertidae

Key to Genera of Lacertidae

1a. Postnasal scale single; median ventral scales meet in straight line .... ***Podarcis*** (p. 82)
1b. Nearly always two postnasal scales; median ventral scales overlap .... ***Lacerta*** (p. 82)

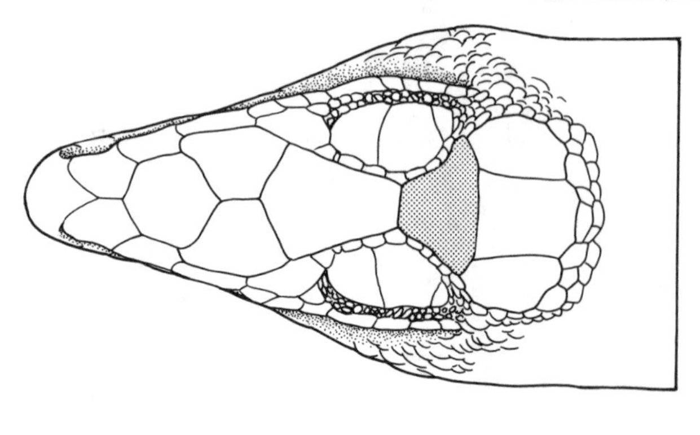

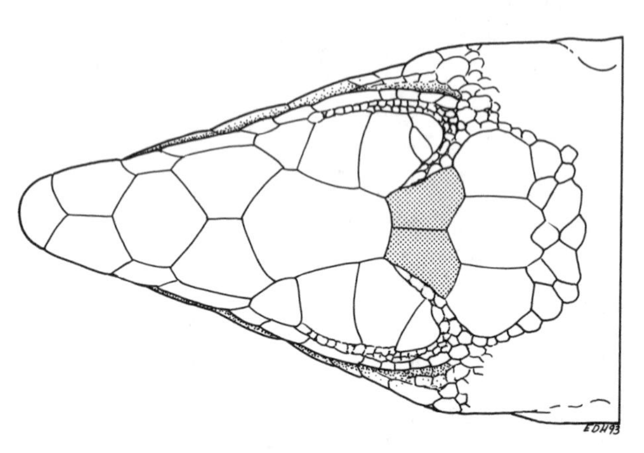

Figure 167. Dorsal view of the heads of two species of the genus *Cnemidophorus*. Upper: *C. hyperythrus*, showing the presence of a single frontoparietal scale (shaded). Drawn from a preserved specimen (KU 62692). Lower: *C. dixoni*, showing the presence of paired frontoparietal scales. Drawn from a preserved specimen (DAK 1127).

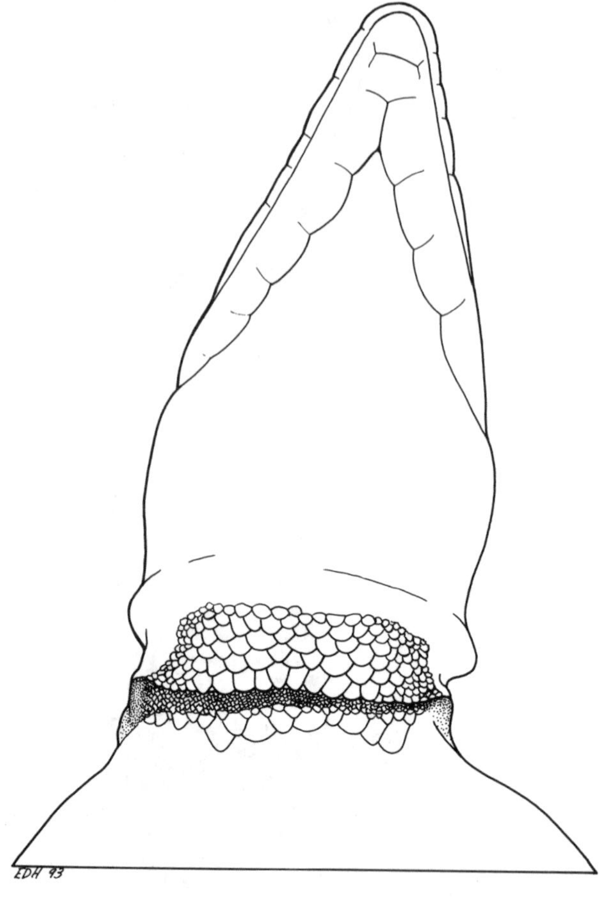

Figure 168. Ventral view of the head and neck of *Cnemidophorus dixoni*, showing the enlarged scales in front of the gular fold. Drawn from a preserved specimen (DAK 1127).

Lacerta

Lacerta viridis (A)

The range of this introduced species is restricted to Topeka, Kansas.

Podarcis

Key to Species of *Podarcis*

1a. Venter light and unmarked; dorsal scales between hindlimbs distinctly keeled; introduced into New York, Topeka, and Philadelphia (the latter probably extirpated) .... ***P. sicula* (A)**
CC 247, pl 12

1b. Venter with at least a few faint spots (often densely spotted); dorsal scales between hindlimbs at most faintly keeled; introduced into Cincinnati .... ***P. muralis* (A)**
CC 247, pl 12

Teiidae

Key to Genera of Teiidae

1a. Ventral scales in 10–12 rows .... ***Ameiva*** (p. 82)
1b. Ventral scales in eight rows .... ***Cnemidophorus*** (p. 82)

Ameiva

Ameiva ameiva (A)

CC 260, pl 16

Cnemidophorus

Key to Species of *Cnemidophorus*

1a. With a dark-bordered, brown (in life) middorsal stripe and yellowish (in life) flanks with light stripes often broken into spots; range restricted to Hialeah and Miami, Florida .... ***C. lemniscatus* (A)**
CC 259, pl 16

1b. Pattern variable, but not as above 2

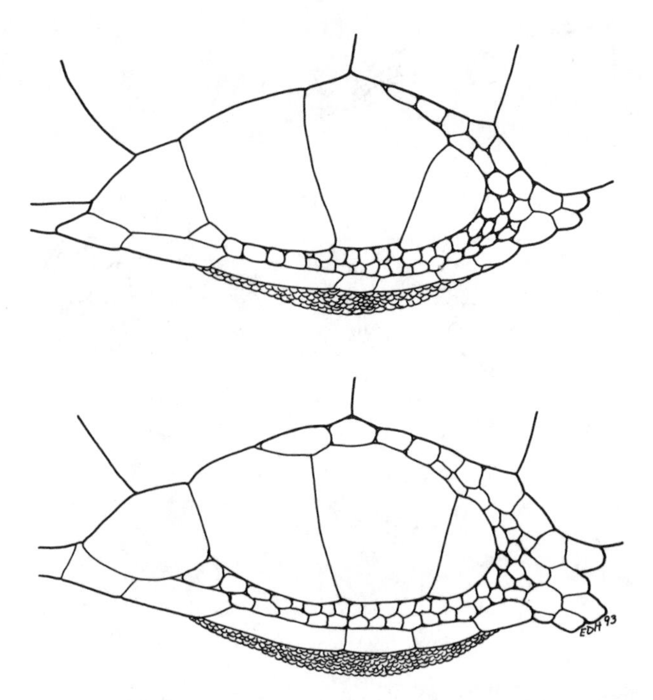

Figure 169. Dorsal view of the orbits of two species of the genus *Cnemidophorus*. Upper: *C. dixoni*, showing the small scales bordering the medial edge of the supraocular scales not extending to or past the middle of the second supraocular scale. Drawn from a preserved specimen (DAK 1127). Lower: *C. tesselatus*, showing the small scales bordering the medial edge of the supraocular scales extending well past the middle of the second supraocular scale. Drawn from a preserved specimen (KU 119255).

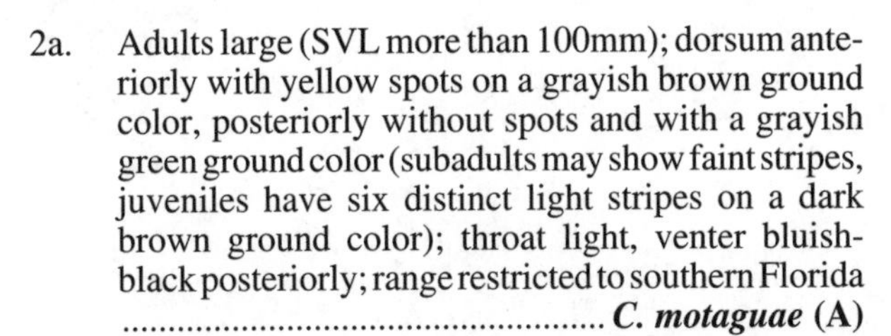

2a. Adults large (SVL more than 100mm); dorsum anteriorly with yellow spots on a grayish brown ground color, posteriorly without spots and with a grayish green ground color (subadults may show faint stripes, juveniles have six distinct light stripes on a dark brown ground color); throat light, venter bluish-black posteriorly; range restricted to southern Florida .... ***C. motaguae* (A)**

2b. Adult size usually smaller; pattern variable, but never with both spots anteriorly and unicolored posteriorly or, if with stripes, not a dark brown ground color; range not restricted to Florida 3

3a. With a single frontoparietal scale (Fig. 167) .... ***C. hyperythrus***
S 111, pl 34 **CAAR** 655 (1998)

3b. With paired frontoparietal scales (Fig. 167) 4

4a. Scales in front of gular fold distinctly enlarged (Fig. 168) 5

4b. Scales in front of gular fold not or barely enlarged 18

5a. Small scales surrounding supraocular scales rarely extend forward past the seam between the 2nd and 3rd supraocular scales (never past the middle of the 2nd supraocular) (Fig. 169); scales on rear of forearm enlarged medially (Fig. 170)........... ***C. dixoni***
CC 256 **CAAR** 398 (1986) (under *C. tesselatus*)

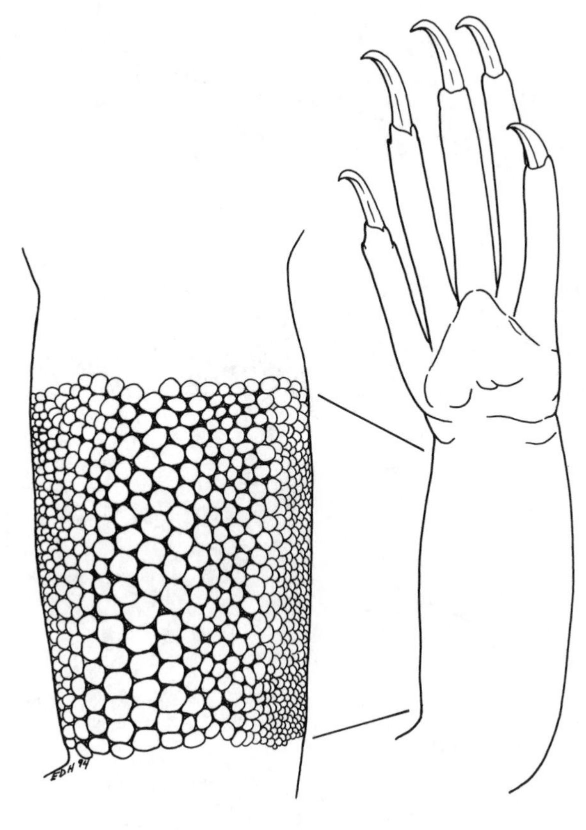

Figure 170. Ventral view of the front leg (right) of *Cnemidophorus dixoni*, showing the enlarged medial scales on the forearm (left). Drawn from a preserved specimen (DAK 1127).

5b. Small scales surrounding supraocular scales usually extend forward to the middle of the 2nd supraocular scale (Fig. 169), or forearm scales all granular 6

6a. Light spots or bars present in areas between stripes 7

6b. No light spots or bars present in areas between stripes 16

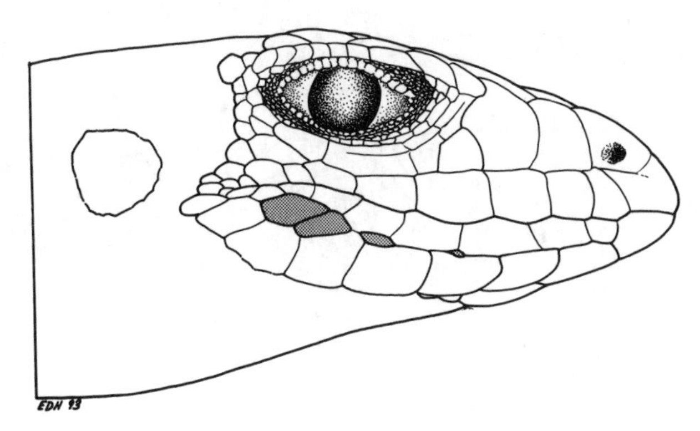

Figure 171. Lateral view of the head of *Cnemidophorus velox*, showing the presence of fewer than five scales (shaded) between chin shields and lower labials. Drawn from a preserved specimen (KU 39343).

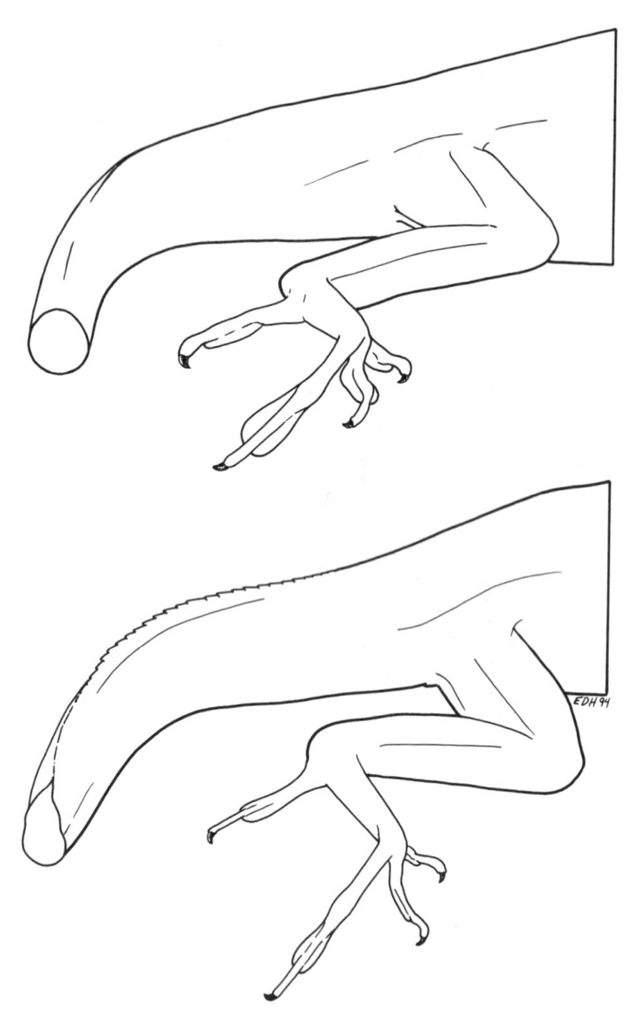

Figure 172. Lateral view of hindquarters and cross-sections of tails of two species of the genus *Anolis*. Upper: *A. carolinensis*, showing a round tail. Drawn from a preserved specimen (KU 22651). Lower: *A. sagrei*, showing a laterally compressed tail. Drawn from a preserved specimen (KU 92689).

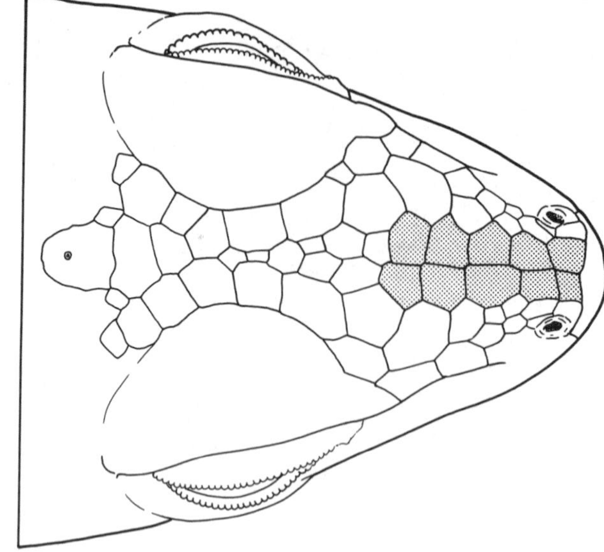

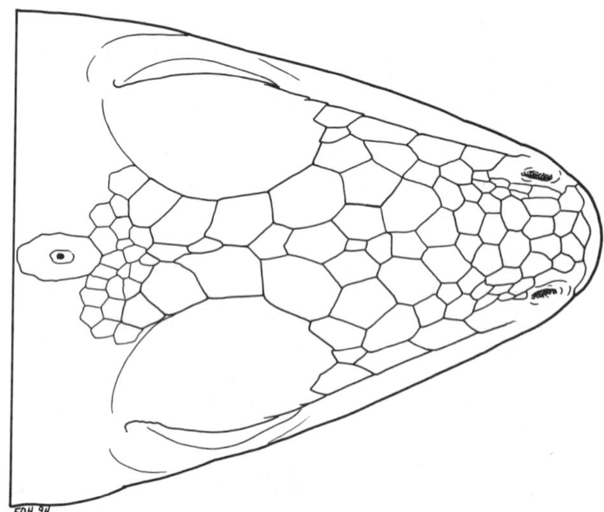

Figure 174. Dorsal view of the heads of two species of the genus *Anolis*. Upper: *A. distichus*, showing the presence of two paramedian rows of scales (shaded). Drawn from a preserved specimen (KU 68977). Lower: *A. cybotes*, showing the absence of two paramedian rows of scales. Drawn from a preserved specimen (KU 254354).

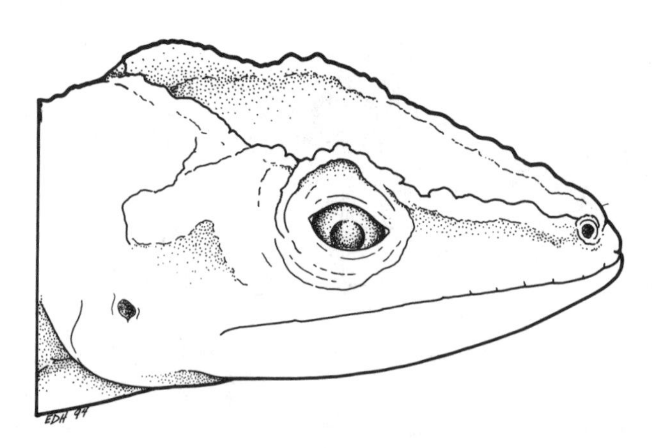

Figure 173. Dorsolateral view of the head of *Anolis equestris*, showing the bony casque. Drawn from a preserved specimen (KU 220258).

7a. Scales on rear of forearm granular, not enlarged .... ***C. tesselatus***
CC 256, pl 18 **S** 108, pl 32 **CAAR** 398 (1986)

7b. Scales on rear of forearm at least slightly enlarged medially (Fig. 170) 8

8a. With 7–8 stripes (at least vestigially present) 9

8b. With six stripes (at least vestigially present) 12

9a. Scales on rear of forearm barely enlarged .... ***C. laredoensis***
CC 253

9b. Scales on rear of forearm distinctly enlarged 10

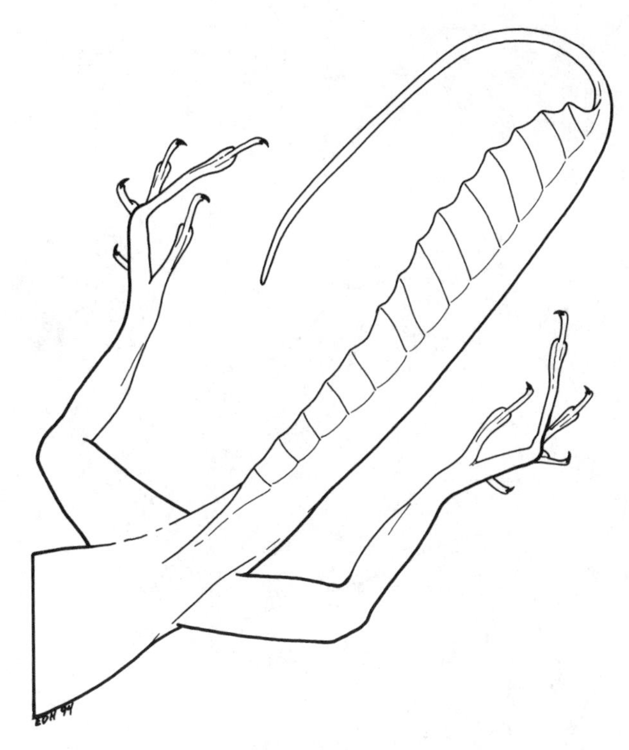

Figure 175. Dorsal view of the hindlimbs and tail of *Anolis cristatellus*, showing the prominent crest on the tail of males. Drawn from a preserved specimen (KU 45943).

10a. From southeastern New Mexico, Texas, and southern Oklahoma south into México; venter in males with at least some blue color 11

10b. From extreme southwestern New Mexico and Arizona south into México; venter white; dorsal stripes often obscure or lost in older individuals 12

11a. Stripes end before hindlimbs; rump rusty red (in life) .... ***C. septemvittatus***
CC 258, pl 18

11b. Stripes extend onto rump .... ***C. gularis***
CC 255, pl 18 **S** 116, pl 33

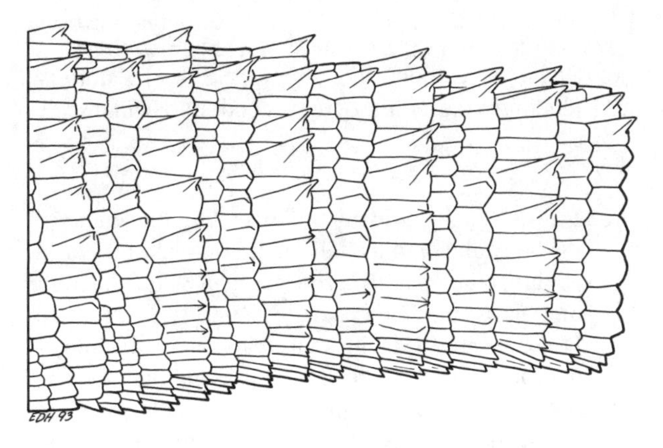

Figure 176. Lateral view of a section of the tail of *Ctenosaura pectinata*, showing the whorls of enlarged spines. Drawn from a preserved specimen (KU 63388).

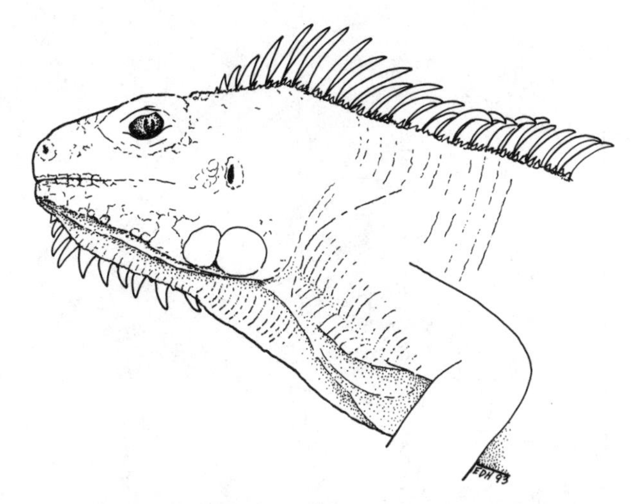

Figure 177. Lateral view of the head and neck of *Iguana iguana*, showing the comblike row of enlarged middorsal scales increasing in length on the neck and the presence of enlarged scales on the lower jaw below the ear. Drawn from a photograph in Capula (1989).

12a. With 85 or more granules across midbody 13

12b. With fewer than 85 granules across midbody 14

13a. From extreme southwestern New Mexico and southeastern Arizona south into Sonora; 100 or more dorsal granules (from back of head to base of tail); in life, reddish dorsal color usually restricted to head and neck .... ***C. burti***
S 118, pl 34

13b. Disjunct populations in southcentral Arizona; 85–100 dorsal granules (from back of head to base of tail); in life, reddish dorsal color extends along body where it is abruptly demarcated from gray or bluish sides .... ***C. xanthonotus***
S 118, pl 34 (as *C. burti xanthonotus*)

14a. With 3–5 granules between paravertebral stripes; usually two preanal scales .... ***C. flagellicaudus***
S 114, pl 33

14b. With 6–8 granules between paravertebral stripes; usually three preanal scales 15

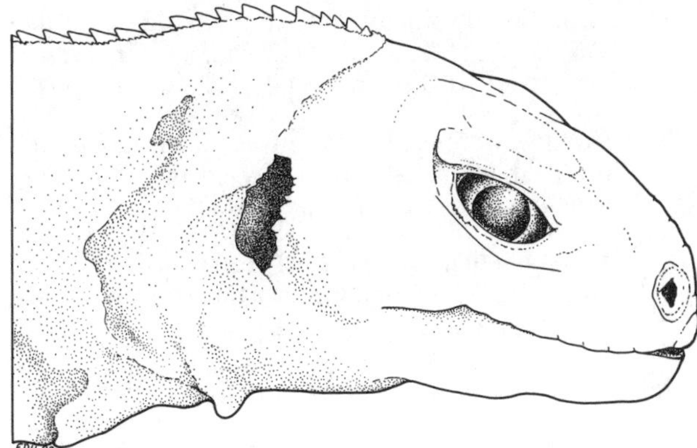

Figure 178. Lateral view of the head and neck of *Dipsosaurus dorsalis*, showing the enlarged row of middorsal scales forming a low crest on the neck and the absence of enlarged scales on the lower jaw below the ear. Drawn from a preserved specimen (KU 6971).

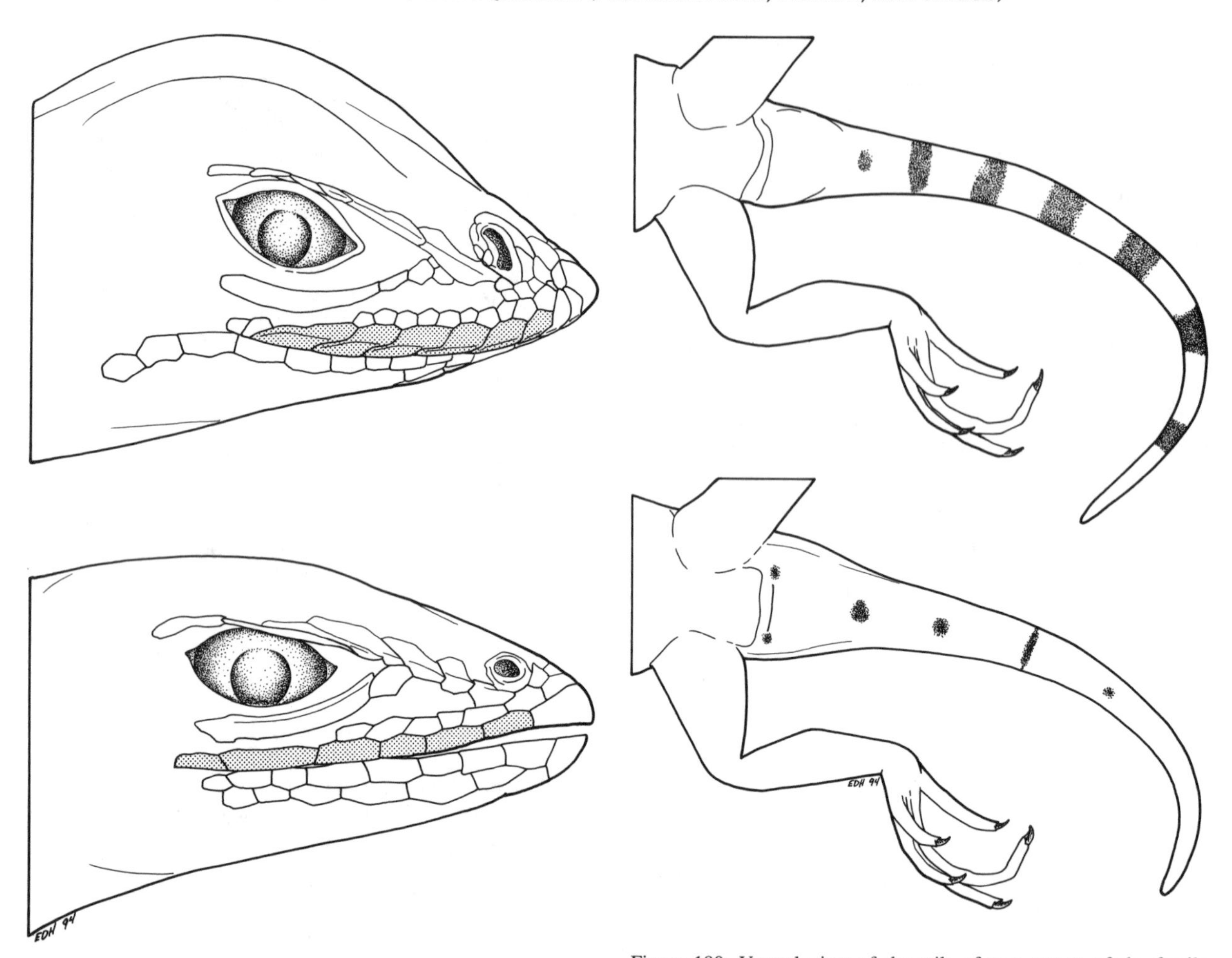

Figure 179. Lateral view of the heads of two genera of the family Phrynosomatidae. Upper: *Uma notata,* showing diagonal seams between supralabial scales (shaded). Drawn from a preserved specimen (KU 154465). Lower: *Urosaurus graciosus*, showing vertical seams between supralabial scales (shaded). Drawn from a preserved specimen (KU 72745).

Figure 180. Ventral view of the tails of two genera of the family Phrynosomatidae. Upper: *Cophosaurus texanus*, showing the presence of broad bands. Drawn from a preserved specimen (KU 203230). Lower: *Holbrookia lacerata*, showing the bands as narrow, reduced, or absent. Drawn from a preserved specimen (KU 477).

15a. Stripes sharply defined .... ***C. sonorae***
S 115, pl 33

15b. Stripes less distinctly defined, often fading on the neck .... ***C. exsanguis***
CC 252, pl 18 **S** 111, pl 33 **CAAR** 516 (1991)

16a. With five or fewer scales between chin shields and lower labial scales (Fig. 171) and three or more preanal scales .... ***C. velox***
S 115, pl 32 **CAAR** 656 (1998)

16b. With more than six scales between chin shields and lower labial scales or fewer than three preanal scales 17

17a. Scales on rear of forearm distinctly enlarged .... ***C. uniparens***
CC 253, pl 18 **S** 112, pl 31

17b. Scales on rear of forearm granular or barely enlarged .... ***C. sexlineatus***
CC 250, pl 18 **S** 109, pl 31 **CAAR** 628 (1996)

18a. Small scales bordering supraocular scales extend to or past the middle of the second supraocular scale .... ***C. neomexicanus***
CC 256, pl 18 **S** 117, pl 32 **CAAR** 109 (1971)

18b. Small scales bordering supraocular scales not extending to the middle of the second supraocular scale 19

19a. Stripes usually evident and extending the length of the body and onto the tail; scales on rear of forearm slightly enlarged medially; less than 80 dorsal granules across midbody 20

19b. Dorsum and sides with spots, bars, or a reticulum of dark markings—if stripes are present, they fade or break up into spots on the lower back and base of the tail (especially middorsally), and the throat and venter are often very dark; scales on rear of forearm

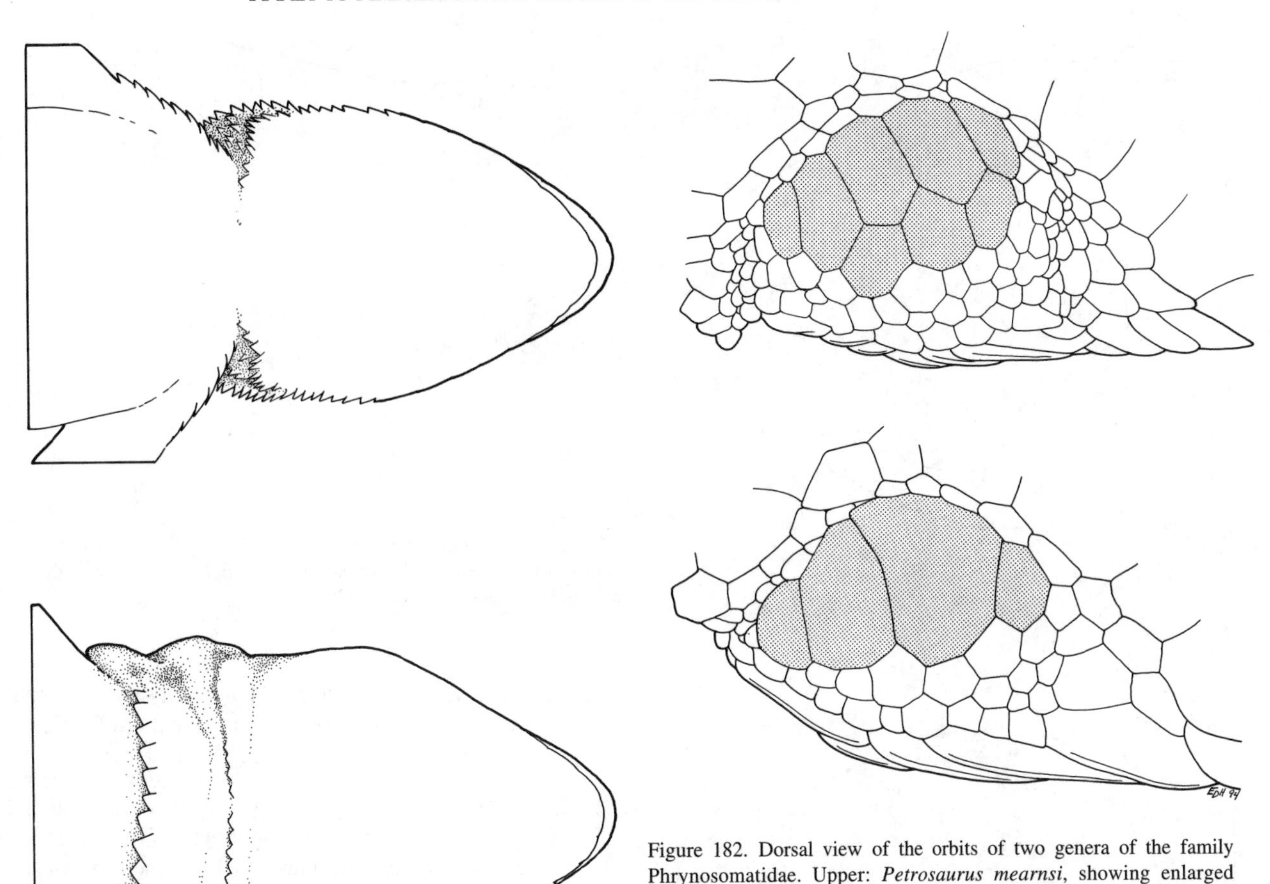

Figure 181. Ventral view of the heads and necks of two genera of the family Phrynosomatidae. Upper: *Sceloporus undulatus*, showing the lack of a transverse gular fold. Drawn from a preserved specimen (KU 214667). Lower: *Uta stansburiana*, showing the presence of a distinct transverse gular fold. Drawn from a preserved specimen (KU 194129).

Figure 182. Dorsal view of the orbits of two genera of the family Phrynosomatidae. Upper: *Petrosaurus mearnsi*, showing enlarged supraoculars (shaded) in two or more rows. Drawn from a preserved specimen (KU 61561). Lower: *Uta stansburiana*, showing enlarged supraoculars (shaded) in a single row. Drawn from a preserved specimen (KU 194095).

granular; more than 80 dorsal granules across midbody 23

20a. Ground color of dorsum and venter very pale, 7–8 stripes barely discernible or absent; from dunes of white sand in Otero County, New Mexico .... ***C. gypsi***
CC 255 (as western subspecies) **S** 113 (as *C. inornatus,* in part)

20b. Dorsal ground color dark, stripes distinct 21

21a. With six dorsal stripes, rarely with any trace of a middorsal stripe; venter blue in both sexes (very dark in males); from central and north-central Arizona .... ***C. pai***
CC 255 (as western subspecies) **S** 113 (as *C. inornatus,* in part)

21b. Usually 7–8 dorsal stripes; if six dorsal stripes, often with a trace of a middorsal stripe 22

22a. Usually with seven stripes; range restricted to Cochise County, Arizona .... ***C. arizonae***
CC 255 (as western subspecies) **S** 113 (as *C. inornatus,* in part)

22b. Usually with 7–8 stripes; from New Mexico and Texas south into México .... ***C. inornatus***
CC 255, pl 18 **S** 113, pl 32

23a. Sides with dark vertical bars, at most with traces of dorsal stripes; throat and chest light in color with some dark spots; from southern New Mexico and southwestern Texas south into México .... ***C. marmoratus***
CC 259, pl 18 **S** 110, pl 31 (as *C. tigris marmoratus*)

23b. Sides often without dark bars, stripes sometimes noticeable (but tend to fade or break up into spots on lower back); from the Great Basin and California south into México, barely into northwestern and extreme southwestern New Mexico (but New Mexico specimens lack dark vertical bars on sides, often retain stripes, and have dark throats and chests) .... ***C. tigris***
S 110, pl 31

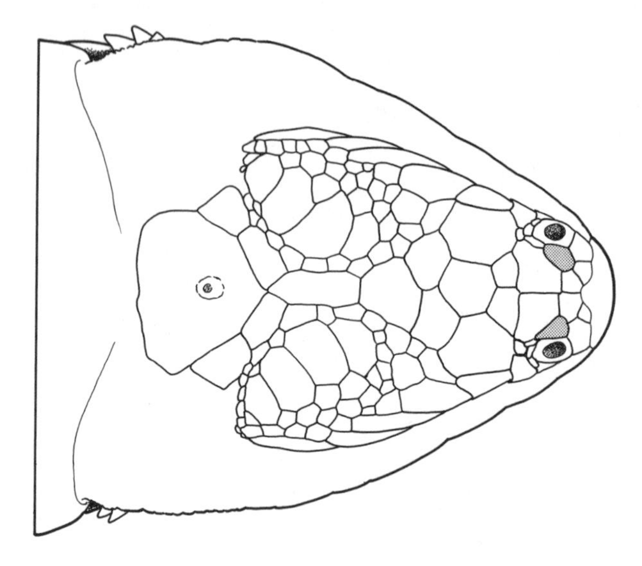

Figure 184. Dorsolateral view of the anterior half of *Uta stansburiana*, showing the presence and position of the dark axillary spot. Drawn from a preserved specimen (KU 194129).

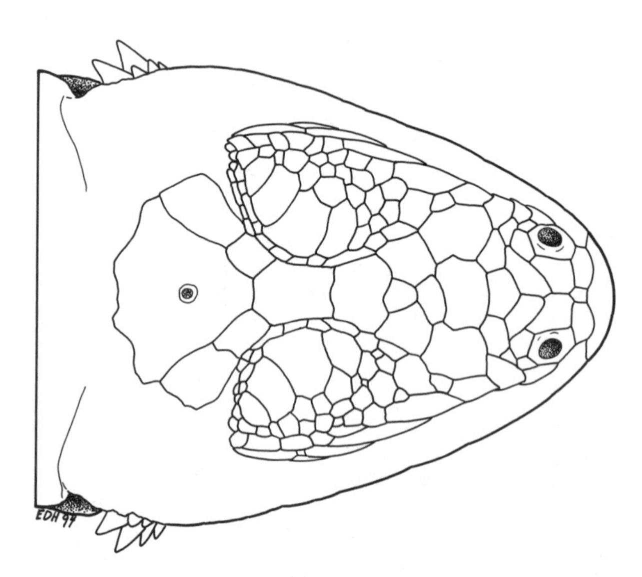

Figure 183. Dorsal view of the heads of two genera of the family Phrynosomatidae. Upper: *Uta stansburiana*, showing the presence of two supranasal scales (shaded). Drawn from a preserved specimen (KU 45918). Lower: *Urosaurus ornatus*, showing the absence of supranasal scales. Drawn from a preserved specimen (KU 88351).

POLYCHROTIDAE

Anolis

Key to Species of *Anolis*

1a. Ventral scales keeled 2
1b. Ventral scales smooth 4

2a. Tail round in section (Fig. 172); dewlap pink (in life) 3
2b. Tail laterally compressed (Fig. 172); dewlap rusty-red and white-edged (in life) .... ***A. sagrei*** (A)
CC 211, pl 13, p 211

3a. Frontal ridges prominent and distinctly higher than canthal ridges; range restricted to southern Florida .... ***A. porcatus*** (A)
CAAR 541 (1992)
3b. Frontal ridges at most barely higher than canthal ridges; range extends throughout the southeastern U.S. north to the Carolinas and Tennessee and west to eastern Texas and extreme southeastern Oklahoma .... ***A. carolinensis***
CC 210, pl 13, p 211

Figure 185. Dorsal view of the head of *Phrynosoma solare*, showing the presence of four occipital spines continuous with the temporal spines. Drawn from a preserved specimen (KU 13192).

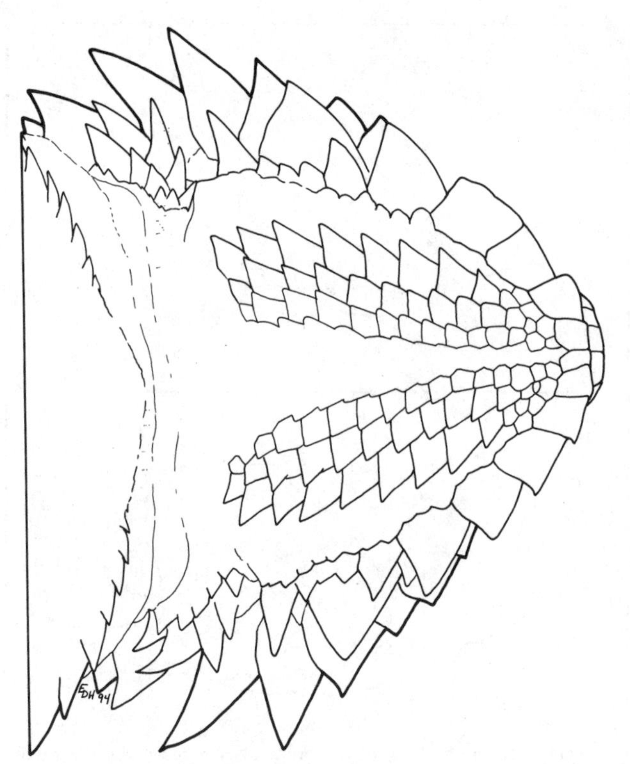

Figure 186. Ventral view of the head and neck of *Phrynosoma coronatum*, showing the presence of three rows of enlarged paramedian scales on each side of the throat. Drawn from a preserved specimen (KU 31337).

4a. Spinelike scales form a middorsal row; size large (SVL to more than 110 mm) 5
4b. No middorsal row of spinelike scales; size smaller (SVL always less than 80 mm) 6

5a. Head with a distinct bony casque (canthal and occipital ridges prominent) (Fig. 173); middorsal spines low; dewlap pinkish-white (in life) .... ***A. equestris*** (A)
CC 210, pl 13

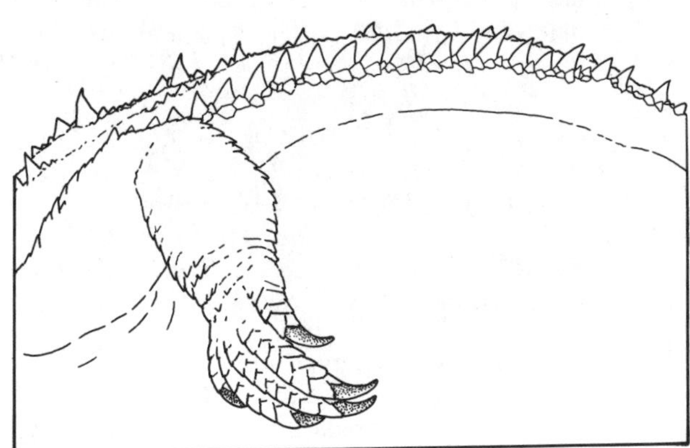

Figure 187. Ventrolateral view of the body of *Phrynosoma hernandesi*, showing the presence of an enlarged row of scales along the abdomen. Drawn from a preserved specimen (KU 27818).

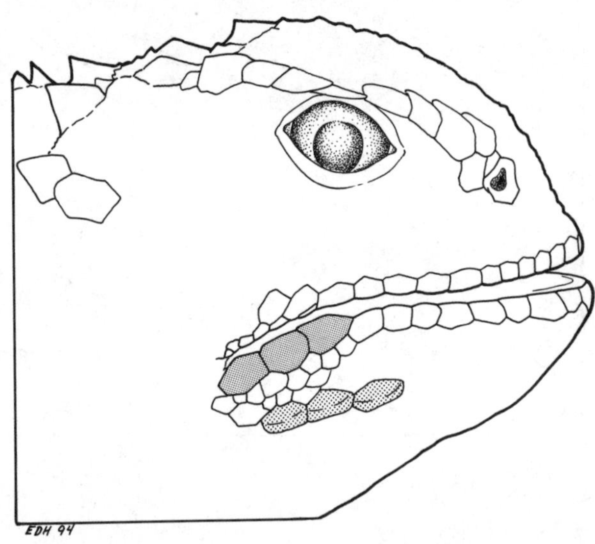

Figure 188. Lateral view of the head of *Phrynosoma hernandesi*, showing posterior chin shields (light shading) smaller than the posterior infralabials (dark shading). Drawn from a preserved specimen (KU 27818).

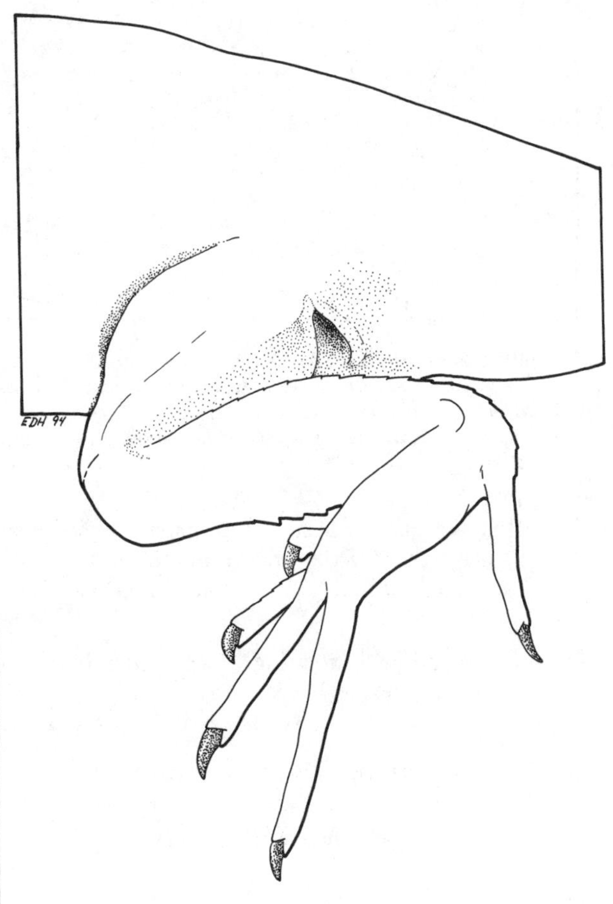

Figure 189. Lateral view of the pelvic area of *Sceloporus marmoratus*, showing the presence and position of the postfemoral pocket. Drawn from a preserved specimen (KU).

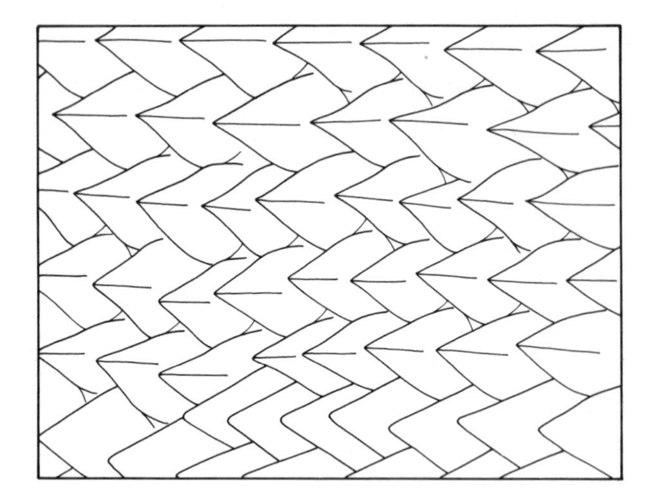

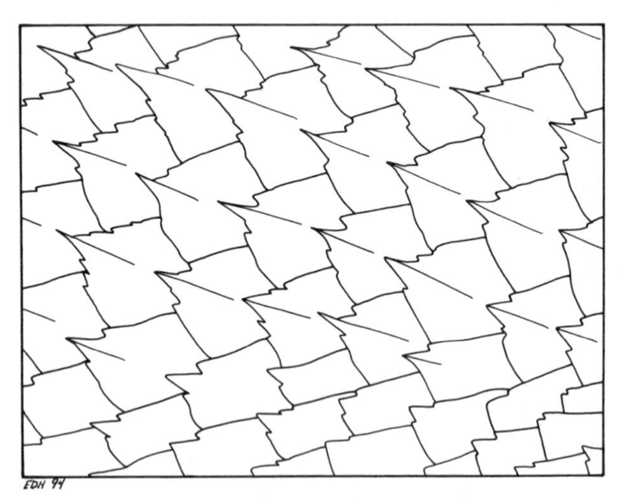

Figure 190. Lateral view of the body scales of two species of the genus *Sceloporus*. Upper: *S. scalaris*, showing lateral scales rows parallel to the body axis. Drawn from a preserved specimen (KU 102937). Lower: *S. clarkii*, showing lateral scale rows ascending diagonally. Drawn from a preserved specimen (KU 218953).

5b. Bony casque on head less prominent; middorsal spines higher; dewlap orange (in life) .... ***A. garmani*** (A)
CC 213, pl 13

6a. Tail round in section (Fig. 172); dewlap blue and yellow or blue and black (in life) .... ***A. chlorocyanus*** (A)

This recently introduced species is currently restricted to southern Florida.

6b. Tail laterally compressed (Fig. 172) 7

7a. Snout with two paramedian series of paired scales (Fig. 174); tail crossbanded; small (SVL less than 60 mm); dewlap yellow, often tinged with orange (in life) .... ***A. distichus*** (A)
CC 214, pl 13, p 215 **CAAR** 108 (1971)

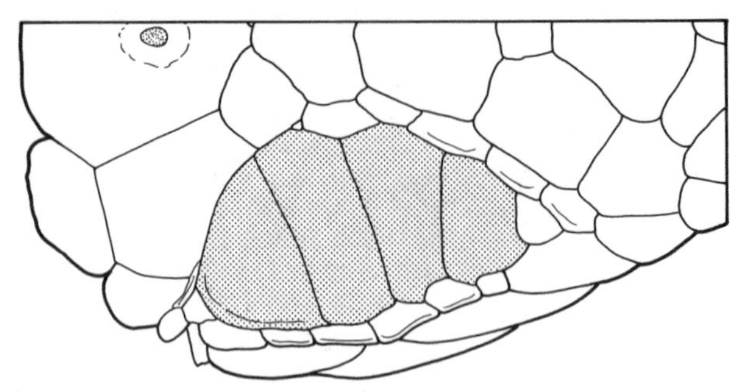

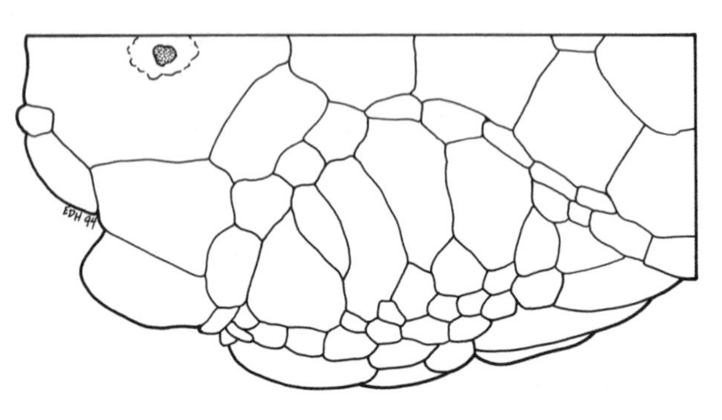

Figure 191. Dorsal view of the orbits of two species of the genus *Sceloporus*. Upper: *S. clarkii*, showing the enlarged supraoculars (shaded) in a single row, with the posterior two scales not separated from median head scales by smaller scales. Drawn from a preserved specimen (KU 218953). Lower: *S. jarrovii*, showing the enlarged supraoculars in more than a single row, irregular in position, and separated from median head scales by smaller scales. Drawn from a preserved specimen (KU 74435).

7b. Snout without paramedian series of paired scales (Fig. 174); tail not crossbanded; larger (SVL often more than 65 mm) 8

8a. Chin distinctly mottled; venter with scattered spots; males with a high crest on tail (Fig. 175); dewlap largely mustard yellow (in life) .... ***A. cristatellus*** (A)
CC 213, pl 13

8b. Chin at most faintly mottled; venter usually immaculate; males without a crest on tail; dewlap usually pale yellow (in life) .... ***A. cybotes*** (A)
CC 214, pl 13, p 211

IGUANIDAE

Key to Genera of Iguanidae

1a. Rostral scale divided .... ***Sauromalus*** (p. 91)
1b. Rostral scale not divided 2

2a. With whorls of enlarged spines on tail (Fig. 176) .... ***Ctenosaura*** (p. 91)
2b. Without whorls of enlarged spines on tail 3

3a. Enlarged middorsal scales comblike, largest on neck; often with an enlarged scale on lower jaw below ear (Fig. 177) .... ***Iguana*** (p. 91)

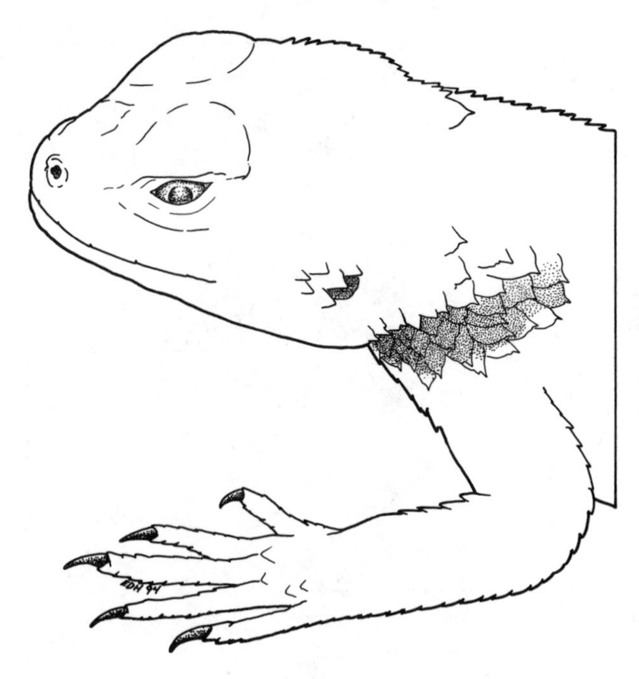

Figure 192. Dorsolateral view of the head and shoulder of *Sceloporus magister*, showing the presence and position of the shoulder patch. Drawn from a preserved specimen (KU 176455).

3b. Enlarged middorsal scales form a low crest, not higher on neck; without an enlarged scale on lower jaw below ear (Fig. 178) .... ***Dipsosaurus*** (p. 91)

Sauromalus

Sauromalus obesus
S 73, pl 20

Ctenosaura

Ctenosaura pectinata (A)
CC 217, pl 12

Iguana

Iguana iguana (A)
CC 217, pl 12

Dipsosaurus

CAAR 242 (1992)

Dipsosaurus dorsalis
S 74, pl 24 **CAAR** 242 (1992)

Phrynosomatidae

Key to Genera of Phrynosomatidae

1a. Head with bony spines or a posteriorly projecting ridge .... ***Phrynosoma*** (p. 92)
1b. Head without bony spines or a projecting ridge 2

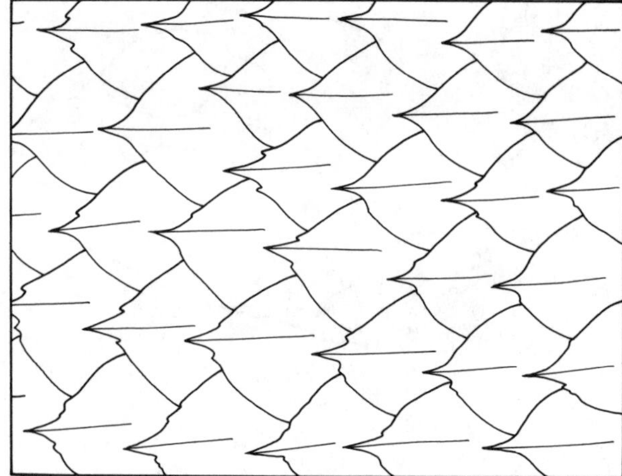

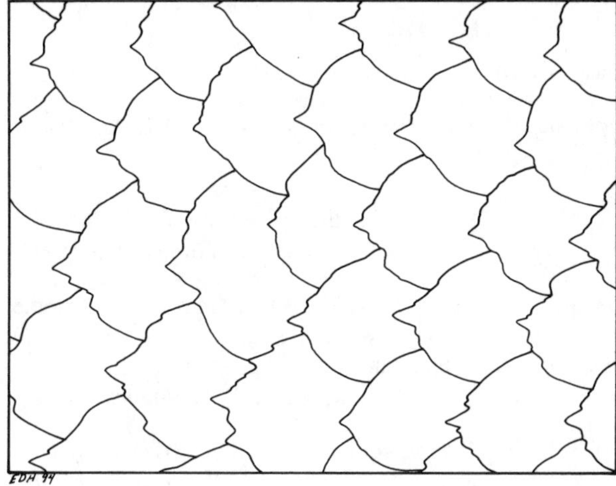

Figure 193. Dorsal view of the body scales of two species of the genus *Sceloporus*. Upper: *S. magister*, showing the dorsal scales deeply notched posteriorly. Drawn from a preserved specimen (KU 62893). Lower: *S. orcutti*, showing the dorsal scales with shallow posterior notches. Drawn from a preserved specimen (KU 49690).

2a. Seams between supralabial scales diagonal (Fig. 179); median postmental scale present 3
2b. Seams between supralabial scales vertical (Fig. 179); median postmental scale absent 6

3a. Ear opening distinct 4
3b. Ear opening absent 5

4a. Ear opening larger than interparietal scale; scales project over ear opening; prominent fringe of scales present on sides of toes (Fig. 148) ...... ***Uma*** (p. 93)
4b. Ear opening smaller than interparietal scale; no scales projecting over ear or forming fringes on sides of toes .... ***Callisaurus*** (p. 94)

5a. Broad black bands present under tail (Fig. 180) .... ***Cophosaurus*** (p. 94)

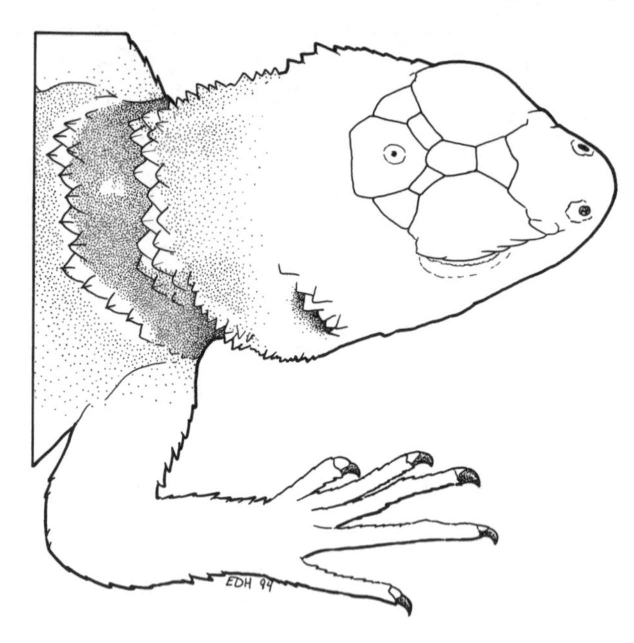

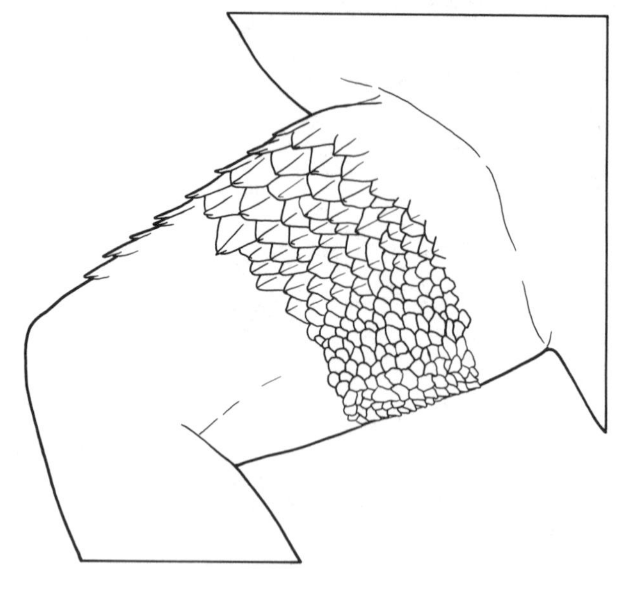

Figure 194. Dorsolateral view of the head and shoulder of *Sceloporus jarrovii*, showing the presence and location of a broad, white-bordered black collar on the neck. Drawn from a preserved specimen (KU 126856).

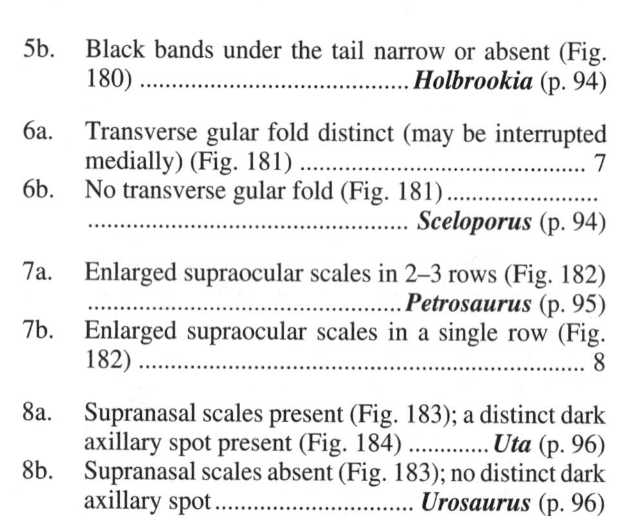

5b. Black bands under the tail narrow or absent (Fig. 180) .... ***Holbrookia*** (p. 94)

6a. Transverse gular fold distinct (may be interrupted medially) (Fig. 181) 7

6b. No transverse gular fold (Fig. 181) .... ***Sceloporus*** (p. 94)

7a. Enlarged supraocular scales in 2–3 rows (Fig. 182) .... ***Petrosaurus*** (p. 95)

7b. Enlarged supraocular scales in a single row (Fig. 182) 8

8a. Supranasal scales present (Fig. 183); a distinct dark axillary spot present (Fig. 184) .... ***Uta*** (p. 96)

8b. Supranasal scales absent (Fig. 183); no distinct dark axillary spot .... ***Urosaurus*** (p. 96)

Phrynosoma

Key to Species of *Phrynosoma*

1a. Four large occipital spines are continuous with temporal spines (Fig. 185) .... ***P. solare***
S 100, pl 21 **CAAR** 162 (1974)

1b. Only two large occipital spines, or spines reduced or absent 2

2a. Three or more rows of enlarged paramedian scales present on throat (Fig. 186) .... ***P. coronatum***
S 97, pl 21 **CAAR** 428 (1988)

2b. Fewer than three rows of enlarged paramedian scales present on throat 3

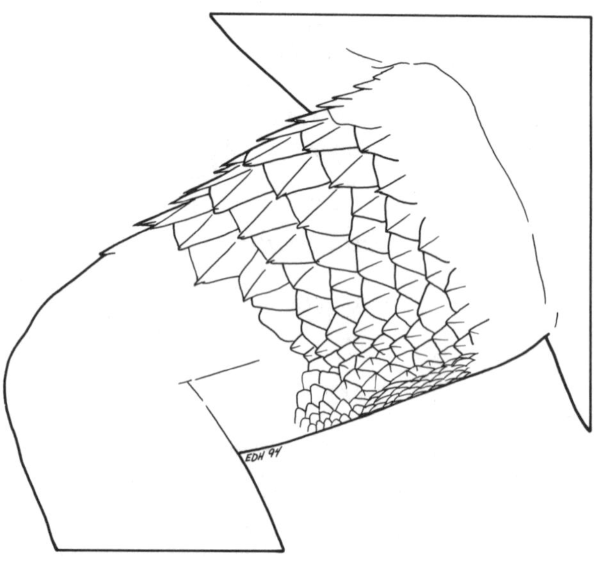

Figure 195. Dorsal view of the thighs of two species of the genus *Sceloporus*. Upper: *S. grammicus*, showing the granular scales on the rear edge. Drawn from a preserved specimen (KU 1003). Lower: *S. olivaceus*, showing the non-granular, keeled, overlapping scales on the rear edge. Drawn from a preserved specimen (KU 15041).

3a. At least one row of enlarged scales present on sides of abdomen (Fig. 187) 4

3b. No enlarged scale rows on sides of abdomen .... ***P. modestum***
CC 243, pl 14, p 242 **S** 99, pl 21 **CAAR** 630 (1996)

4a. Posterior chin shields smaller than posterior infralabial scales (Fig. 188) .... ***P. douglasii***
CC 243, pl 14, p 242 **S** 96, pl 21

4b. Posterior chin shields much larger than posterior infralabial scales 5

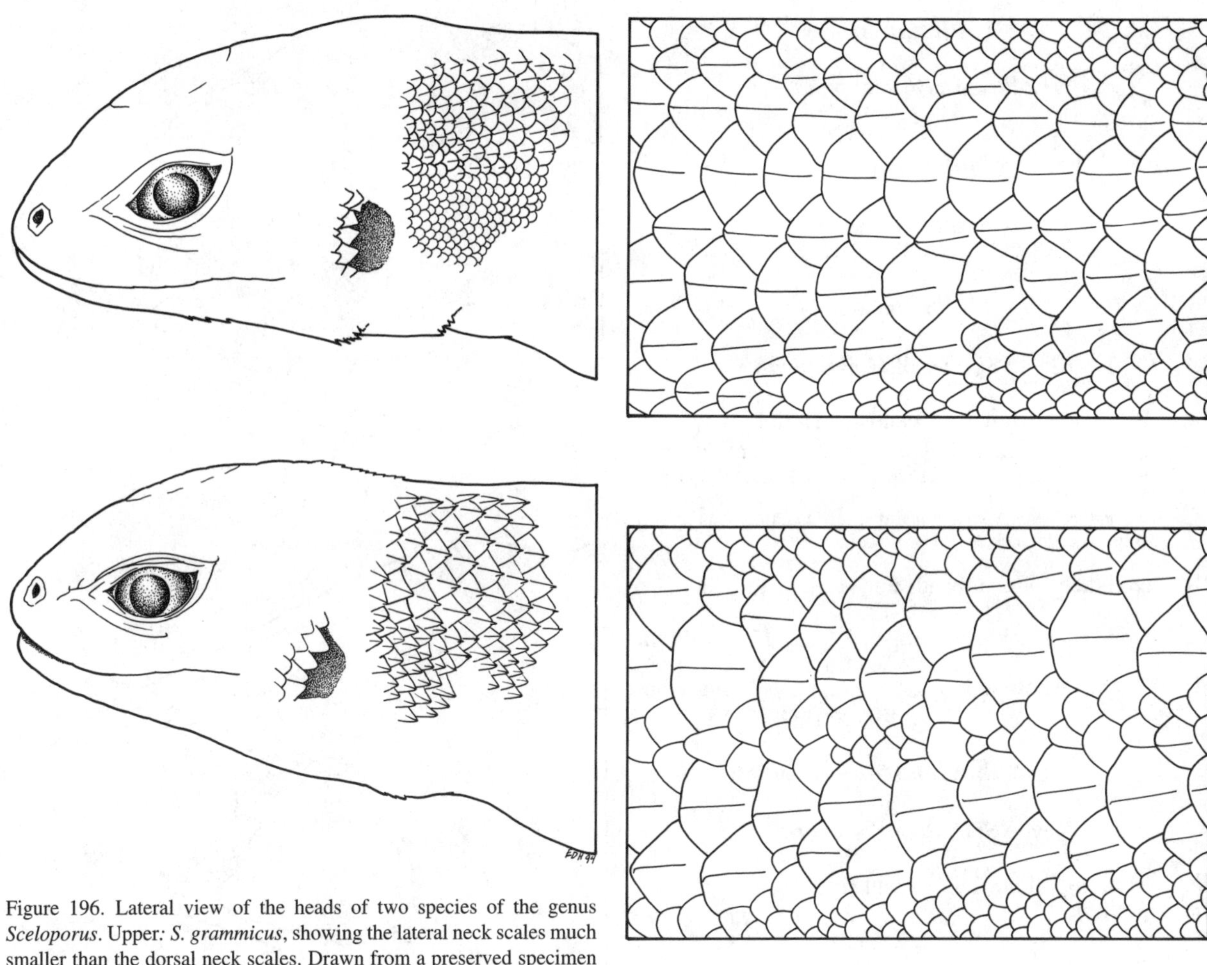

Figure 196. Lateral view of the heads of two species of the genus *Sceloporus*. Upper: *S. grammicus*, showing the lateral neck scales much smaller than the dorsal neck scales. Drawn from a preserved specimen (KU 11001). Lower: *S. vandenburgianus*, showing the lateral neck scales nearly as large as the dorsal neck scales. Drawn from a preserved specimen (KU 193854).

5a. Vertebral line dark; spines long .... ***P. mcallii***
S 102, pl 21 **CAAR** 281 (1981)

5b. No dark vertebral line; spines short 6

6a. One enlarged scale row present on sides of abdomen; ventral scales smooth .... ***P. platyrhinos***
S 98, pl 21 **CAAR** 517 (1991)

6b. Two enlarged scale rows present on sides of abdomen; ventral scales keeled .... ***P. cornutum***
CC 241, pl 14, p 242 **S** 101, pl 21 **CAAR** 469 (1990)

Uma

Key to Species of *Uma*

1a. Lateral black streaks on throat connect medially to form crescents .... ***U. scoparia***
S 78, pl 23 **CAAR** 155 (1974)

1b. Lateral black streaks on throat not connected medially .... ***U. notata***
S 78, pl 23 **CAAR** 197 (1977)
and **CAAR** 126 (1973, as *U. inornata*)

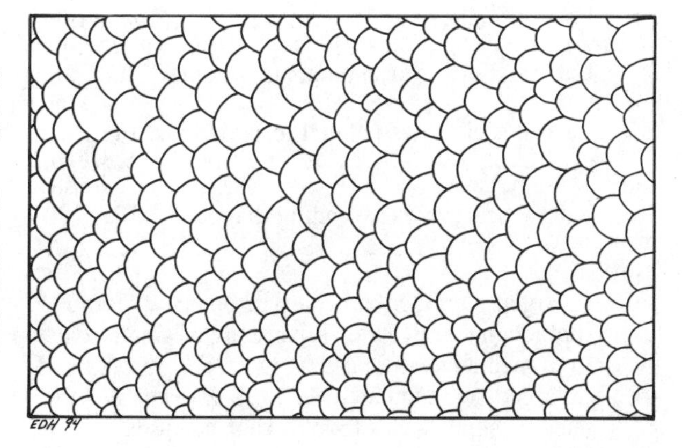

Figure 197. Dorsal view of the middorsal body scales of three species of the genus *Urosaurus*. Upper: *U. graciosus*, showing a broad band of enlarged scales abruptly differentiated from the smaller lateral scales. Drawn from a preserved specimen (KU 72745). Middle: *U. ornatus*, showing enlarged scales arranged in two rows and abruptly differentiated from smaller scales. Drawn from a preserved specimen (KU 61558). Lower: *U. microscutatus*, showing slightly enlarged middorsal scales that gradually merge with smaller lateral scales. Drawn from a preserved specimen (BWMC 56).

Callisaurus

Callisaurus draconoides
S 77, pl 22

Cophosaurus

Cophosaurus texanus
CC 222, pl 14, p 223 **S** 76, pl 22

Holbrookia

Key to Species of *Holbrookia*

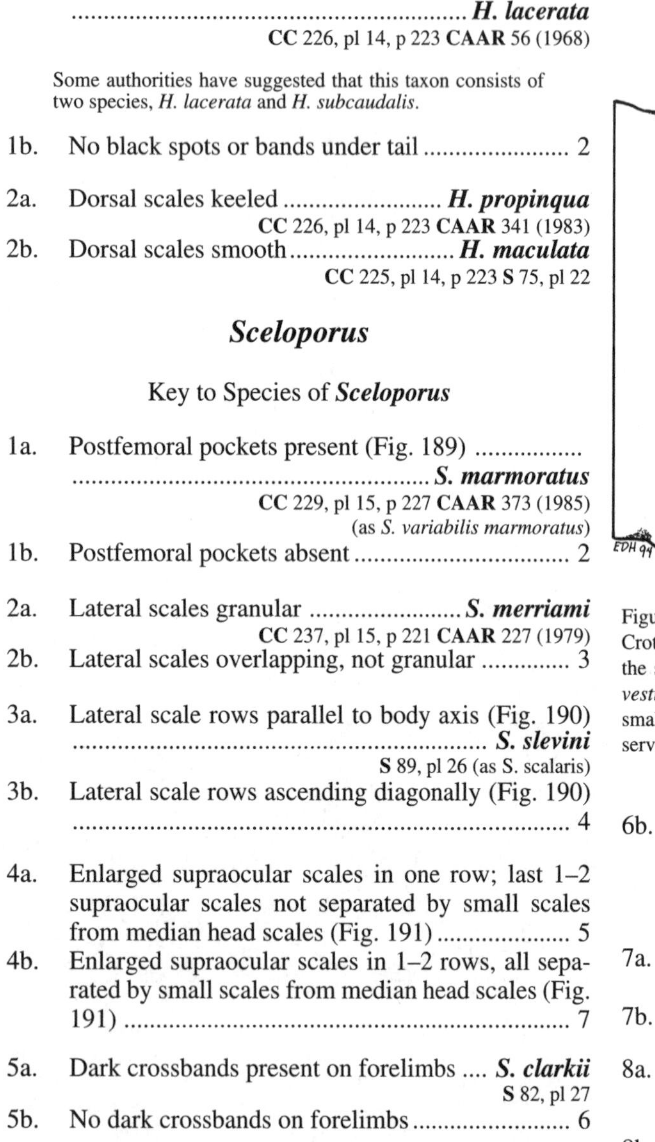

1a. Black spots or narrow bands under tail .... ***H. lacerata***
CC 226, pl 14, p 223 **CAAR** 56 (1968)

Some authorities have suggested that this taxon consists of two species, *H. lacerata* and *H. subcaudalis*.

1b. No black spots or bands under tail 2

2a. Dorsal scales keeled .... ***H. propinqua***
CC 226, pl 14, p 223 **CAAR** 341 (1983)

2b. Dorsal scales smooth.......................... ***H. maculata***
CC 225, pl 14, p 223 **S** 75, pl 22

Sceloporus

Key to Species of *Sceloporus*

1a. Postfemoral pockets present (Fig. 189) .... ***S. marmoratus***
CC 229, pl 15, p 227 **CAAR** 373 (1985) (as *S. variabilis marmoratus*)

1b. Postfemoral pockets absent 2

2a. Lateral scales granular .... ***S. merriami***
CC 237, pl 15, p 221 **CAAR** 227 (1979)

2b. Lateral scales overlapping, not granular 3

3a. Lateral scale rows parallel to body axis (Fig. 190) .... ***S. slevini***
S 89, pl 26 (as S. scalaris)

3b. Lateral scale rows ascending diagonally (Fig. 190) 4

4a. Enlarged supraocular scales in one row; last 1–2 supraocular scales not separated by small scales from median head scales (Fig. 191) 5

4b. Enlarged supraocular scales in 1–2 rows, all separated by small scales from median head scales (Fig. 191) 7

5a. Dark crossbands present on forelimbs .... ***S. clarkii***
S 82, pl 27

5b. No dark crossbands on forelimbs 6

6a. Distinct black shoulder patch present (Fig. 192); dorsal scales deeply notched posteriorly on either side of the median spine (Fig. 193)...... ***S. magister***
CC 231, pl 16 **S** 81, pl 27 **CAAR** 290 (1982)

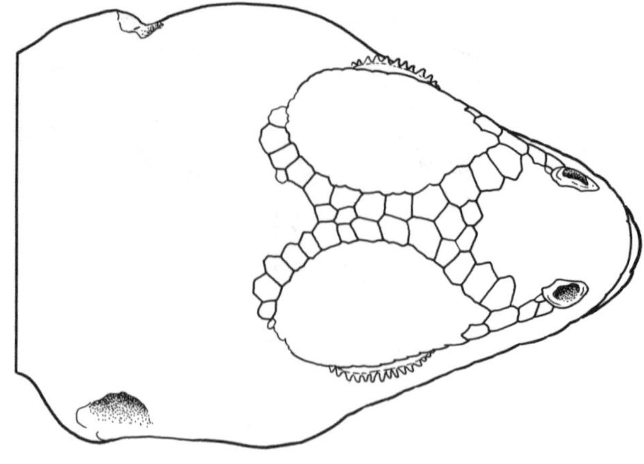

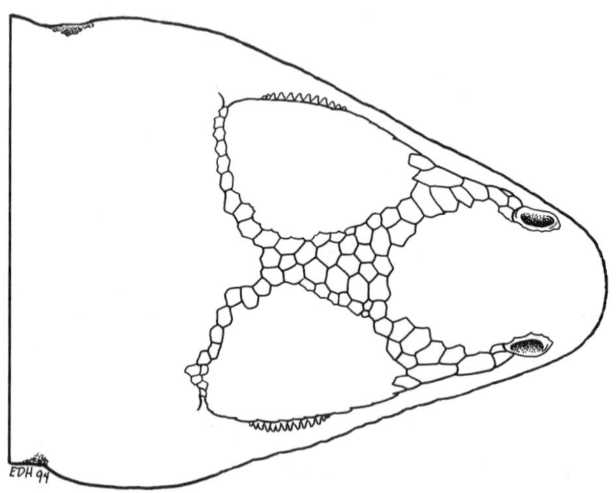

Figure 198. Dorsal view of the heads of two genera of the family Crotaphytidae. Upper: *Crotaphytus*, showing two rows of scales between the supraorbital semicircles. Drawn from a preserved specimen of *C. vestigium* (KU 126910). Lower: *Gambelia*, showing several rows of small scales between the supraorbital semicircles. Drawn from a preserved specimen of *G. wislizenii* (KU 121748).

6b. Black shoulder patch indistinct or absent; dorsal scales with a shallow notch posteriorly on either side of the median spine (Fig. 193) .... ***S. orcutti***
S 84, pl 27 **CAAR** 265 (1980)

7a. Broad light-bordered black collar present (Fig. 194) 8

7b. Broad light-bordered black collar absent 10

8a. Enlarged supraocular scales in one row .... ***S. jarrovii***
S 90, pl 26

8b. Supraocular scales in two rows or irregular......... 9

9a. Tail with crossbands; supraocular scales in two rows .... ***S. poinsettii***
CC 230, pl 16 **S** 83, pl 27

9b. Crossbands on tail indistinct or absent; supraocular scales irregular .... ***S. serrifer***
CC 230, pl 16

10a. Scales on rear of thigh granular (Fig. 195) 11
10b. Scales on rear of thigh overlapping and keeled, not granular (Fig. 195) 14

11a. Lateral neck scales much smaller than and well differentiated from dorsal neck scales (Fig. 196) .... ***S. grammicus***
CC 229, pl 15, p 227
11b. Lateral neck scales barely smaller than dorsal neck scales (Fig. 196) 12

12a. Pattern consisting of a very pale ground color with faint, slightly darker dorsolateral lines extending posteriorly from above the ear to the tail; males without blue (in life) markings on venter and throat .... ***S. arenicolus***
CC 237, pl 15 **S** 87 **CAAR** 386 (as *S. graciosus arenicolus*)
12b. Pattern with four longitudinal rows of spots that may sometimes merge to form stripes; usually with two dorsolateral pale lines extending posteriorly from eye to tail; often with an irregular black spot on each shoulder; males with blue (in life) patches on sides of venter and on throat 13

13a. Blue belly patches of males usually separated by light areas; venter of females light; from extreme western Dakotas, Wyoming, western Colorado, and northwestern New Mexico west into the Great Basin and much of California (but not southern California) .... ***S. graciosus***
CC 237, pl 15 **S** 87, pl 26 **CAAR** 386 (1986)
13b. Blue belly patches of males extensive, often connected with throat patch; venter of females dusky; from southern California and northern Baja California .... ***S. vandenburgianus***
CC 237 **S** 87 **CAAR** 386 (1986)
(as *S. graciosus vandenburgianus*)

14a. Posterior surface of thigh immaculate; row of enlarged supraocular scales covers ¾ or more of supraocular area .... ***S. olivaceus***
CC 231, pl 16 **CAAR** 143 (1973)
14b. Posterior surface of thigh with a dark longitudinal line or blotches; row of enlarged supraocular scales covers ⅔ or less of supraocular area 15

15a. Scales on posterior surface of thigh abruptly differentiated from larger scales on dorsal surface of thigh .... ***S. occidentalis***
S 85, pl 26, p 132 **CAAR** 631 (1996)
15b. Scales on posterior surface of thigh gradually merging with slightly larger scales on dorsal surface of thigh 16

16a. With a distinct dark lateral stripe; dorsal pattern usually indistinct .... ***S. woodi***
CC 236, pl 15 **CAAR** 196 (1977)

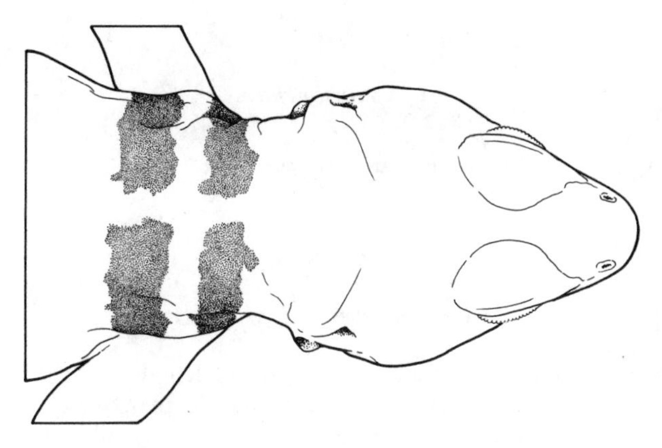

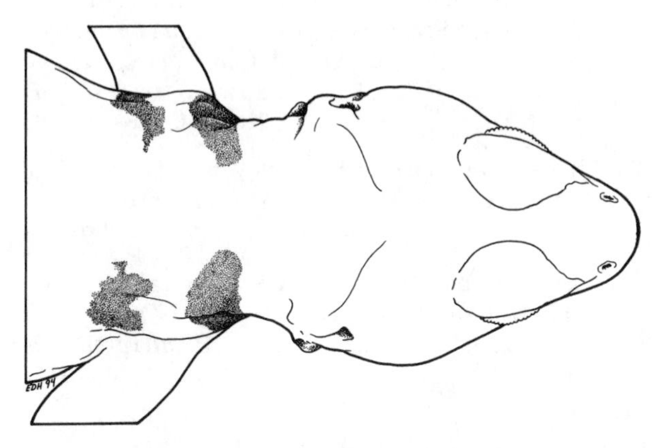

Figure 199. Dorsal view of the heads and shoulders of two species of the genus *Crotaphytus*. Upper: *C. bicinctores*, showing the presence and position of two robust neck collars narrowly separated at midline (12 pale scales or fewer). Drawn from preserved specimens (KU 126913 and 126915). Lower: *C. vestigium,* showing the presence and position of two neck collars widely separated at midline (13 pale scales or more). Drawn from preserved specimens (KU 1346, 121464, and 121465).

16b. Dark lateral stripe indistinct or with two dark lateral stripes; dorsal pattern often with distinct stripes or dark chevrons 17

17a. With distinct light dorsolateral stripes; males unmarked below; usually with notches on femoral pore scales .... ***S. virgatus***
S 88, pl 26 **CAAR** 72 (1968)
17b. Light dorsolateral stripes often indistinct; males with dark ventrolateral marks; without notches on femoral pore scales .... ***S. undulatus***
CC 233, pl 15 **S** 86, p 132

Some authorities have suggested that this taxon consists of three species, *S. undulatus*, *S. consobrinus,* and *S. tristichus*.

Petrosaurus
CAAR 494 (1990)

Petrosaurus mearnsi
S 95, pl 25 **CAAR** 495 (1990)

Uta

Uta stansburiana
CC 240, pl 14, p 221 **S** 93, pl 25

Some authorities have suggested that this taxon consists of two species, *U. stansburiana* and *U. stejnegeri.*

Urosaurus

Key to Species of *Urosaurus*

1a. Postfemoral pocket present; tail length less than twice SVL; enlarged middorsal scales in two rows or not abruptly demarcated from adjacent lateral scales (Fig. 197) 2

1b. Postfemoral pocket absent; tail length at least twice SVL; middorsal area with a broad band of enlarged scales abruptly demarcated from adjacent lateral scales (Fig. 197) .... ***U. graciosus***
S 92, pl 25 **CAAR** 448 (1988)

2a. One or two median rows of very small vertebral scales bordered by rows of much larger scales (Fig. 197) .... ***U. ornatus***
CC 239, pl 15 **S** 91, pl 25

2b. Slightly enlarged middorsal scales merging gradually with smaller lateral scales (Fig. 197) .... ***U. microscutatus***
S 94, pl 25

CROTAPHYTIDAE

Key to Genera of Crotaphytidae

1a. One or two rows of large scales present between supraorbital semicircles (Fig. 198) .... ***Crotaphytus*** (p. 96)

1b. Several rows of small scales present between supraorbital semicircles (Fig. 198) .... ***Gambelia*** (p. 97)

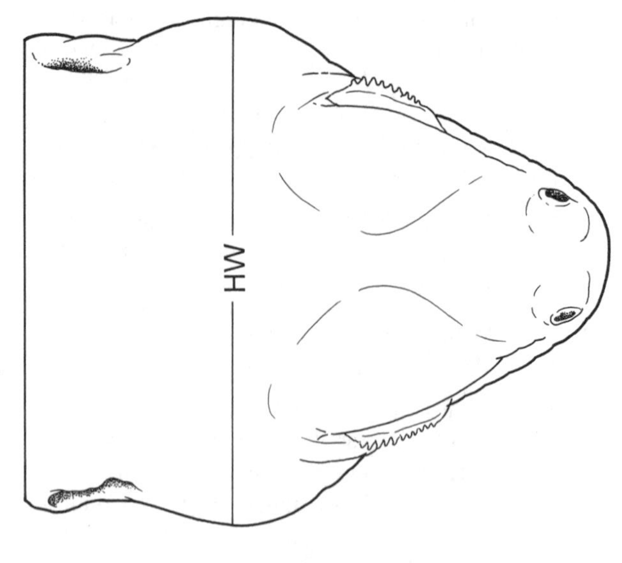

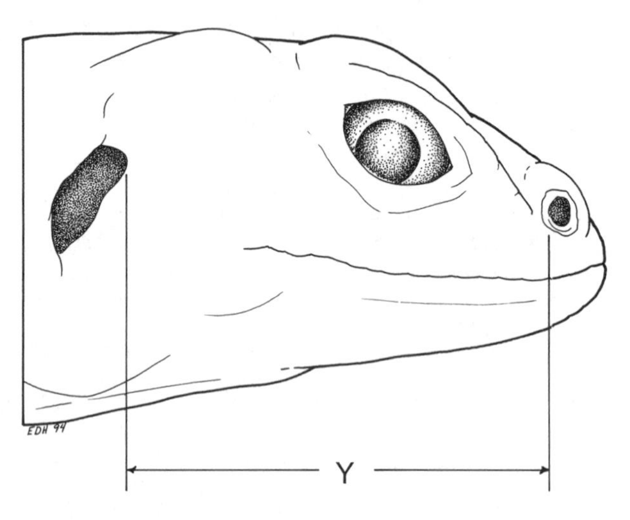

Figure 200. Views of the head of *Gambelia sila*. Upper: Dorsal view, showing how to measure head width (HW). Lower: Lateral view, showing how to measure distance from ear to nostril (Y) for comparison with HW. Drawn from a preserved specimen (KU 121537).

Crotaphytus

Key to Species of *Crotaphytus*

1a. One or two black collars and postfemoral pockets present 2

1b. No black collar or postfemoral pockets .... ***C. reticulatus***
CC 219, pl 17 **CAAR** 185 (1976)

2a. Tail round (or nearly so) in cross-section; interior throat lining black 3

2b. Tail compressed laterally; interior throat lining white 4

3a. Anterior collar marking incomplete ventrally (does not pass through gular fold) .... ***C. collaris***
CC 219, pl 17 **S** 80, pl 24

3b. Anterior collar marking complete ventrally .... ***C. nebrius***
S 80 (as *C. collaris nebrius*)

4a. Two black collars present, separated middorsally by twelve or fewer pale scales (Fig. 199); usually six or more internasal scales .... ***C. bicinctores***
S 80 (as *C. insularis bicintores*)

4b. Two black collars (rear collar occasionally reduced or absent), separated middorsally by thirteen or more pale scales (Fig. 199); usually five or fewer internasal scales .... ***C. vestigium***
S 80 (as *C. insularis vestigium*)

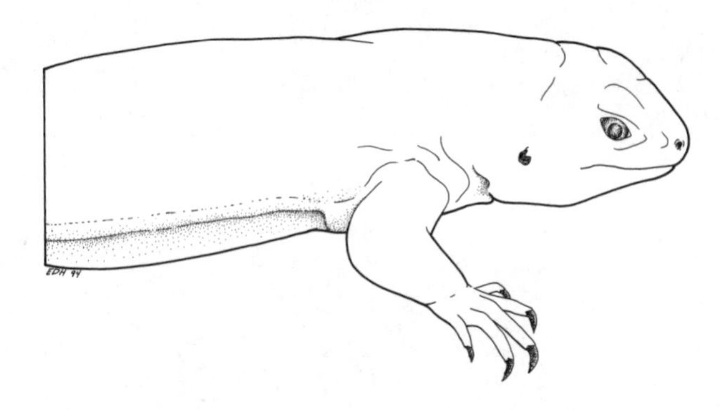

Figure 201. Lateral view of the head and body of *Leiocephalus schreibersii*, showing the presence and position of the lateral skin fold behind the front leg. Drawn from a preserved specimen (KU 93354).

Gambelia

Key to Species of ***Gambelia***

1a. Throat darkly mottled or blotched; width of head at least equal to distance from nostril to ear (Fig. 200) .... ***G. sila***
S 79, pl 24 **CAAR** 612 (1995)

1b. Throat with dark longitudinal streaks; width of head less than distance from nostril to ear 2

2a. Dorsal spotting extends onto the temporal region of the head and often onto the snout; dorsal ground color very pale .... ***G. wislizenii***
CC 220, pl 17 **S** 79, pl 24

2b. Dorsal spotting does not extend onto the top of the head or snout; dorsal ground color tan to dark brown .... ***G. copeii***
CC 220 (as western subspecies) (part)
S 79 (as *G. wislizenii*) (part)

CORYTOPHANIDAE

Basiliscus

Basiliscus vittatus (A)
CC 216, pl 12

TROPIDURIDAE

Leiocephalus

Key to Species of ***Leiocephalus***

1a. Distinct lateral fold of skin bearing scales smaller than those above and below the fold (Fig. 201) .... ***L. schreibersii*** (A)
CC 245, pl 12 **CAAR** 613 (1995)

1b. No distinct lateral fold of skin 2

2a. Dorsal surface of tail with transverse bands; preauricular scales enlarged; face mask absent or very faint .... ***L. carinatus*** (A)
CC 245, pl 12

2b. Dorsal surface of tail without transverse bands; preauricular scales small; dark face mask distinct in males (reduced in many females to a dark-bordered temporal rectangle) .... ***L. personatus*** (A)

This recently introduced species is currently restricted to southern Florida.

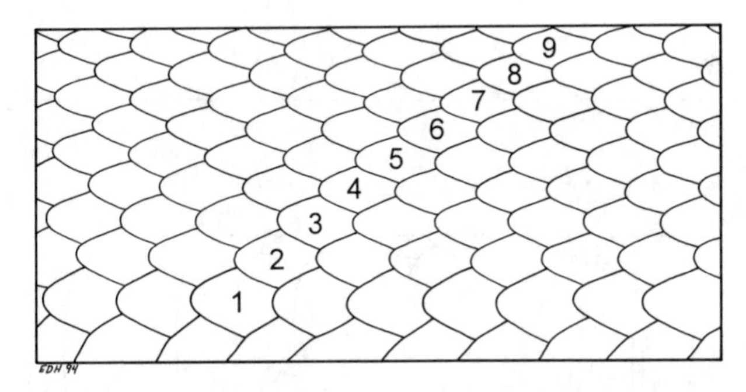

Figure 202. Generalized drawings of a snake. Upper: Dorsolateral view of the shape and configuration of a snake (*Storeria dekayi*). Drawn from a photograph by Suzanne L. Collins. Lower: Lateral view of the body of a snake, showing dorsal scales rows one to nine (lowest row is the edge of the ventral scales).

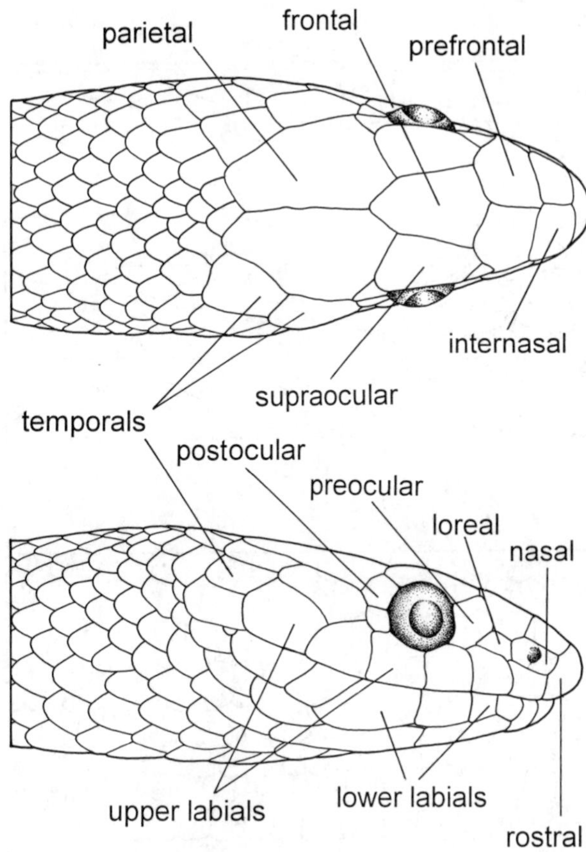

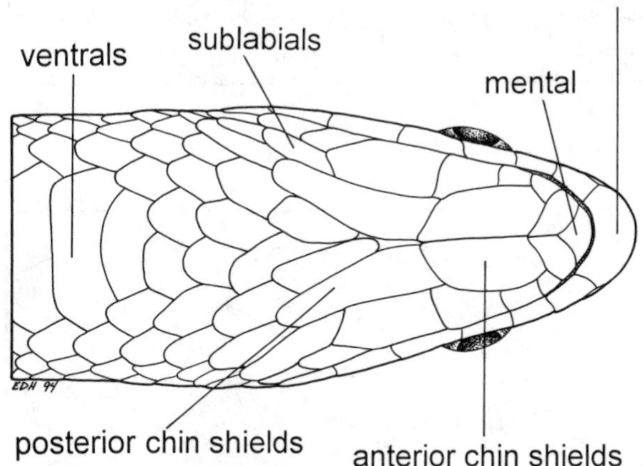

Figure 203. Dorsal (upper), lateral (middle), and ventral (lower) views of the head of a snake, with individual scales identified as used in the key. Drawn from a preserved specimen of *Opheodrys aestivus* (KU).

SERPENTES
(Snakes)

KEY TO FAMILIES OF SERPENTES

1a. Ventral scales not transversely elongated (Fig. 205); eyes covered by scales 2

1b. Ventral scales transversely elongated (if barely larger than dorsal scales, divided into two rows with a midventral sulcus) (Fig. 205); eyes not covered by scales 3

2a. Twenty scale rows at midbody .... **Typhlopidae** (p. 101)

2b. Fourteen scale rows at midbody .... **Leptotyphlopidae** (p. 101)

3a. Loreal pit present between eye and nostril (Fig. 206) .... **Viperidae** (p. 101)

Some authorities place the pit-vipers, subfamily Crotalinae (including all New World vipers), into the separate family Crotalidae.

3b. No loreal pit 4

4a. Scales on top of head behind eyes small (Fig. 207); chin shields not enlarged.............. **Boidae** (p. 104)

4b. Scales on top of head behind eyes enlarged (Fig. 207); chin shields enlarged 5

5a. Body and tail laterally compressed, with a midventral keel; ventral scales in two rows .... **Hydrophiidae** (p. 104)

Some authorities place species of Hydrophiidae in the family Elapidae (below). Species in both families possess a pair of permanently erect, grooved fangs in the anterior part of the upper jaws.

5b. Body and tail not laterally compressed; ventral scales in a single row 6

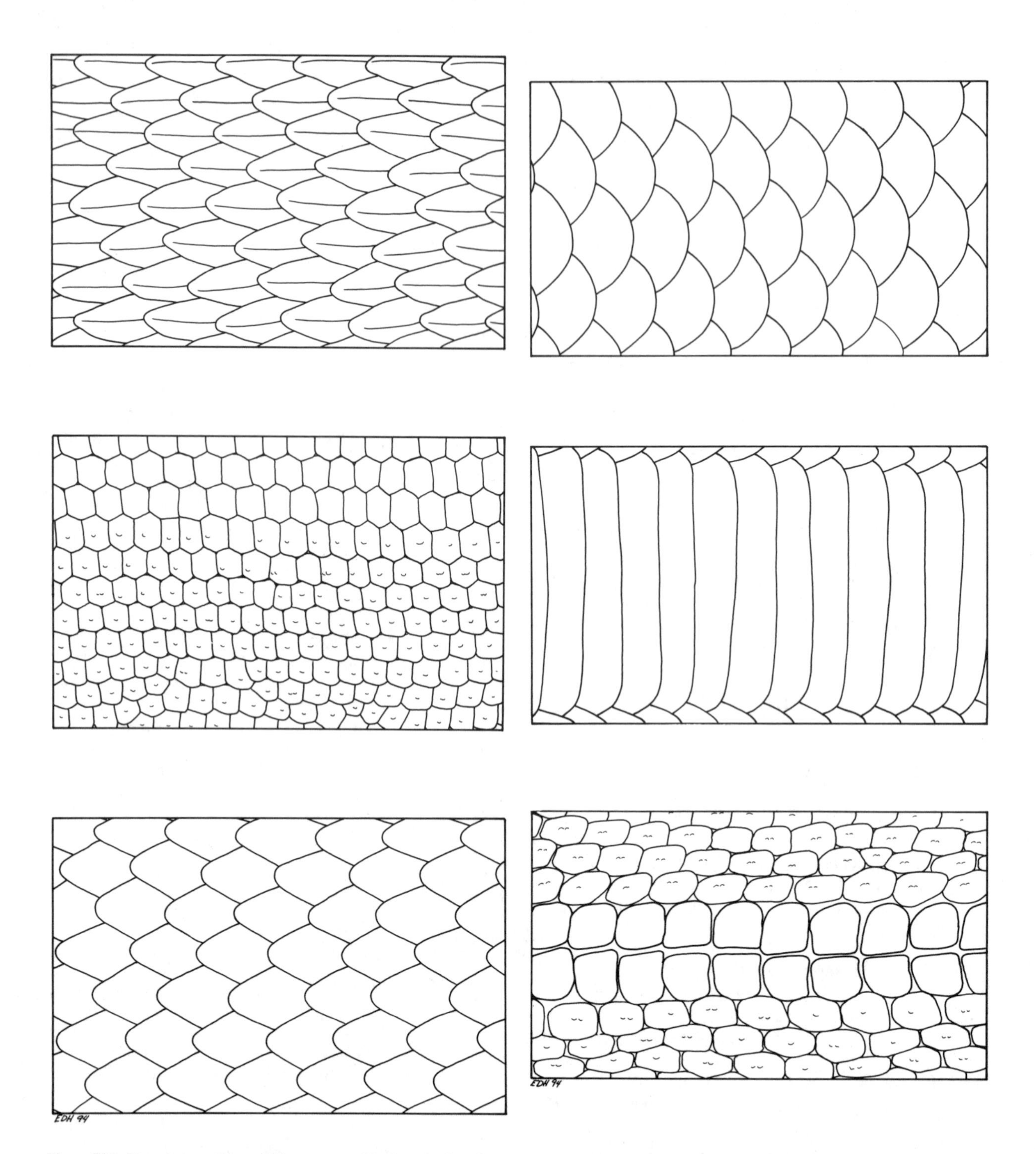

Figure 204. Dorsal view of three different types of body scales found on snakes. Upper: Elongate, overlapping, keeled scales. Drawn from a preserved specimen of *Sistrurus catenatus* (KU 220803). Middle: Nonoverlapping, weakly keeled scales that are hexagonal or quadrangular in shape. Drawn from a preserved specimen of *Pelamis platurus* (KU 183980). Lower: Elongate, overlapping, smooth scales. Drawn from a preserved specimen of *Lampropeltis triangulum* (KU 204843).

Figure 205. Ventral view of three different types of belly scales found on snakes. Upper: Overlapping scales of equal size; not transversely elongate. Drawn from a preserved specimen of *Leptotyphlops humilus* (KU 61352). Middle: Overlapping, transversely elongate scales. Drawn from a preserved specimen of *Sistrurus catenatus* (KU 220803). Lower: Two rows of slightly enlarged scales. Drawn from a preserved specimen of *Pelamis platurus* (KU 183930).

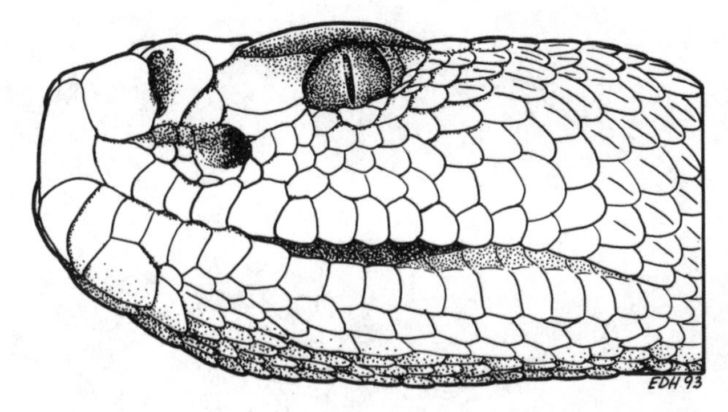

Figure 206. Lateral view of the head of *Crotalus atrox*, showing the size and position of the loreal pit, between and below the nostril and eye.

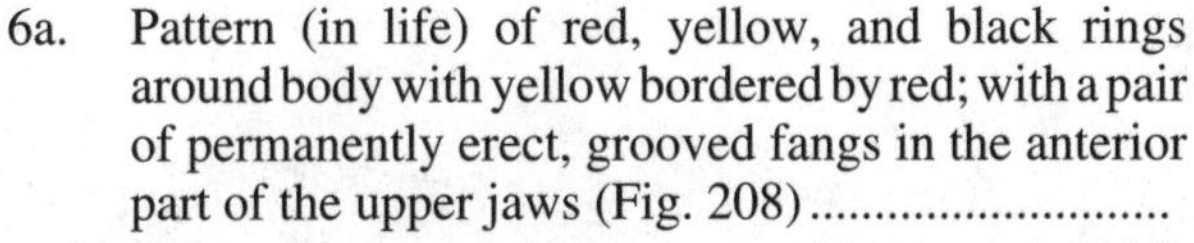

6a. Pattern (in life) of red, yellow, and black rings around body with yellow bordered by red; with a pair of permanently erect, grooved fangs in the anterior part of the upper jaws (Fig. 208) .... **Elapidae** (p. 104)

Some authorities place American members of the family Elapidae in the family Micruridae.

6b. Uniformly pigmented or with pattern involving stripes, blotches, crossbands, or rings (if rings around body are red, yellow, and black , yellow is never bordered by red); without a pair of permanently erect, grooved fangs in the anterior part of the upper jaws .... **Colubridae** (p. 105)

Some authorities divide the family Colubridae into three families, Natricidae, Xenodontidae, and Colubridae.

Typhlopidae

Ramphotyphlops

***Ramphotyphlops braminus* (A)**
CC 284

Leptotyphlopidae
CAAR 230 (1979)

Leptotyphlops
CAAR 230 (1979)

Key to Species of ***Leptotyphlops***

1a. Supraocular scales present (Fig. 209) ...... ***L. dulcis***
CC 284, pl 33, p 285 **S** 125, pl 36 **CAAR** 231 (1979)

1b. No supraocular scales .... ***L. humilis***
CC 285, p 285 **S** 123, pl 36 **CAAR** 232 (1979)

Viperidae

Key to Genera of Viperidae

1a. Tail pointed, without rattles; scales weakly keeled .... ***Agkistrodon*** (p. 101)

1b. Tail blunt, nearly always with at least one segment of a rattle; scales distinctly keeled 2

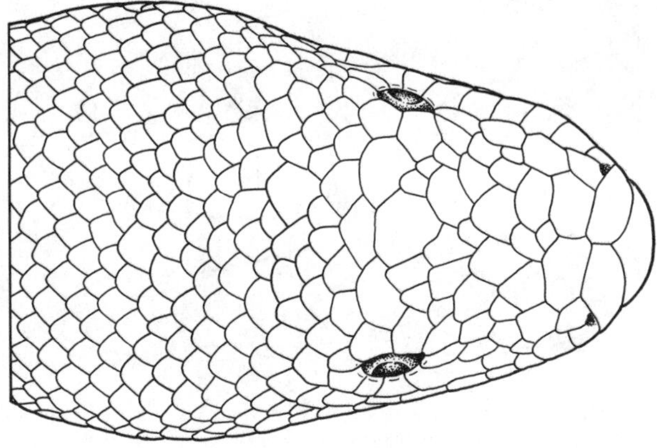

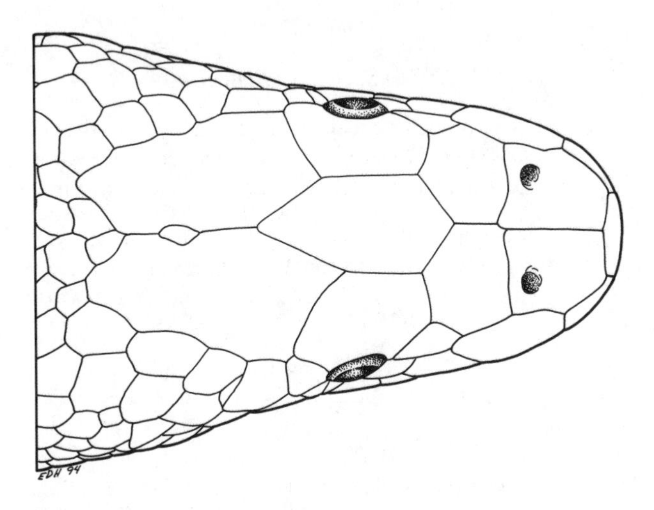

Figure 207. Dorsal view of scales on the heads of two families of snakes: Upper: *Charina trivirgata*, showing the small scales on top of the head behind the eyes, typical of the family Boiidae. Drawn from a preserved specimen (KU 80921). Lower: *Pelamis platurus*, showing the enlarged scales on top of the head behind the eyes, typical of the family Hydrophiidae. Drawn from a preserved specimen (KU 183930).

2a. Area between eyes with large scales (Fig. 210) .... ***Sistrurus*** (p. 102)

2b. Area between eyes with small scales (Fig. 210) .... ***Crotalus*** (p. 102)

Agkistrodon

Key to Species of ***Agkistrodon***

1a. Loreal scale present (Fig. 211); 23 scale rows at midbody .... ***A. contortrix***
CC 399, pl 34

1b. Loreal scale absent (Fig. 211); 25 scale rows at midbody .... ***A. piscivorus***
CC 403, pl 34

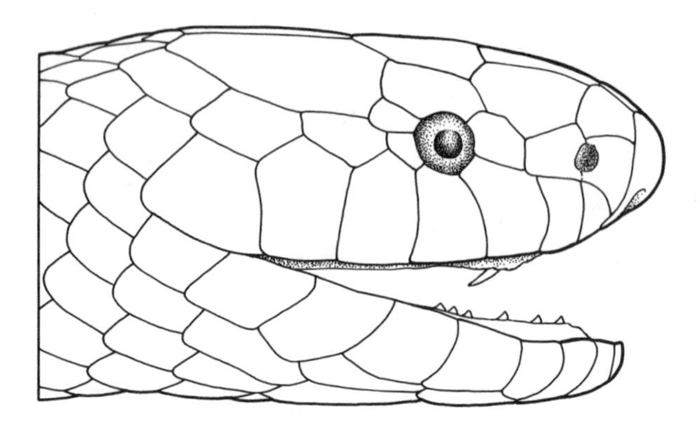

Figure 208. Lateral view of the head of *Micrurus tener*, showing the permanently erect front fang in the upper jaw, typical of the family Elapidae. Drawn from a preserved specimen (KU 61344).

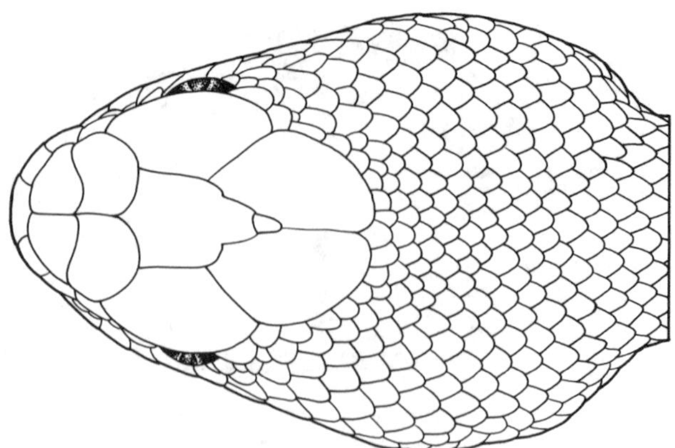

Sistrurus

Key to Species of *Sistrurus*

1a. Upper preocular scale not in contact with postnasal scale (Fig. 212); tail long and slender, rattle barely wider than eye .... ***S. miliarius***
CC 407, pl 35 **CAAR** 220 (1978)

1b. Upper preocular scale in contact with postnasal scale (Fig. 212); tail short and thick, rattle at least twice as wide as eye .... ***S. catenatus***
CC 405, pl 35, p404, 405 **S** 178, pl 45 **CAAR** 332 (1983)

Crotalus

Key to Species of *Crotalus*

1a. Supraocular scales hornlike (Fig. 213) .... ***C. cerastes***
S 186, pl 44

1b. Supraocular scales not hornlike 2

2a. Tip of snout and canthus rostralis raised into a sharp ridge (Fig. 214); a light vertical line present on rostral scale and mental scale .... ***C. willardi***
S 183, pl 45

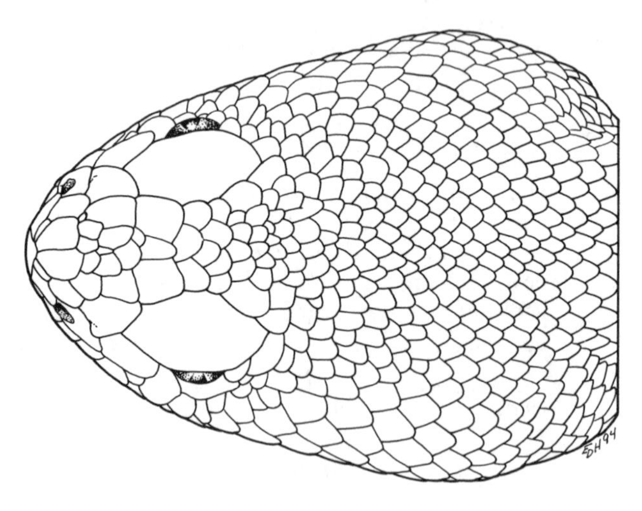

Figure 210. Dorsal view of the heads of two genera of the family Viperidae. Upper: *Sistrurus catenatus*, showing the presence of large scales in the area between the eyes. Drawn from a preserved specimen (KU 220803). Lower: *Crotalus viridis*, showing the presence of small scales in the area between the eyes. Drawn from a preserved specimen (KU 192386).

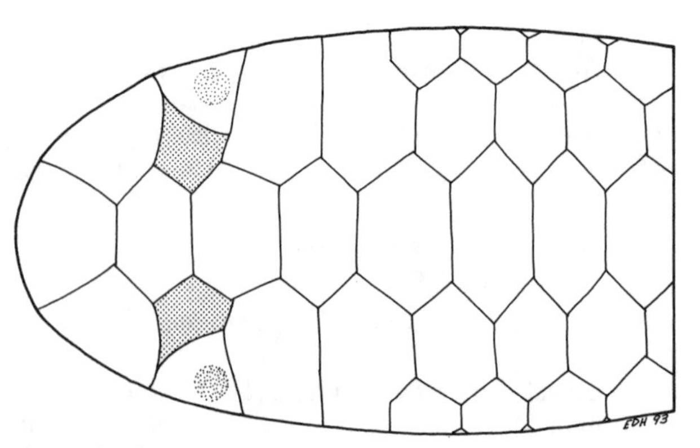

Figure 209. Dorsal view of the head of *Leptotyphlops dulcis*, showing the presence and position of the supraocular scales (shaded). Adapted from a photograph in Tennant (1984).

2b. No sharp ridge on snout and canthus rostralis; no light vertical line on rostral scale and mental scale 3

3a. Prenasal scales curved under postnasal scales (Fig. 215); usually with a dorsal pattern of widely spaced crossbands .... ***C. lepidus***
CC 415, pl 36 **S** 182, pl 45

3b. Prenasal scales not curved under postnasal scales; pattern variable, but not with widely spaced crossbands 4

4a. Dorsal pattern with two parallel rows of small dark spots .... ***C. pricei***
S 179, pl 45 **CAAR** 266 (1981)

4b. Dorsal pattern variable, but not with two parallel rows of small dark spots 5

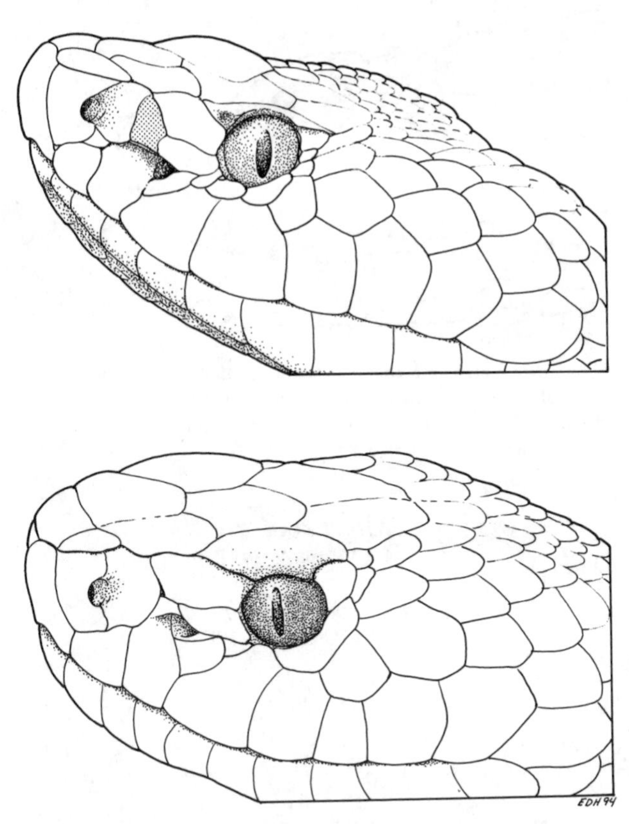

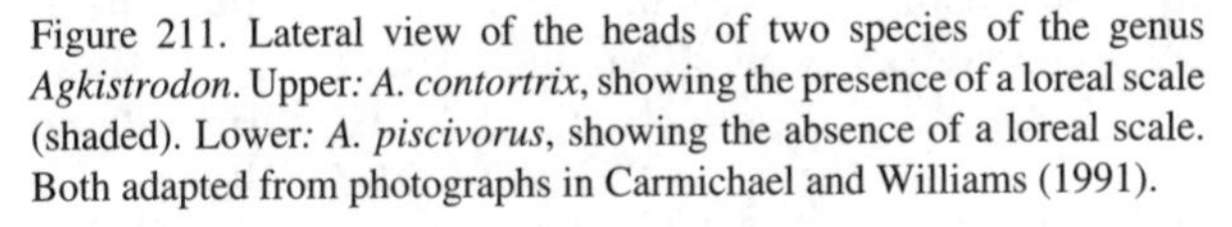

Figure 211. Lateral view of the heads of two species of the genus *Agkistrodon*. Upper: *A. contortrix*, showing the presence of a loreal scale (shaded). Lower: *A. piscivorus*, showing the absence of a loreal scale. Both adapted from photographs in Carmichael and Williams (1991).

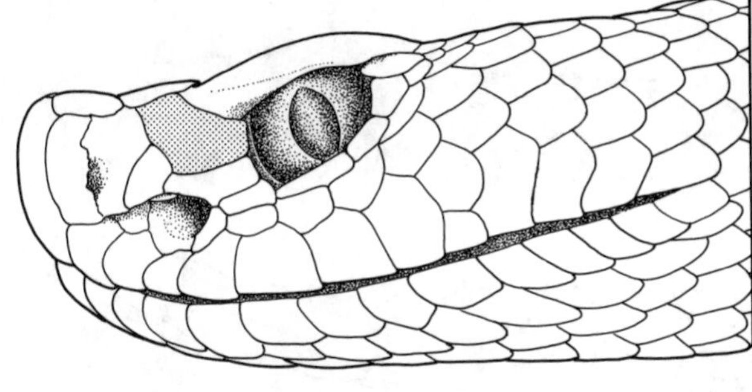

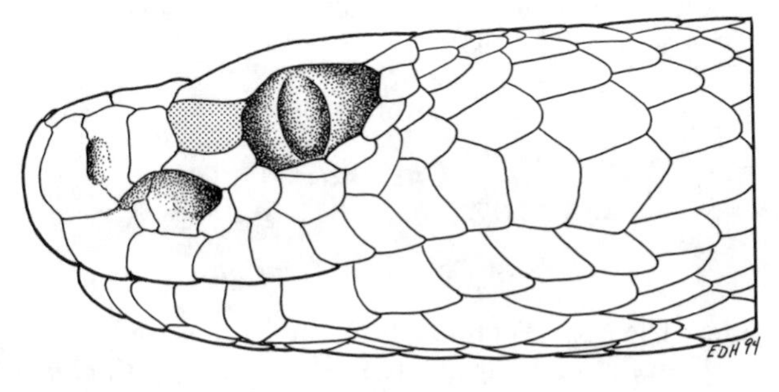

Figure 212. Lateral view of the heads of two species of the genus *Sistrurus*. Upper: *S. catenatus*, showing the upper preocular scale (shaded) in contact with the postnasal scale. Drawn from a preserved specimen (KU 220803). Lower: *S. miliarius*, showing the upper preocular scale (shaded) not in contact with the postnasal scale. Drawn from a preserved specimen (KU 29999).

5a. Rostral scale separated from prenasal scales by small scales (Fig. 216); supraocular scales with one or more grooves .... ***C. mitchellii***
S 188, pl 44 **CAAR** 388 (1986)

5b. Rostral scale in contact with prenasal scales; supraocular scales without grooves 6

6a. Vertical light lines present on the posterior edges of the prenasal scales and 1st upper labial scales .... ***C. adamanteus***
CC 411, pl 35 **CAAR** 252 (1980)

6b. No light lines on the posterior edges of the prenasal scales and 1st upper labial scales 7

7a. Three or more internasal scales in contact with rostral scale (Fig. 217) .... ***C. viridis***
CC 413, pl 36, p 411 **S** 184, pl 44

7b. Two internasal scales in contact with rostral scale 8

8a. Dorsum black or with crossbands or chevrons narrower than interspaces (Fig. 218); tail black .... ***C. horridus***
CC 409, pl 35, p 411 **CAAR** 253 (1980)

8b. Dorsum variable, but not uniformly black; if banded, the bands are wider than the interspaces; tail black or not 9

9a. Scales on snout and between supraocular scales enlarged (Fig. 219); dark tail rings usually narrower than light rings; lower ½ of proximal rattle usually light .... ***C. scutulatus***
CC 413, pl 36, p 411 **S** 189, pl 44 **CAAR** 291 (1982)

9b. Scales on snout and between supraocular scales not enlarged; dark and light tail rings about equal in width or dark rings wider; entire proximal rattle dark 10

10a. Tail with distinct black and white rings 11

10b. Tail rings absent or indistinct, and not black and white 12

11a. First lower labial scales usually divided transversely (Fig. 220); with faint or no dark spots in dorsal diamond-shaped marks .... ***C. exsul***
S 187, pl 44 (as *Crotalus ruber*)

11b. First lower labial scales usually not divided; with distinct dark spots in dorsal diamond-shaped marks .... ***C. atrox***
CC 411, pl 36, p 411 **S** 185, pl 44

12a. Tail colored like body; head small, rattle wide (head length less than 2½ times width of proximal rattle) .... ***C. tigris***
S 181, pl 45

12b. Tail black; head length more than 2½ times width of proximal rattle ....***C. molossus***
CC 415, pl 36 **S** 180, pl 45 **CAAR** 242 (1980)

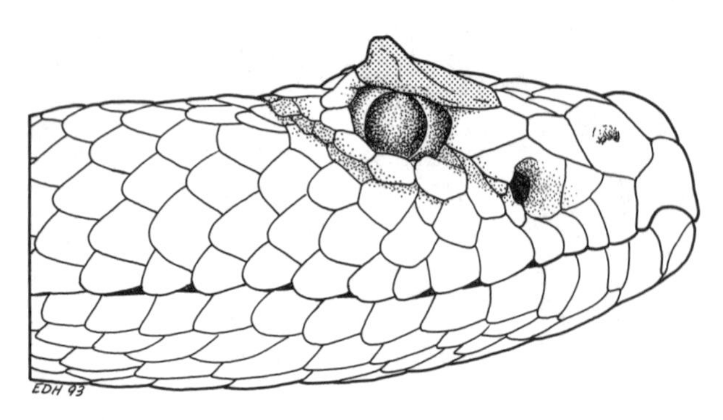

Figure 213. Lateral view of the head of *Crotalus cerastes*, showing the presence of a horn-like supraocular scale (shaded). Drawn from a preserved specimen (KU 5340).

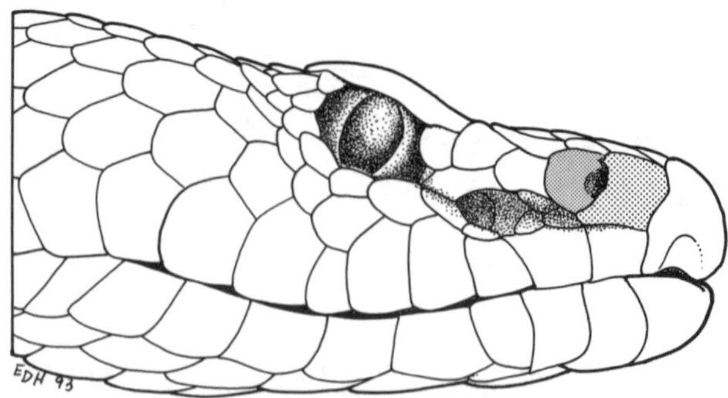

Figure 215. Lateral view of the head of *Crotalus lepidus*, showing the prenasal scale (light shading) curved under the postnasal scale (dark shading). Drawn from a preserved specimen (KU 157866).

BOIDAE

Charina

CAAR 205 (1977) and 294 (1982, as *Lichanura*)

Key to Species of ***Charina***

1a. Three scales present between the eyes, the middle one largest (Fig. 221) .... ***C. bottae***
S 127, pl 36 **CAAR** 205 (1977)

1b. Several small scales of about equal size present between eyes (Fig. 221) .... ***C. trivirgata***
S 126, pl 36 **CAAR** 294 (1982) (as *Lichanura trivirgata*)

HYDROPHIIDAE

Some authorities consider this family to be confamilial with the Family Elapidae.

Pelamis

CAAR 255 (1980)

Pelamis platurus
S pl 48 **CAAR** 255 (1980)

ELAPIDAE

Key to Genera of Elapidae

1a. Band on neck red (in life) (Fig. 222); from west of the 105th meridian .... ***Micruroides*** (p. 104)

1b. Band on neck black (Fig. 222); from east of the 105th meridian .... ***Micrurus*** (p. 104)

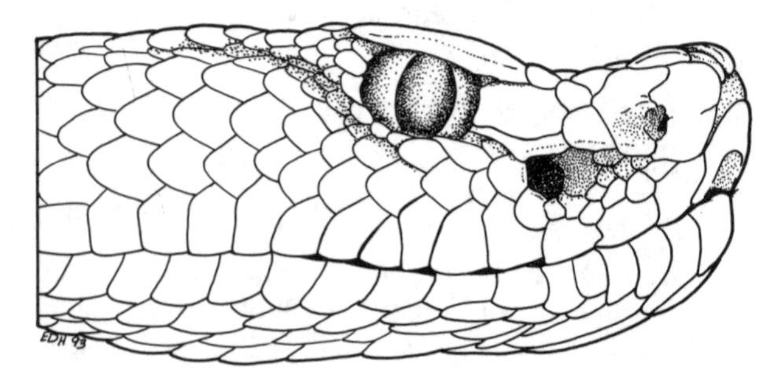

Figure 214. Lateral view of the head of *Crotalus willardi*, showing the tip of the snout and the canthus rostralis raised to form a sharp, longitudinal ridge. Drawn from a preserved specimen (KU 193276).

Micruroides

CAAR 163 (1974)

Micruroides euryxanthus
S 176, pl 37 **CAAR** 163 (1974)

Micrurus

Key to species of ***Micrurus***

1a. Red bands with few black dots or spots; black neck band does not involve parietal scales (Fig. 223) .... ***Micrurus fulvius***
CC 394, pl 30, p 372 **CAAR** 316 (1983)

1b. Red bands with widely scattered black spots; black neck band extends forward to involve posterior tips of parietal scales (Fig. 223) .... ***Micrurus tener***
CC 394, pl 36 **CAAR** 316 (1983) (as *M. fulvius tener*)

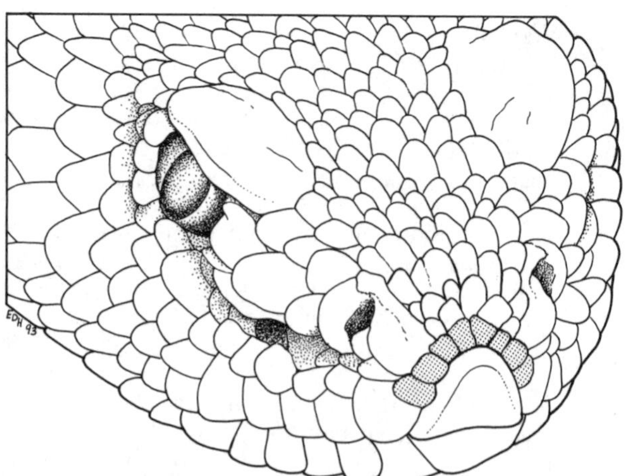

Figure 216. Anterior view of the snout of *Crotalus mitchellii*, showing the rostral scale and prenasal scales separated by a row of small scales (shaded). Drawn from a preserved specimen (KU 5335).

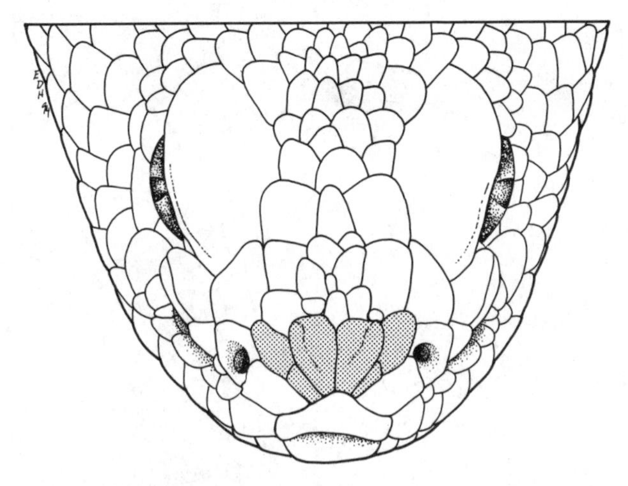

Figure 217. Dorsal view of the snout of *Crotalus viridis*, showing the presence of three or more internasal scales (shaded) in contact with the rostral scale. Drawn from a preserved specimen (KU 192386).

COLUBRIDAE

Key to Genera of Colubridae

1a. Rostral scale turned upward and keeled (Fig. 224) .... ***Heterodon*** (p. 111)
1b. Rostral scale not turned upward and keeled 2

2a. Anal plate divided (Fig. 225) 3
2b. Anal plate not divided (Fig. 225) 39

3a. At least some dorsal scales keeled (Fig. 204) 4
3b. Dorsal scales smooth (Fig. 204) 15

4a. Either the loreal scale or preocular scale absent (Fig. 226) 5
4b. Both loreal and preocular scales present (Fig. 226) 7

5a. Dorsal scale rows number nineteen .... ***Farancia*** (part) (p. 112)
5b. Dorsal scale rows number 15–17 6

6a. Prefrontal scale in contact with eye (Fig. 227) .... ***Virginia*** (part) (p. 112)
6b. Prefrontal scale not in contact with eye (Fig. 227) .... ***Storeria*** (p. 112)

7a. Dorsal scale rows number seventeen 8
7b. Dorsal scale rows number nineteen or more 10

8a. Less than 60 subcaudals .. ***Seminatrix*** (part) (p. 113)
8b. More than 80 subcaudals 9

9a. Seven upper labial scales and 7–8 lower labial scales; dorsum uniformly pigmented (green in life) .... ***Opheodrys*** (p. 113)
9b. Eight or nine upper labial scales and 10–11 lower labial scales; dorsum dark with a light spot on each scale .... ***Drymobius*** (p. 113)

10a. Keels on dorsal scales faint, seldom extending to tips of scales 11
10b. Dorsal scales strongly keeled, keels extending to tips of all dorsal scales except sometimes those in lower rows 13

11a. Subocular scales present (Fig. 228) .... ***Bogertophis*** (p. 113)
11b. Subocular scales absent (Fig. 228) 12

12a. Dorsal scale rows number thirty or more .... ***Senticolis*** (p. 113)
12b. Dorsal scale rows number less than thirty .... ***Elaphe*** (p. 113)

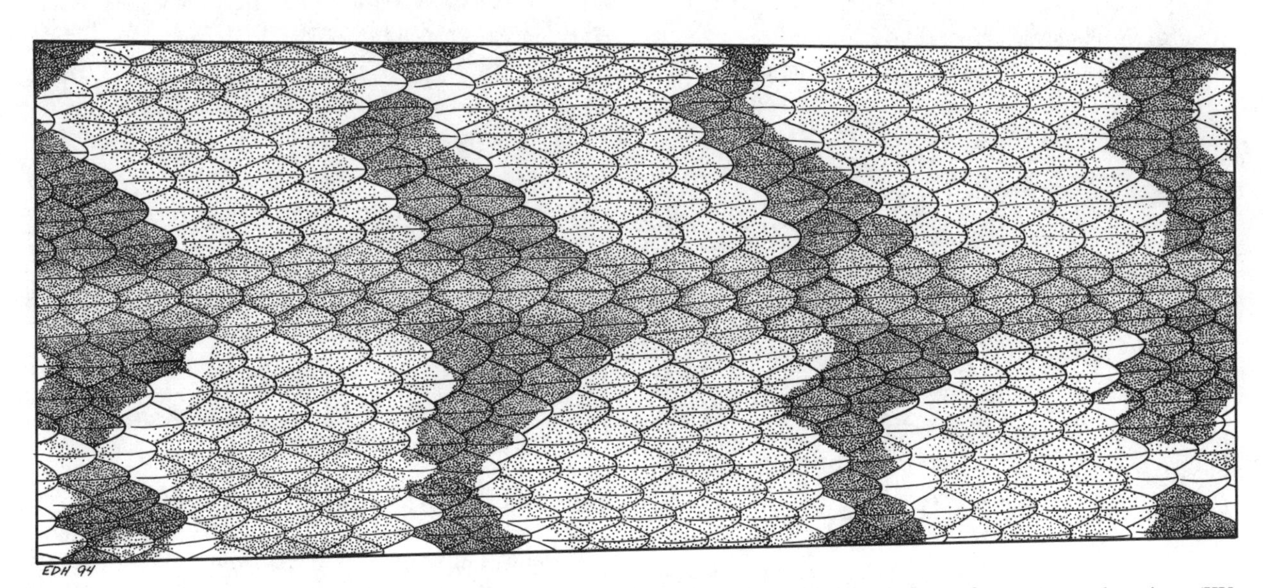

Figure 218. Dorsal view of the body of *Crotalus horridus*, showing the chevronlike pattern of dark bands. Drawn from a preserved specimen (KU 221491).

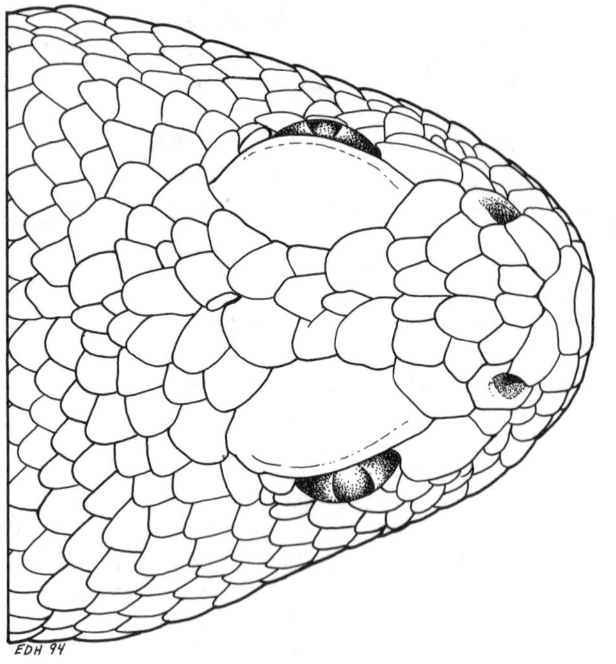

Figure 219. Dorsal view of the head of *Crotalus scutulatus*, showing head scales on snout and between eyes slightly larger than those on rear and sides of head. Drawn from a preserved specimen (KU 13782).

13a. Dorsal scale rows number 21 or more .... ***Nerodia*** (p. 114)
13b. Dorsal scale rows number nineteen 14

14a. Seven or more upper labial scales present; usually two preocular scales present .... ***Regina*** (p. 115)
14b. Six or fewer upper labial scales present; one preocular scale present .... ***Clonophis*** (p. 115)

15a. Dorsal scale rows number nineteen or more 16
15b. Dorsal scale rows number less than nineteen 21

16a. Preocular scale absent (Fig. 226) .... ***Farancia*** (part) (p. 112)

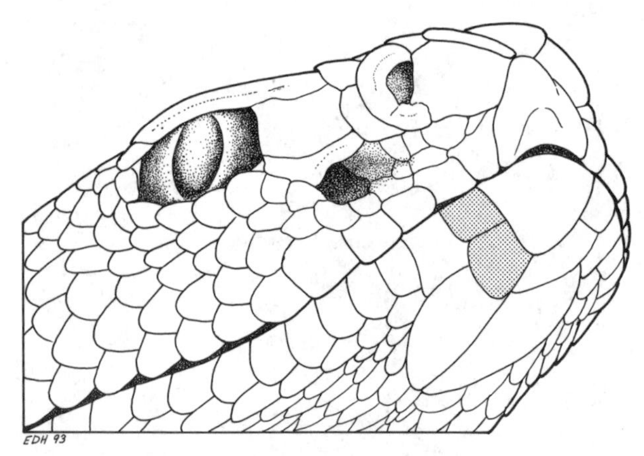

Figure 220. Ventrolateral view of the chin of *Crotalus exsul*, showing the first lower labial (shaded) divided transversely. Drawn from a preserved specimen (KU 61322).

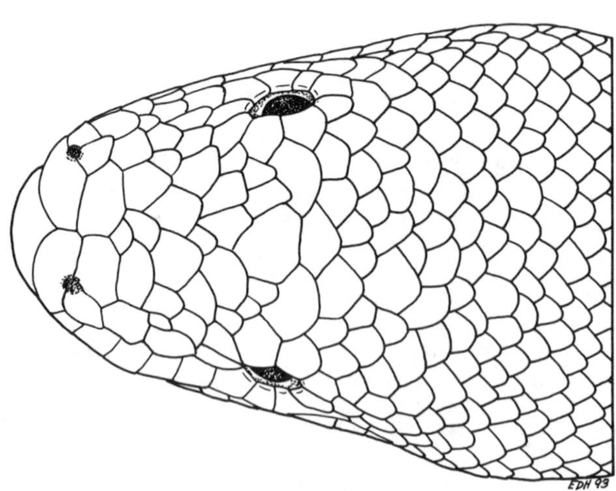

Figure 221. Dorsal view of the heads of two species of the genus *Charina*. Upper: *C. bottae*, showing the presence of three scales between the eyes, with the center scale much larger. Drawn from a preserved specimen (KU 6650). Lower: *C. trivirgata*, showing the presence of many scales of more or less equal size between the eyes. Drawn from a preserved specimen (KU 80921).

16b. Preocular and loreal scales present (Fig. 226) 17

17a. Pupil round (Fig. 229) 18
17b. Pupil elliptical (Fig. 229) 19

18a. Dorsal scale rows number nineteen .... ***Coniophanes*** (p. 115)
18b. Dorsal scale rows number 25 or more 11

19a. Two or more loreal scales, 3–4 postocular scales, 2–3 anterior temporal scales, and usually nine or more upper labial scales ..... ***Trimorphodon*** (part) (p. 116)
19b. One loreal scale, two postocular scales, one anterior temporal scale, and 7–8 upper labials 20

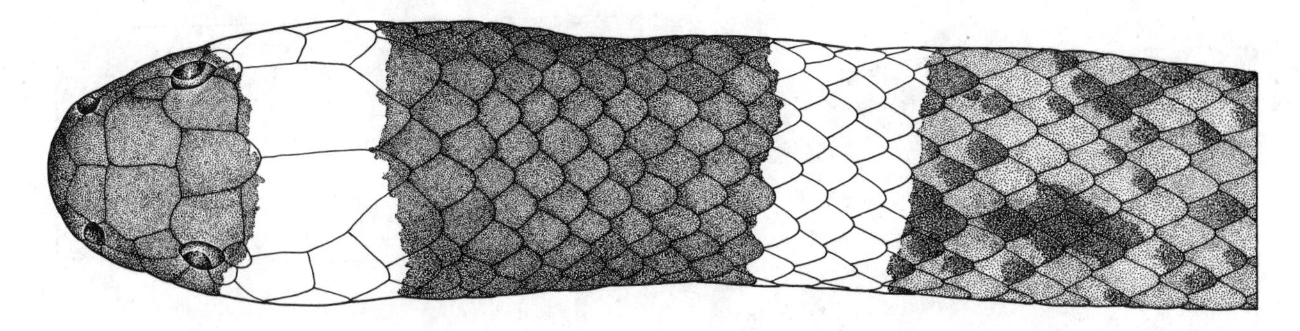

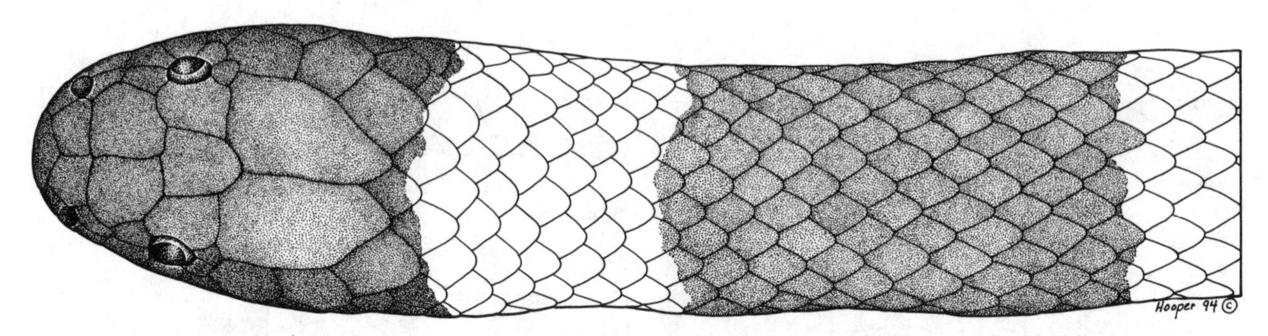

Figure 222. Dorsal view of the heads and anterior bodies of two genera of the family Elapidae. Upper: *Micrurus tener*, showing the black band (darkest shading) on rear of head and neck. Drawn from a preserved specimen (DAK 1125). Lower: *Micruroides euryxanthus*, showing red band (lightest shading) reaching posterior part of neck. Drawn from a preserved specimen (KU).

20a. Dorsum with small blotches; sides with smaller spots .... ***Hypsiglena*** (p. 116)
20b. Dorsum with large blotches; without lateral spots .... ***Leptodeira*** (p. 116)

21a. Either loreal or preocular scale absent 22
21b. Both a loreal scale and at least one preocular scale present 28

22a. Dorsal scale rows number thirteen 23
22b. Dorsal scale rows number more than thirteen 24

23a. Dorsum uniformly colored; one postocular scale present .... ***Carphophis*** (p. 116)
23b. Dorsum with dark crossbars; two postocular scales present .... ***Chilomeniscus*** (p. 116)

24a. Six upper labial scales present; prefrontal scale in contact with eye .... ***Virginia*** (part) (p. 112)
24b. Seven or more upper labial scales present; prefrontal scale not in contact with eye 25

25a. Dorsal scale rows number fifteen 26
25b. Dorsal scale rows number seventeen 27

26a. Snout flat, shovel-like (Fig. 230) .... ***Chionactis*** (part) (p. 116)
26b. Snout normal; top of head flattened .... ***Tantilla*** (p. 116)

27a. Rostral scale in contact with frontal (Fig.231); tail short and thick; head not long and pointed .... ***Ficimia*** (p. 118)
27b. Rostral scale not in contact with frontal; tail long and slender (> ½ SVL); head long and pointed .... ***Oxybelis*** (p. 118)

28a. One preocular scale present 29
28b. Two or three preocular scales present 35

29a. Dorsal scale rows number seventeen 30
29b. Dorsal scale rows number less than seventeen 32

30a. Rostral scale turned upward .... ***Gyalopion*** (part) (p. 118)
30b. Rostral scale not turned upward 31

31a. Dark line through eye to angle of jaw (Fig. 232); head darker than body; usually with seven upper labial scales .... ***Rhadinaea*** (p. 118)
31b. No dark line through eye; head and body black; usually with eight upper labial scales .... ***Seminatrix*** (part) (p. 113)

32a. Posterior and anterior chin shields about equal in length (Fig. 233); more than 65 subcaudals .... ***Liochlorophis*** (part) (p. 113)
32b. Posterior chin shields much shorter than anterior chin shields; 64 or fewer subcaudals 33

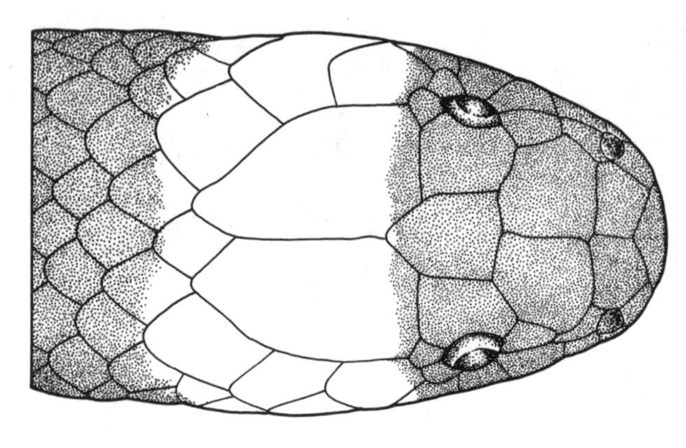

Figure 223. Dorsal view of the heads of two species of the genus *Micrurus*. Upper: *M. fulvius*, showing the black neck band not reaching the parietal scales. Drawn from a preserved specimen (KU). Lower: *M. tener*, showing the black neck band reaching the parietal scales. Drawn from a preserved specimen (DAK 1125).

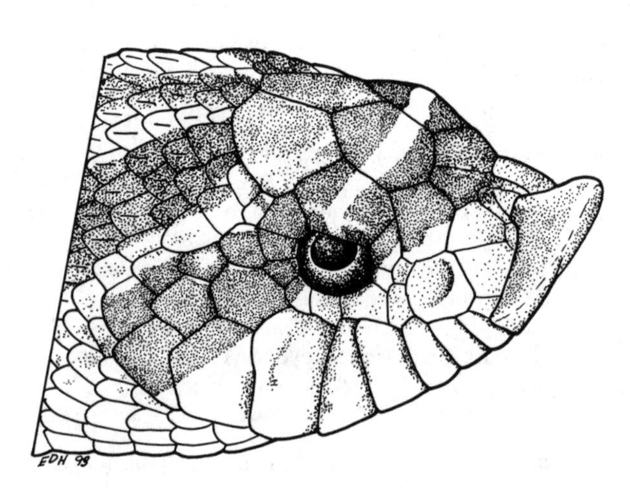

Figure 224. Dorsolateral view of the head of *Heterodon nasicus*, showing the upturned and keeled rostral scale. Drawn from a photograph by Suzanne L. Collins.

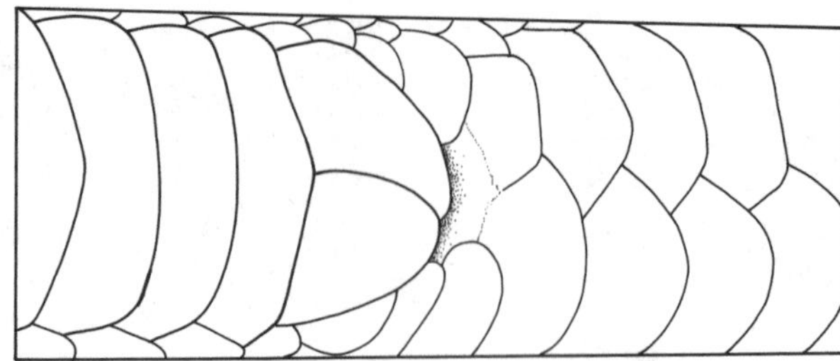

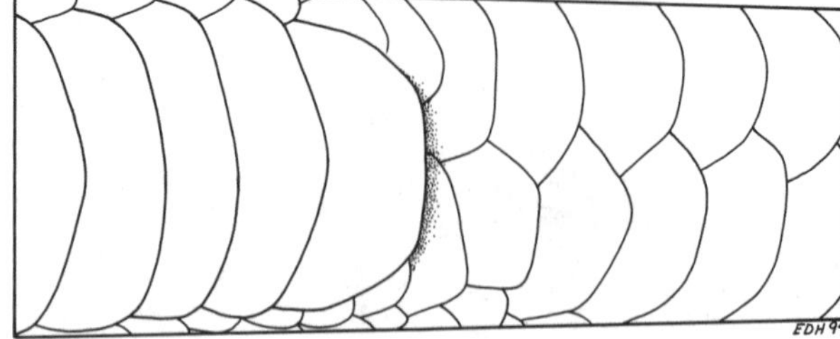

Figure 225. Ventral view of the cloacal area of two snakes. Upper: Anal plate divided. Drawn from a preserved specimen of *Opheodrys aestivus* (KU). Lower: Anal plate not divided.

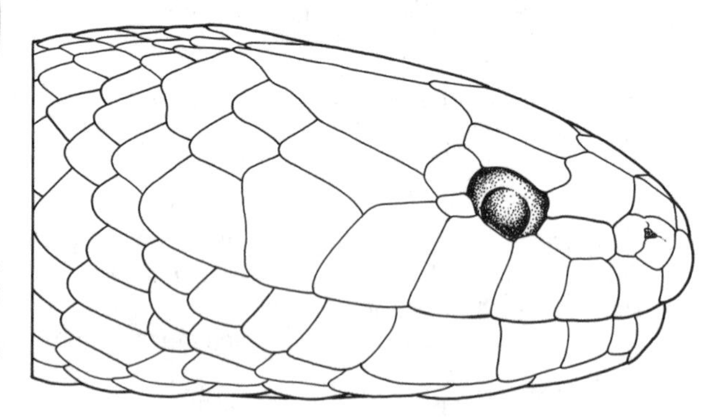

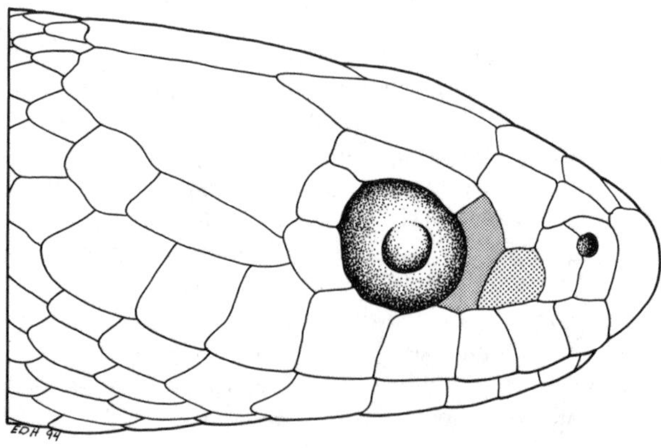

Figure 226. Lateral view of the heads of two genera of the family Colubridae. Upper: *Farancia abacura*, showing the presence of only one scale (loreal or preocular) between the eye and nostril. Drawn from a preserved specimen (KU 21572). Lower: *Seminatrix pygaea*, showing the presence of both the loreal scale (light shading) and the preocular scale (dark shading) between the eye and nostril. Drawn from a preserved specimen (KU 61122).

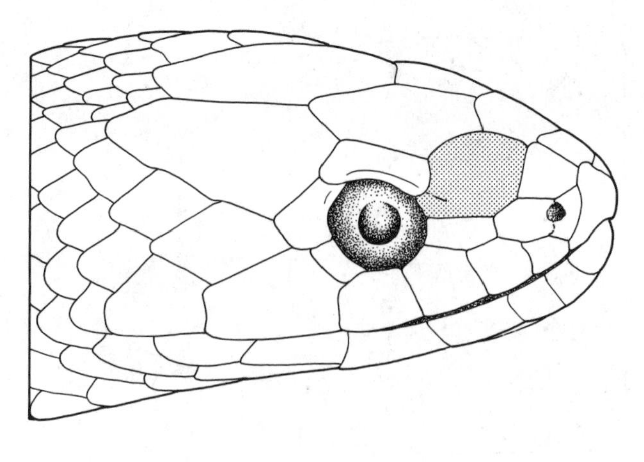

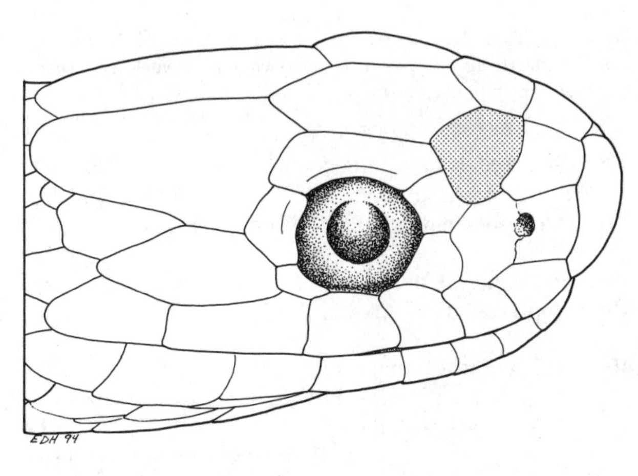

Figure 227. Lateral view of the heads of two genera of the family Colubridae. Upper: *Virginia valeriae*, showing the prefrontal scale in contact with the eye. Drawn from a preserved specimen (KU 192254). Lower: *Storeria occipitomaculata*, showing the prefrontal scale not in contact with the eye. Drawn from a preserved specimen (KU 129470).

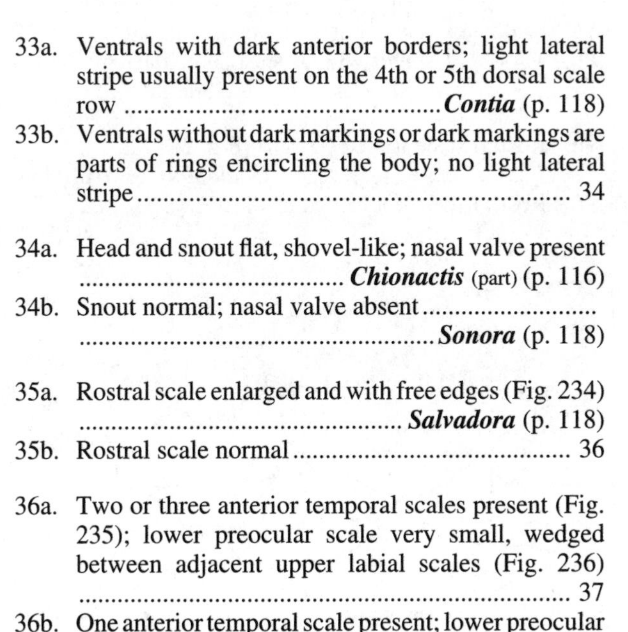

33a. Ventrals with dark anterior borders; light lateral stripe usually present on the 4th or 5th dorsal scale row .... ***Contia*** (p. 118)

33b. Ventrals without dark markings or dark markings are parts of rings encircling the body; no light lateral stripe 34

34a. Head and snout flat, shovel-like; nasal valve present .... ***Chionactis*** (part) (p. 116)

34b. Snout normal; nasal valve absent .... ***Sonora*** (p. 118)

35a. Rostral scale enlarged and with free edges (Fig. 234) .... ***Salvadora*** (p. 118)

35b. Rostral scale normal 36

36a. Two or three anterior temporal scales present (Fig. 235); lower preocular scale very small, wedged between adjacent upper labial scales (Fig. 236) 37

36b. One anterior temporal scale present; lower preocular scale not as above 38

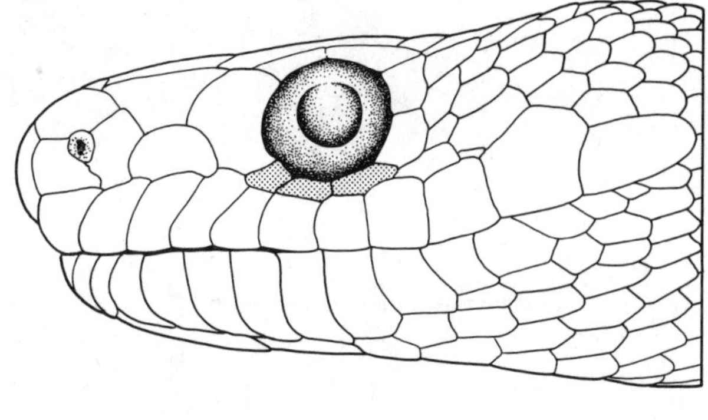

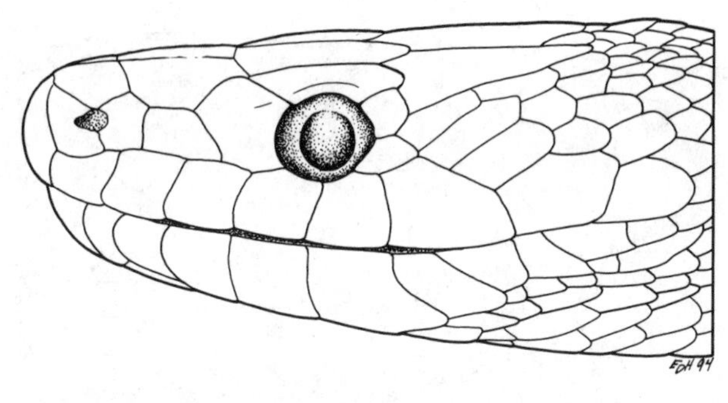

Figure 228. Lateral view of the heads of two genera of the family Colubridae. Upper: *Bogertophis subocularis*, showing the presence of subocular scales (shaded). Drawn from a preserved specimen (KU 8436). Lower: *Senticolis triaspis*, showing the absence of subocular scales. Drawn from a preserved specimen (KU 87744).

37a. Fifteen dorsal scale rows present at posterior end of body .... ***Coluber*** (p. 118)

37b. Thirteen or fewer dorsal scale rows present at posterior end of body .... ***Masticophis*** (p. 119)

38a. Nasal plate divided (Fig. 237); light colored ring on neck (may occasionally be broken dorsally) and/or with black spots on venter .... ***Diadophis*** (p. 119)

38b. Nasal plate entire; without a ring on the neck or black spots on venter .... ***Liochlorophis*** (part) (p. 113)

39a. At least some dorsal scales keeled 40

39b. Dorsal scales smooth 44

40a. Dorsal scale rows number 27 or more; usually with four prefrontal scales (Fig. 238) ***Pituophis*** (p. 119)

40b. Dorsal scale rows number less than 27; with two prefrontal scales 41

41a. Pupil elliptical; subocular scales present; rostral scale enlarged and with free edges .... ***Phyllorhynchus*** (part) (p. 119)

41b. Pupil round; subocular scales absent; rostral scale normal 42

42a. Eight or more lower labial scales present .... ***Thamnophis*** (p. 119)

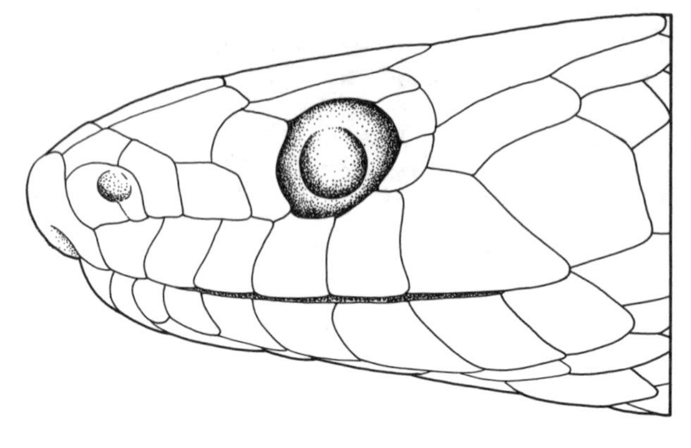

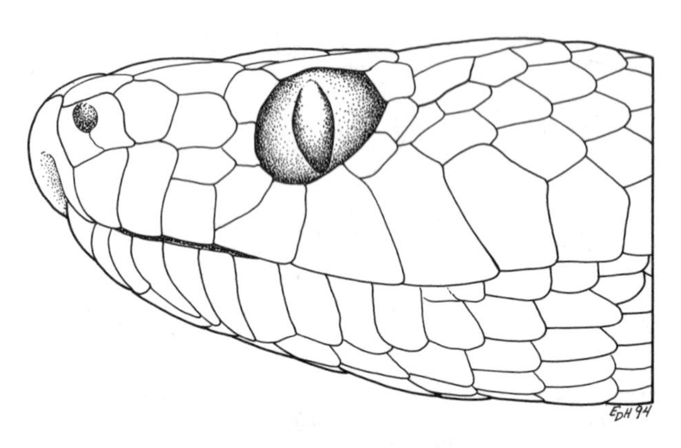

Figure 229. Lateral view of the heads of two genera of the family Colubridae. Upper: *Coniophanes imperialis*, showing the round pupil of the eye. Drawn from a preserved specimen (KU 23262). Lower: *Trimorphodon biscutatus*, showing the elliptical pupil of the eye. Drawn from a preserved specimen (KU 191934).

42b. Less than eight lower labial scales present 43

43a. Double row of black spots present on venter .... ***Tropidoclonion*** (p. 122)
43b. Venter immaculate or with tiny spots not in rows .... ***Virginia*** (part) (p. 112)

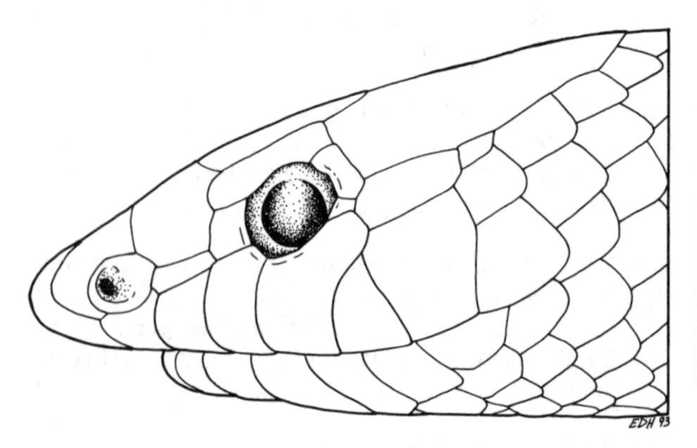

Figure 230. Lateral view of the head of *Chionactis occipitalis*, showing the flat, shovellike snout. Drawn from a preserved specimen (KU 6620).

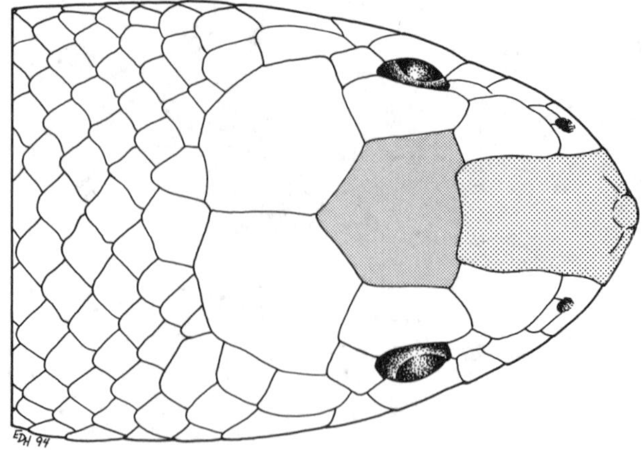

Figure 231. Dorsal view of the head of *Ficimia streckeri*, showing the rostral scale (light shading) in contact with the frontal scale (dark shading). Drawn from a preserved specimen (KU).

44a. Subcaudals not divided (Fig. 239) .... ***Rhinocheilus*** (p. 122)
44b. Most subcaudals divided (Fig. 239) 45

45a. Pupil elliptical 46
45b. Pupil round 47

46a. Subocular scales present .... ***Phyllorhynchus*** (part) (p. 119)
46b. Subocular scales absent.. .... ***Trimorphodon*** (part) (p. 116)

47a. Parietal scales in contact with upper labial scales (Fig. 240) .... ***Stilosoma*** (p. 122)
47b. Parietal scales not in contact with upper labial scales 48

48a. Middle of venter immaculate 49
48b. Middle of venter with at least some dark markings 51

49a. Rostral scale turned upward (Fig. 241) .... ***Gyalopion*** (part) (p. 118)
49b. Rostral scale not turned upward 50

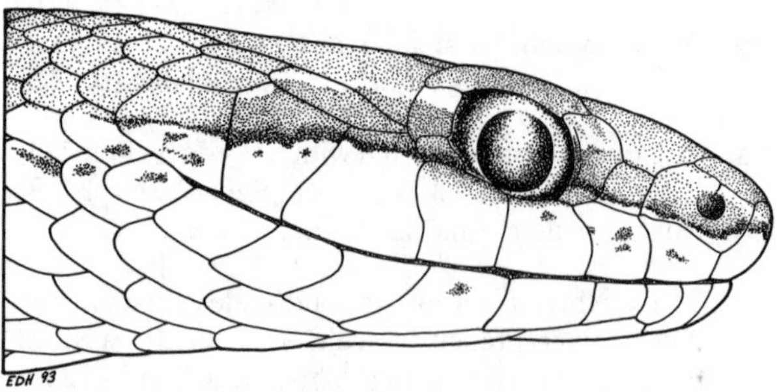

Figure 232. Lateral view of the head of *Rhadinaea flavilata*, showing the dark line running from the snout through the eye to the angle of the jaw. Drawn from a preserved specimen (KU).

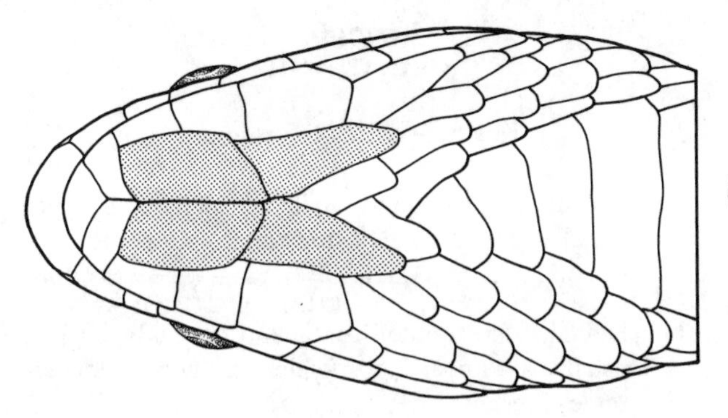

Figure 233. Ventral view of the head of *Liochlorophis vernalis*, showing the posterior and anterior chin shields (shaded) about equal in length. Drawn from a preserved specimen (KU 68791).

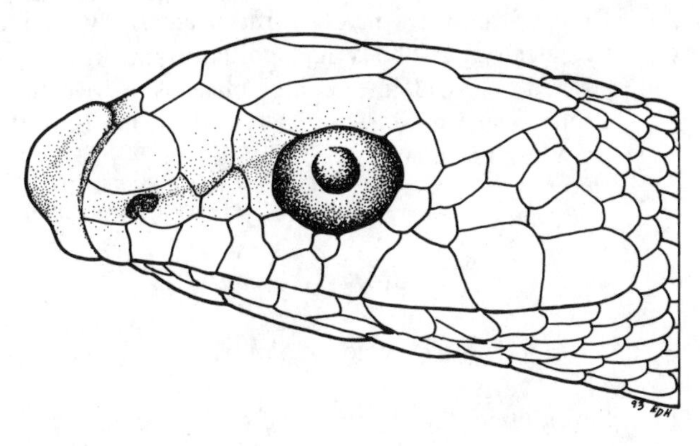

Figure 234. Lateral view of the head of *Salvadora hexalepis*, showing the enlarged rostral scale with free edges. Drawn from a photograph in Shaw and Campbell (1974).

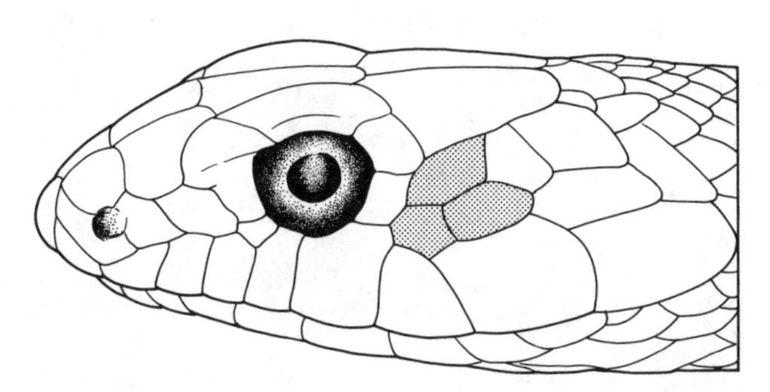

Figure 235. Lateral view of the head of *Masticophis flagellum*, showing the presence of three anterior temporal scales (shaded). Drawn from a photograph by Suzanne L. Collins.

50a. Upper labial scales number seven or less; eight lower labial scales present .... ***Cemophora*** (p. 122)
50b. Upper labial scales number eight; 12–15 lower labial scales present .... ***Arizona*** (p. 122)

51a. Dorsal scale rows number seventeen .... ***Drymarchon*** (p. 122)
51b. Dorsal scale rows number more than seventeen .... ***Lampropeltis*** (p. 123)

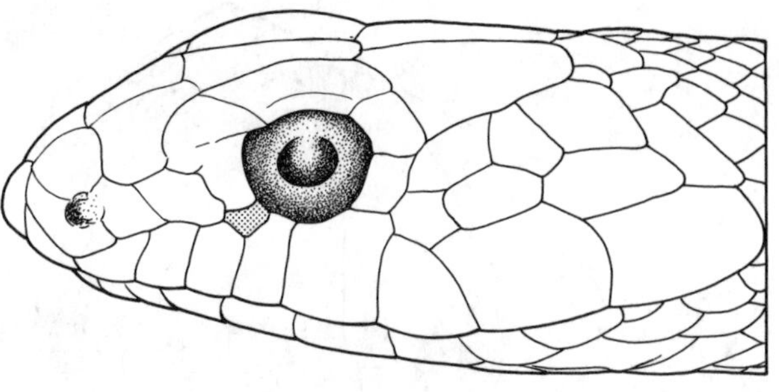

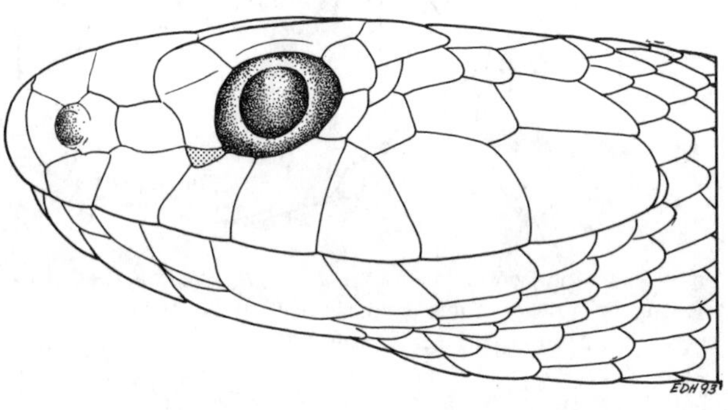

Figure 236. Lateral view of the heads of two genera of the family Colubridae. Upper: *Masticophis flagellum* (drawn from a photograph by Suzanne L. Collins). Lower: *Coluber constrictor* (adapted from a photograph in Ortenburger, 1928). Both show the presence and position of the lower preocular scale (shaded).

Heterodon

CAAR 315 (1983)

Key to Species of *Heterodon*

1a. Prefrontal scales in contact; underside of tail lighter than venter (Fig. 242) .... ***H. platirhinos***
CC 327, pl 25, p 326 **CAAR** 282 (1981)
1b. Prefrontal scales separated by small scales; underside of tail like venter (Fig. 242) 2

2a. Dorsal scale rows number 25–27; venter clouded, but light (Fig. 242) .... ***H. simus***
CC 328, pl 25, p 326 **CAAR** 375 (1985)
2b. Dorsal scale rows number 23; venter black with small light patches (Fig. 242) .... ***H. nasicus***
CC 328, pl 25, p 326 **S** 124, pl 40

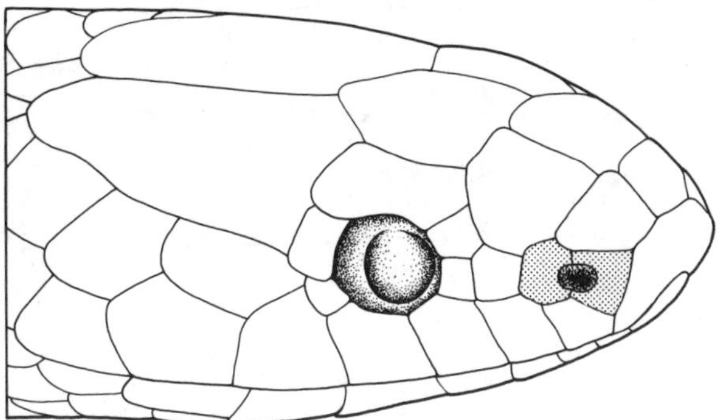

Figure 237. Lateral view of the head of *Diadophis punctatus*, showing the divided nasal scale (shaded). Drawn from a preserved specimen (KU).

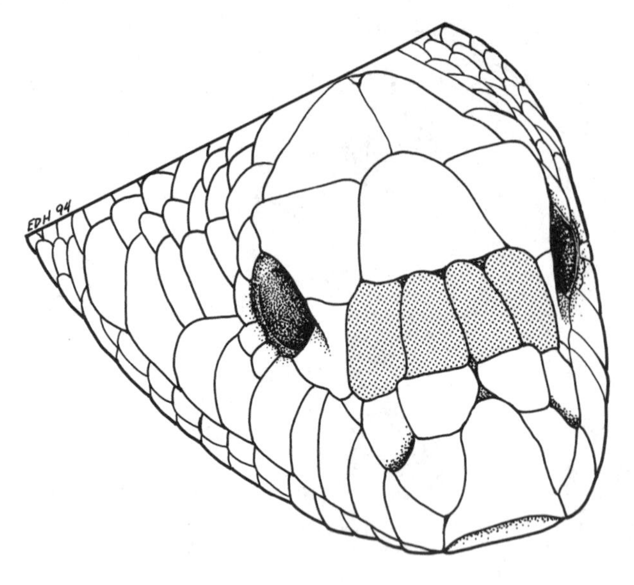

Figure 238. Anterodorsal view of the head of *Pituophis melanoleucus*, showing the presence of four prefrontal scales (shaded). Drawn from a photograph in Carmichael and Williams (1991).

Farancia

CAAR 292 (1982)

Key to Species of *Farancia*

1a. One internasal scale present (Fig. 243) .... ***F. abacura***
CC 336, pl 25 **CAAR** 314 (1983)

1b. Two internasal scales present (Fig. 243) .... ***F. erytrogramma***
CC 336, pl 25 **CAAR** 293 (1982)

Some authorities have suggested that this taxon consists of two species, *F. erytrogramma* and *F. seminola*.

Virginia

CAAR 529 (1991)

Key to Species of *Virginia*

1a. Dorsal scales strongly keeled; usually five upper labial scales and one postocular scale present .... ***V. striatula***
CC 326, pl 22 **CAAR** 599 (1994)

1b. Dorsal scales smooth or at most weakly keeled; usually six upper labial scales and two postocular scales present 2

2a. Dorsal scale rows number fifteen or seventeen; from the eastern U.S. west into the Great Plains .... ***V. valeriae***
CC 325, pl 22 **CAAR** 552 (1992)

2b. Dorsal scale rows number fifteen anteriorly, and seventeen at midbody and posteriorly; range restricted to mountains and high plateaus of western Pennsylvania and adjacent Maryland into northeastern West Virginia .... ***V. pulchra***
CC 325, pl 22 **CAAR** 552 (1992)
(as *V. valeriae pulchra*)

Storeria

Key to Species of *Storeria*

1a. Seven upper labial scales and one preocular scale present; dorsal scale rows usually number seventeen .... ***S. dekayi***
CC 308, pl 22, p 307 **CAAR** 306 (1982)

1b. Six upper labial scales and two preocular scales present; dorsal scale rows usually number fifteen .... ***S. occipitomaculata***
CC 308, pl 22 **S** 153, pl 41

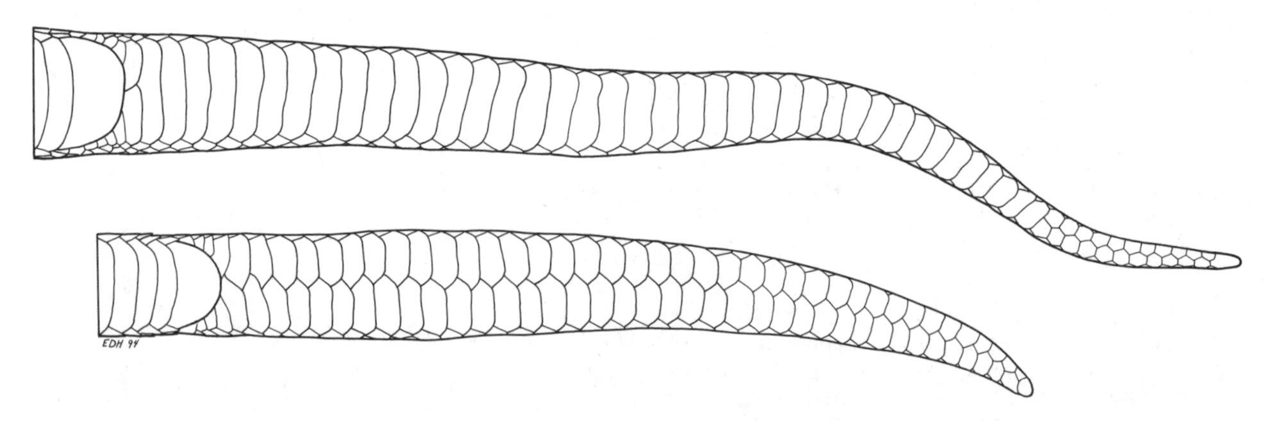

Figure 239. Ventral view of the undersides of the tails of two genera of the family Colubridae. Upper: *Rhinocheilus lecontei*, showing undivided subcaudal scales. Drawn from a preserved specimen (KU). Lower: *Phyllorhynchus decurtatus*, showing divided subcaudal scales. Drawn from a preserved specimen (KU 189210).

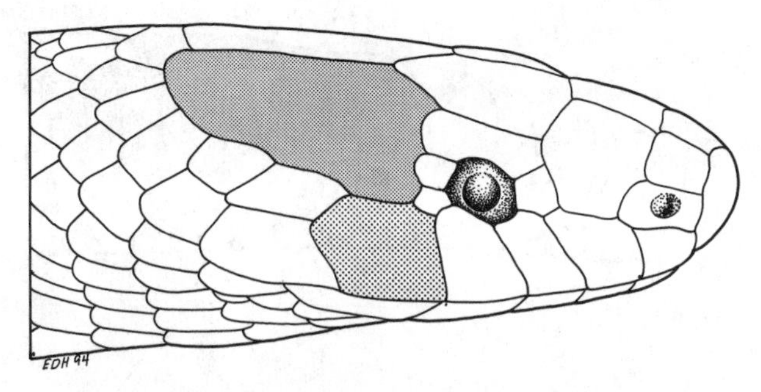

Figure 240. Lateral view of the head of *Stilosoma extenuatum*, showing the parietal scale (dark shading) in contact with an upper labial scale (light shading). Drawn from a preserved specimen (KU 68943).

Seminatrix

Seminatrix pygaea
CC 305, pl 22

Opheodrys

Opheodrys aestivus
CC 347, pl 25

Liochlorophis

Liochlorophis vernalis
CC 347, pl 25 **S** 137, pl 39 (as *Opheodrys vernalis*)

Drymobius
CAAR 170 (1975)

Drymobius margaritiferus
CC 349, pl 32 **CAAR** 172 (1975)

Bogertophis
CAAR 497 (1990)

Key to Species of *Bogertophis*

1a. Keeled middorsal scale rows number five or fewer; dorsal pattern indistinct or composed of light netlike dorsal and lateral streaks .... ***B. rosaliae***
S 198, pl 48 (as *Elaphe rosaliae*) **CAAR** 498 (1990)

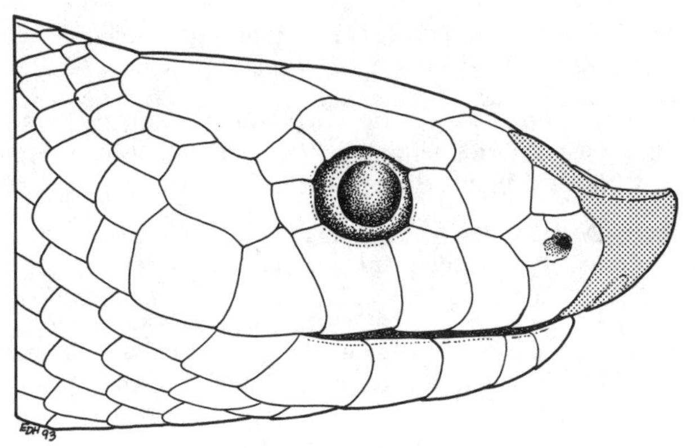

Figure 241. Lateral view of the head of *Gyalopion canum*, showing the upturned (but not keeled) rostral scale (shaded). Drawn from a preserved specimen (KU 129027).

1b. Keeled middorsal scale rows number seven or more; dorsal pattern composed of 21–28 distinct dark H-shaped marks .... ***B. subocularis***
CC 361, pl 32
S 145, pl 38 **CAAR** 268 (1981) (as *Elaphe subocularis*)

Senticolis
CAAR 525 (1991)

Senticolis triaspis
S 144, pl 38 (as *Elaphe triaspis*) **CAAR** 525 (1991)

Elaphe

Key to Species of *Elaphe*

1a. V-shaped mark on head; postocular stripe extends onto neck (Fig. 244) 2
1b. No V-shaped mark on head; postocular stripe stops at posterior margin of mouth 3

2a. Dorsal markings often reddish or orange in life, very pronounced and distinctly outlined in black; distribution largely east of the Mississippi River (except in southeastern Louisiana), absent from Illinois .... ***E. guttata***
CC 355, pl 28, p 353
2b. Dorsal markings pale brown, tan, or gray in life, often indistinct, outlines of blotches dark, rarely black; distribution from southwestern Illinois west to Colorado and southwest through Texas to New Mexico and México (a disjunct population exists in western Colorado and adjacent Utah) .... ***E. emoryi***
CC 355, pl 28, p 353
S 143, pl 38 (as *E. guttata emoryi*)

3a. Ventrals less than 220 4
3b. Ventrals more than 220 5

4a. Number of large blotches on body averages 41; top of head usually brown; from west and south of Lake Michigan .... ***E. vulpina***
CC 357, p 353 **CAAR** 470 (1990)
4b. Number of large blotches on body averages 34; top of head often reddish-orange; from southern Ontario through eastern Michigan into north-central Ohio .... ***E. gloydi***
CC 357, pl 28, p 353 (as *E. vulpina gloydi*)
CAAR 470 (1990) (as *E. vulpina gloydi*)

5a. Eight upper labial scales present; if pattern of longitudinal stripes present, not from Texas and adjacent México .... ***E. obsoleta***
CC 357, pl 28, p 353
5b. Usually nine upper labial scales present; pattern of four longitudinal stripes present; from Texas south into adjacent regions of México .... ***E. bairdi***
CC 361, pl 28

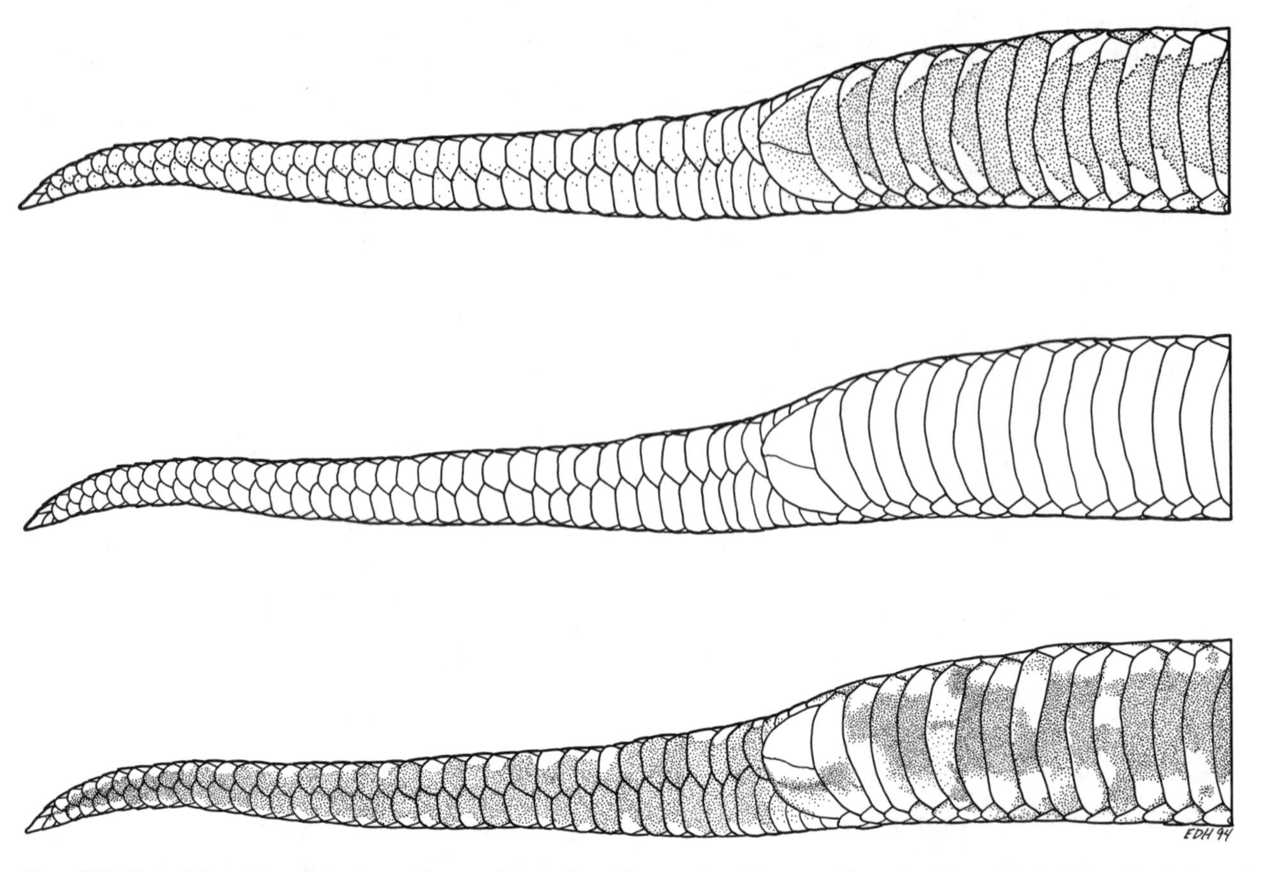

Figure 242. Ventral view of the tails and posterior areas of the bodies of three species of the genus *Heterodon*. Upper: *H. platirhinos*. Drawn from a preserved specimen (KU 220709). Middle: *H. simus*. Drawn from a preserved specimen (KU 82134). Lower: *H. nasicus*. Drawn from a preserved specimen (KU 207105).

Nerodia

Key to Species of *Nerodia*

1a. At least one subocular scale present 2
1b. No subocular scales 3

2a. Dark venter with light markings; from coastal Alabama westward .... ***N. cyclopion***
CC 288, pl 21, p 288
2b. Venter light and largely unmarked; from coastal Alabama eastward .... ***N. floridana***
CC 289, pl 21, p 288

3a. Dorsal pattern with four alternating rows of dark blotches (rarely fused to form crossbars) (Fig. 245); range restricted to central Texas 4
3b. Dorsal pattern variable, but not as above; range not restricted to central Texas 5

4a. Distinct dark spots in rows on either side of venter; from the Brazos River system .... ***N. harteri***
CC 299, pl 21 **CAAR** 330 (1983)
4b. Venter immaculate or at most lightly marked, if spots are present, they are tiny and indistinct; from the Colorado and Concho river systems .... ***N. paucimaculata***
CC 299, pl 21 **CAAR** 330 (1983) (as *N. harteri paucimaculata*)

Some authorities consider this taxon a subspecies of *N. harteri*.

5a. Anterior temporal scales number 2–4 (Fig. 246); usually 11–13 lower labial scales present .... ***N. taxispilota***
CC 289, pl 21, p 290 **CAAR** 331 (1983)
5b. One anterior temporal scale present; usually ten lower labial scales present 6

6a. Dorsal spots in alternating median and lateral rows, widely separated, but connected at corners .... ***N. rhombifer***
CC 291, pl 21, p 290, 292 **CAAR** 376 (1985)
6b. Dorsal pattern variable, with stripes, crossbands, or blotches, but not medial and lateral rows of spots connected at corners 7

7a. Venter immaculate or markings limited to edges of ventrals .... ***N. erythrogaster***
CC 293, pl 20, p 292 **S** 151, pl 43 **CAAR** 500 (1990)

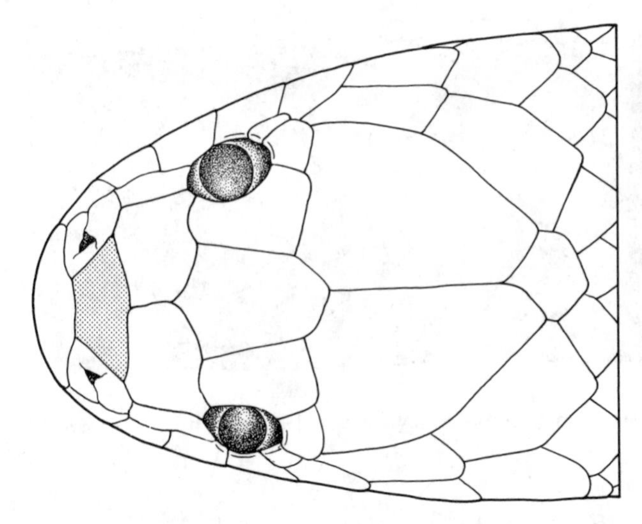

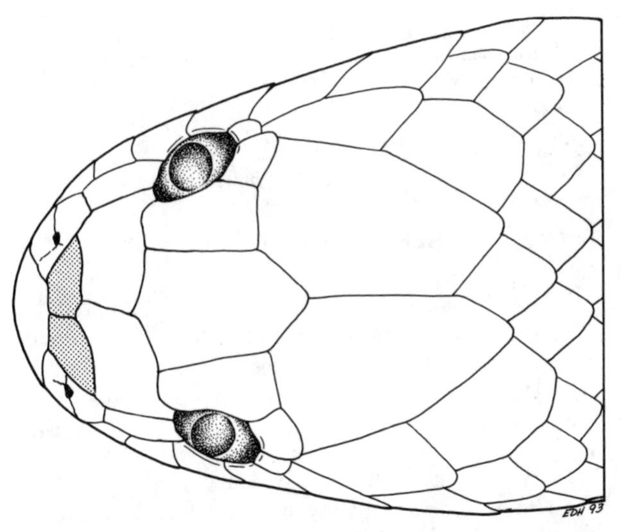

Figure 243. Dorsal view of the heads of two species of the genus *Farancia*. Upper: *F. abacura*, showing the presence of one internasal scale (shaded). Drawn from a preserved specimen (KU 214422). Lower: *F. erytrogramma*, showing the presence of two internasal scales (shaded). Drawn from a preserved specimen (KU 197245).

7b. Venter with conspicuous markings 8

8a. Dorsal pattern usually with dark crossbands on neck and anterior ¼ of body, and with alternating dark dorsal and lateral blotches posteriorly (Fig. 247) (Lake Erie populations may have a very faint or absent pattern, coastal North Carolina populations may be very dark) .... ***N. sipedon***
CC 294, pl 20 **S** 152, pl 43

8b. Dorsal pattern not as above, or if very dark, not from coastal North Carolina 9

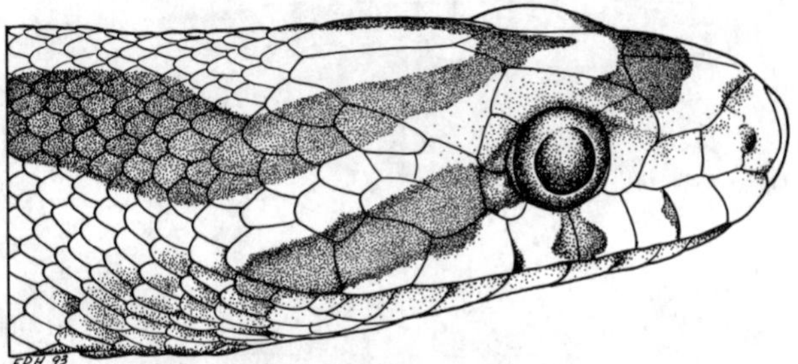

Figure 244. Lateral view of the head of *Elaphe emoryi*, showing the presence of a V-shaped mark on the dorsum of the head and neck and a lateral postocular stripe that extends onto the lower neck. Drawn from a photograph by Suzanne L. Collins.

9a. Dark line present from eye to angle of jaw (Fig. 248); dorsum usually with crossbands (rarely with stripes) .... ***N. fasciata***
CC 297, pl 20, p 288

9b. No dark stripe from eye to angle of jaw; dorsum often with longitudinal stripes (at least anteriorly); from coastal marshes, often in brackish water .... ***N. clarkii***
CC 299, pl 21

Regina

Key to Species of ***Regina***

1a. Anterior dorsal scales smooth; one internasal scale present (Fig. 249) .... ***R. alleni***
CC 304, pl 22

1b. Anterior dorsal scales keeled; two internasal scales present (Fig. 249) 2

2a. Venter immaculate or with a median stripe or row of spots .... ***R. grahamii***
CC 302, pl 22

2b. Venter with two median stripes or rows of spots ... 3

3a. Lower dorsal scale rows keeled; four ventral stripes present (rarely broken into spots) .... ***R. septemvittata***
CC 301, pl 22

3b. Lower dorsal scale rows smooth; two ventral rows of spots present .... ***R. rigida***
CC 302, pl 22

Clonophis

CAAR 364 (1985)

Clonophis kirtlandii

CC 305, pl 22 **CAAR** 364 (1985)

Coniophanes

Coniophanes imperialis

CC 384, pl 33

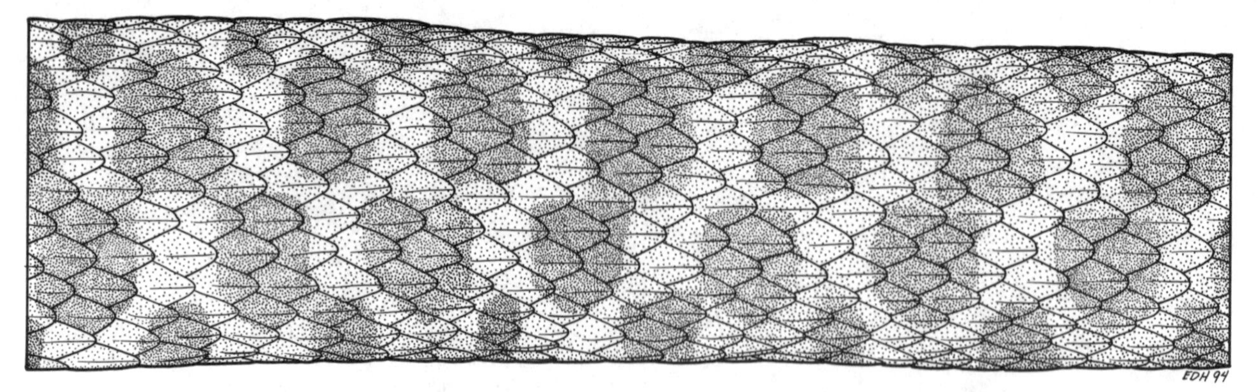

Figure 245. Dorsal view of the body of *Nerodia harteri*, showing the pattern of four alternating rows of dark blotches. Drawn from a preserved specimen (KU 82687).

Trimorphodon

CAAR 352 (1984)

Trimorphodon biscutatus

CC 387, pl 32 **S** 177, pl 39 **CAAR** 353 (1984)

Hypsiglena

Hypsiglena torquata

CC 385, pl 33 **S** 175, pl 39

Leptodeira

Leptodeira septentrionalis

CC 387, pl 32

Carphophis

Key to Species of ***Carphophis***

1a. Pale ventral coloration extends upward only onto the first or first and second dorsal scale rows .... ***C. amoenus***
CC 334, pl 25

1b. Pale ventral coloration extends upward to involve the third dorsal scale rows .... ***C. vermis***
CC 334, pl 25

Chilomeniscus

Chilomeniscus cinctus

S 168, pl 38

Chionactis

Key to Species of ***Chionactis***

1a. Snout convex above (Fig. 250); usually less than 21 dark bands present .... ***C. palarostris***
S 166, pl 38

1b. Snout flat above (Fig. 250); usually 21 or more dark bands present .... ***C. occipitalis***
S 167, pl 38

Evidence indicates that *Chionactis saxatilis,* which is still included in some checklists, is an invalid taxon, and we have chosen to omit it from this key.

Tantilla

CAAR 307 (1982)

Key to Species of ***Tantilla***

1a. Head scarcely if any darker than body, if darker not sharply demarcated from lighter body color; usually six upper labial scales present (if with seven, 6th shorter than 5th); usually one postocular scale present (Fig. 251A) .... ***T. gracilis***
CC 390, pl 33, p 389

1b. Head dark above, sharply demarcated from lighter body color; usually seven upper labial scales present (6th about as high as 5th); usually two postocular scales present 2

2a. From east of the Mississippi River 3

2b. From west of the Mississippi River 5

3a. Rostral scale black; front edge of light collar level with rear of 7th labial; black neck band 3–5 scales wide; postocular labial scales mostly light (Fig. 251B) .... ***T. coronata***
CC 388, pl 33, p 389 **CAAR** 308 (1982)

3b. Without above combination of characters (one or more may be present) 4

4a. Rostral scale black; light collar incomplete dorsally or absent; dark cap on head usually with two posterior notches (Fig. 251C); hemipenes with two basal hooks; from extreme southern Florida ... ***T. oolitica***
CC 390 **CAAR** 256 (1980)

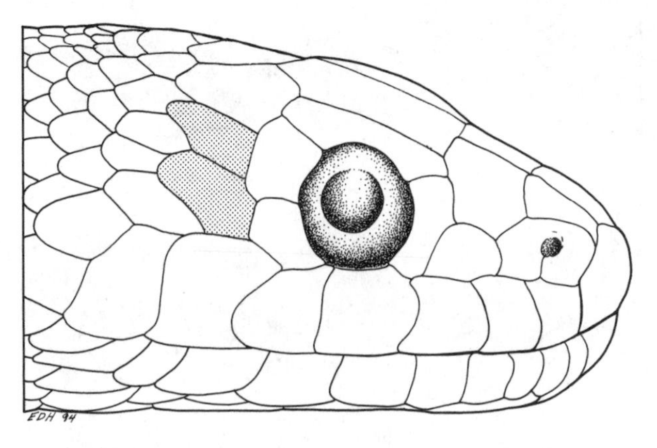

Figure 246. Lateral view of the head of *Nerodia taxispilota*, showing the presence of two anterior temporal scales (shaded). Drawn from a preserved specimen (KU 61081).

4b. Rostral scale with a light spot or if black, all upper labial scales black; light collar usually complete; dark cap on head usually with one posterior notch (Fig. 251D); hemipenes with one basal hook; from peninsular, but not extreme southern Florida .... ***T. relicta***
CC 388, pl 33, p 389 **CAAR** 257 (1980)

Some authorities have suggested that this taxon consists of two species, *T. relicta* and *T. pamlica*.

5a. Head all black or with light collar (sometimes broken) bordered posteriorly by a broad black bar (two or more scale rows wide) or three black spots (Fig. 251E); restricted to the Big Bend area of southern Texas .... ***T. cucullata***
CC 394, pl 33, p 391 (as *T. rubra cucullata*)

5b. Head not all black or if black, with a light collar, when present, followed by body color or a narrow bar no more than one scale row wide 6

6a. Head cap stops no more than one dorsal scale length behind parietals; collar bordered posteriorly by a narrow bar (Fig. 251F) .... ***T. wilcoxi***
S 171, pl 41 **CAAR** 345 (1983)

6b. Head cap extends at least one dorsal scale length past the end of the interparietal suture; light collar, if present, followed by body color (but dark smudges may be present) 7

7a. Head cap extends below angle of jaw 8

7b. Head cap does not extend below angle of jaw 9

8a. Supralabial scales 5 and 6 all white; ¼ or more of the anterior temporal scale white (Fig. 251G) .... ***T. yaquia***
S 174, pl 41, p 217 **CAAR** 198 (1977)

8b. Supralabial scales 5 and 6 dark-edged dorsally; less than ⅕ of the anterior temporal scale white (Fig. 251H) .... ***T. planiceps***
S 172, pl 37, 41, p 217 **CAAR** 319 (1983)

9a. Head cap extends at least three scale rows past the end of the interparietal suture; posterior edge of cap angular; without a light collar (Fig. 251I) .... ***T. nigriceps***
CC 392, pl 33, p 391, 392 **S** 173, pl 41, p 217

9b. Head cap extends < three scale rows past the end of the interparietal suture; posterior edge of cap straight or convex; collar usually present (may be faint).... 10

10a. Mental scale usually not touching anterior chin shields; often only one postocular scale present (Fig. 251J); hemipenes not capitate .... ***T. atriceps***
CC 393, pl 33, p 391 **CAAR** 317 (1983)

10b. Mental scale usually touching anterior chin shields; usually two postocular scales present (Fig. 251K); hemipenes capitate .... ***T. hobartsmithi***
CC 392, p 392 **S** 172, p 217 **CAAR** 318 (1983)

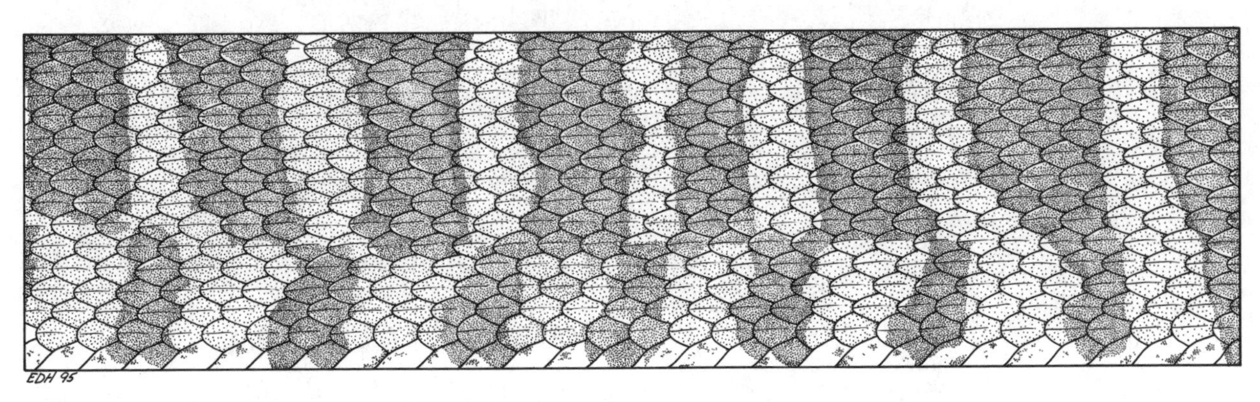

Figure 247. Dorsolateral view of the body pattern of *Nerodia sipedon*, showing crossbands on anterior part of body (right) and alternating dorsal and lateral blotches posteriorly (left). Drawn from a preserved specimen (KU 206318).

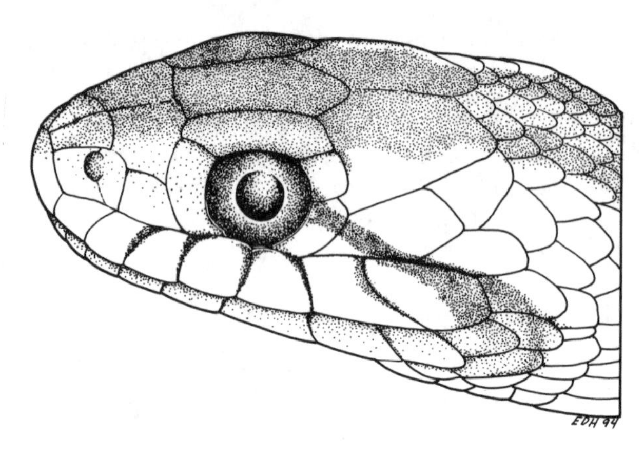

Figure 248. Lateral view of the head of *Nerodia fasciata*, showing a dark line from the eye to the angle of the jaw. Drawn from a photograph in Carmichael and Williams (1991).

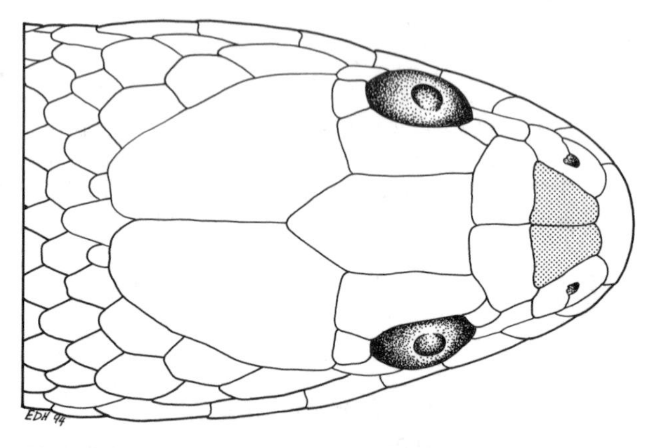

Figure 249. Dorsal view of the heads of two species in the genus *Regina*. Upper: *R. alleni*, showing the presence of a single internasal scale (shaded). Drawn from a preserved specimen (KU 189198). Lower: *R. grahamii*, showing the presence of two internasal scales (shaded). Drawn from a preserved specimen (KU 203615).

Ficimia

CAAR 471 (1990)

Ficimia streckeri

CC 384, pl 33 **CAAR** 181 (1976)

Oxybelis

Oxybelis aeneus

S 139, pl 40 **CAAR** 305 (1982)

Gyalopion

CAAR 182 (1976)

Key to Species of *Gyalopion*

1a. Anal plate divided .... ***G. canum***
CC 385, pl 33 **S** 169, pl 40 **CAAR** 182 (1976)

1b. Anal plate not divided .... ***G. quadrangulare***
S 170, pl 40 **CAAR** 182 (1976)

Rhadinaea

Rhadinaea flavilata

CC 334, pl 25

Contia

Contia tenuis

S 129, pl 39

Sonora

Sonora semiannulata

CC 383, pl 31, p 382 **S** 165, pl 38 **CAAR** 333 (1983)

Salvadora

Key to Species of *Salvadora*

1a. Posterior chin shield in contact or separated by only one scale (Fig. 252); usually eight upper labial scales present; lateral stripe (if present) on scale rows two and three .... ***S. grahamiae***
CC 351, pl 31, p 350 **S** 140, pl 40

1b. Posterior chin shields separated by 2–3 scales; usually 9–10 upper labial scales present; lateral stripe involves the 4th scale row (at least anteriorly) 2

2a. Two upper labial scales in contact with eye (Fig. 253); one loreal scale present .... ***S. deserticola***
CC 352, pl 31, p 350

2b. One or no upper labial scales in contact with eye; usually two loreal scales present .... ***S. hexalepis***
S 138, pl 40

Coluber

CAAR 399 (1986)

Key to Species of *Coluber*

1a. Largely from east of the Rocky Mountains, but specimens from the westernmost portions of the range usually have seven upper labial scales and less than 85 subcaudals .... ***C. constrictor***
CC 339, pl 26, p 337 **S** 136, pl 36 **CAAR** 218 (1978)

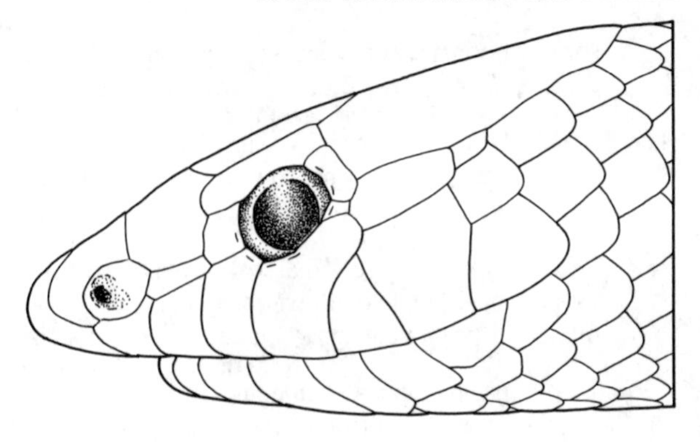

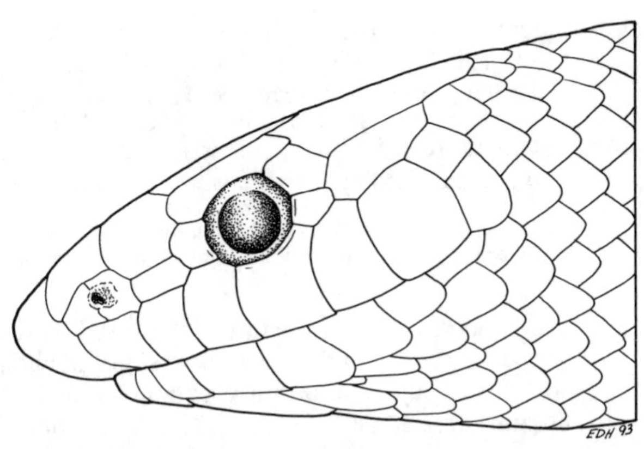

Figure 250. Lateral view of the heads of two species of the genus *Chionactis*. Upper: *C. occipitalis*, showing the flat aspect of the snout. Drawn from a preserved specimen (KU 6620). Lower: *C. palarostris*, showing the convex aspect of the snout. Drawn from a preserved specimen (KU 77973).

1b. From the Rocky Mountains west; usually eight upper labial scales and 85 or more subcaudals present .... ***C. mormon***
S 136, pl 36 **CAAR** 218 (1978) (as *C. constrictor mormon*)

Masticophis

CAAR 144 (1973)

Key to Species of *Masticophis*

1a. Dorsal scale rows number fifteen at midbody...... 2
1b. Dorsal scale rows number seventeen at midbody 3

2a. Dorsal head plates with light edges; no paired light specks on dorsal scales....................... ***M. taeniatus***
CC 344, pl 26 **S** 134, pl 36 **CAAR** 639 (1996)
2b. Dorsal head plates uniformly dark; paired light specks present on dorsal scales, or edges of middorsal scales light .... ***M. schotti***
CC 344, pl 26, p 337 (as *M. taeniatus schotti*) **CAAR** 638 (1996)

3a. At least one well-defined light lateral line present 4
3b. No well-defined light lateral line ...... ***M. flagellum***
CC 343, pl 26 **S** 133, pl 36 **CAAR** 145 (1973)

4a. Single light lateral line extends onto tail................ .... ***M. lateralis***
S 132, pl 36 **CAAR** 343 (1983)
4b. Two or three light lateral lines not extending onto tail ....***M. bilineatus***
S 135, pl 36 **CAAR** 637 (1996)

Diadophis

Diadophis punctatus

CC 331, pl 25 p 331 **S** 128, pl 37

Some authorities have suggested that this taxon consists of three species, *D. punctatus*, *D. acricus*, and *D. amabilis*.

Pituophis

Key to Species of *Pituophis*

1a. Usually a dark line extending from the eye to the angle of the jaw (Fig. 254); from western Wisconsin and Illinois west across the Great Plains to the Pacific Ocean and south into México (isolated populations in eastern Illinois and western Indiana) .... ***P. catenifer***
CC 365, pl 27 **S** 141, pl 39 **CAAR** 474 (1990) (as *P. melanoleucus*, in part)

1b. Head black or, if light, no dark line from the eye to the angle of the jaw; from the southeastern U.S. 2

2a. Dark dorsal blotches near head obscure the ground color and often run together, whereas those near the tail are distinctly separated; from west-central Louisiana and eastern Texas....................... ***P. ruthveni***
CC 365, pl 27 **CAAR** 474 (1990) (as *P. melanoleucus ruthveni*)
2b. Dorsal pattern obscure, absent, or with blotches distinctly separated both anteriorly and posteriorly; from Kentucky and Virginia to the Coastal Plain (west to extreme eastern Louisiana) and into peninsular Florida .... ***P. melanoleucus***
CC 365, pl 27 **CAAR** 474 (1990)

Phyllorhynchus

Key to Species of *Phyllorhynchus*

1a. Dorsal blotches before tail number seventeen or more ....***P. decurtatus***
S 131, pl 40
1b. Dorsal blotches before tail number sixteen or fewer .... ***P. browni***
S 130, pl 40

Thamnophis

Key to Species of *Thamnophis*

1a. Lateral light stripe anteriorly involving the 4th scale row 2
1b. Lateral light stripe lower or absent 8

2a. Slender habitus; tail long (usually more than ¼ total length) 3
2b. Less slender; tail shorter (usually less than ¼ total length) 4

3a. Parietal spots often absent; from east of the Mississippi River .... ***T. sauritus***
CC 321, pl 23 **CAAR** 99 (1970)
3b. Parietal spots present; from Indiana and southern Wisconsin south through the Mississippi Valley and west into the Great Plains .... ***T. proximus***
CC 321, pl 23 **S** 159, pl 42 **CAAR** 98 (1970)

4a. Lateral light stripe anteriorly involving the 2nd scale row 5
4b. Lateral light stripe anteriorly limited to the 3rd and 4th scale rows 7

5a. Seventeen dorsal scale rows present at midbody ... . ***T. brachystoma*** (part)
CC 316, pl 23 **CAAR** 190 (1976)
5b. Nineteen dorsal scale rows present at midbody ... 6

6a. From east of the Mississippi River .... ***T. butleri***
CC 316, pl 23 **CAAR** 258 (1980)
6b. From west of the Mississippi River .... ***T. sirtalis*** (part)
CC 312, pl 23, 24 **S** 157, pl 43 **CAAR** 270 (1981)

Some authorities have suggested that this taxon consists of two species, *T. sirtalis* and *T. dorsalis*.

7a. Eight or nine upper labial scales present... ***T. eques***
S 155, pl 42
7b. Seven upper labial scales present .... ***T. radix***
CC 315, pl 24 **S** 160, pl 42

8a. Lateral light stripe anteriorly limited to 3rd scale row .... ***T. marcianus***
CC 316, pl 24 **S** 154, pl 42
8b. Lateral light stripe anteriorly involving 2nd scale row or absent 9

9a. Seventeen or fewer dorsal scale rows present at midbody 10
9b. Nineteen or more dorsal scale rows present at midbody 11

10a. Seven upper labial scales present...... ***T. ordinoides***
S 164, pl 42 **CAAR** 233 (1979)
10b. Six upper labial scales present .... ***T. brachystoma*** (part)
CC 316, pl 23 **CAAR** 190 (1976)

11a. Seven upper labial scales present.... ***T. sirtalis*** (part)
CC 312, pl 23, 24 **S** 157, pl 43 **CAAR** 270 (1981)

Some authorities have suggested that this taxon consists of two species, *T. sirtalis* and *T. dorsalis*.

11b. Eight upper labial scales present 12

12a. Black blotches present on sides of neck (Fig. 255); no more than nineteen dorsal scale rows; two lateral series of conspicuous dark spots present .... ***T. cyrtopsis***
CC 319, pl 24 **S** 156, pl 42 **CAAR** 245 (1980)
12b. No black blotches on sides of neck; often with more than nineteen dorsal scale rows (if nineteen, lacking two lateral series of conspicuous dark spots)..... 13

13a. Usually only one upper labial scale in contact with eye; usually no light lateral stripes; dorsum often with numerous prominent spots .... ***T. rufipunctatus***
S 158, pl 43 **CAAR** 505 (1990)
13b. Two upper labial scales in contact with eye; a light lateral stripe often present; no dorsal spots or, if any are present, not prominent 14

14a. Internasal scales broader than long, not tapered anteriorly (Fig. 256); 6th and 7th upper labial scales enlarged (often higher than wide) .... ***T. elegans***
CC 319, pl 24 **S** 161, pl 43 **CAAR** 320 (1983)
14b. Internasal scales longer than broad, tapered anteriorly; no enlarged upper labial scales 15

The following species of *Thamnophis* are members of the *T. couchii* complex; some authorities recognize only one species.

15a. No middorsal light stripe (a very faint and narrow stripe may be present on the neck of some individuals); from southern California south into Baja California .... ***T. hammondii***
S 162, p 205 **CAAR** 351 (1984) (as *T. couchii hammondii*)
15b. Middorsal light stripe present (may be faint or narrow), if fading posteriorly, not from southern California 16

16a. Usually eleven lower labial scales present; middorsal light stripe narrow, occasionally fading posteriorly .... ***T. couchii***
S 162, pl 43, p 205 **CAAR** 351 (1984) (as *T. c. couchii*)
16b. Usually ten lower labial scales present; middorsal light stripe variable 17

17a. Usually 21 or more dorsal scale rows present anteriorly; middorsal light stripe narrow and dull .... ***T. gigas***
S 162, p 205 **CAAR** (1984) (as *T. couchii gigas*)
17b. Usually nineteen or fewer dorsal scale rows present anteriorly; middorsal light stripe variable, but often distinct and bright .... ***T. atratus***
S 162, pl 43, p 205 **CAAR** (1984)
(as *T. couchii atratus*, *T. c. aquaticus*, *T. c. hydrophilus*)

Figure 251 (facing page). Dorsolateral views of the head and neck patterns of eleven species of the genus *Tantilla*. (A) *T. gracilis*. Drawn from a preserved specimen (KU 83632). (B) *T. coronata*. Drawn from a preserved specimen (KU 19583). (C) *T. oolitica*. (D) *T. relicta*. Drawn from a preserved specimen (KU 222231). (E) *T. cucullata*. Drawn from a preserved specimen (KU 176979). (F) *T. wilcoxi*. Drawn from a preserved specimen (KU 39970). (G) *T. yaquia*. Drawn from a preserved specimen (KU 93500). (H) *T. planiceps*. Drawn from a preserved specimen (KU 74363). (I) *T. nigriceps*. Drawn from a preserved specimen (KU 206494). (J) *T. atriceps*. Drawn from a preserved specimen (KU 16144). (K) *T. hobartsmithi*. Drawn from a preserved specimen (KU 72771).

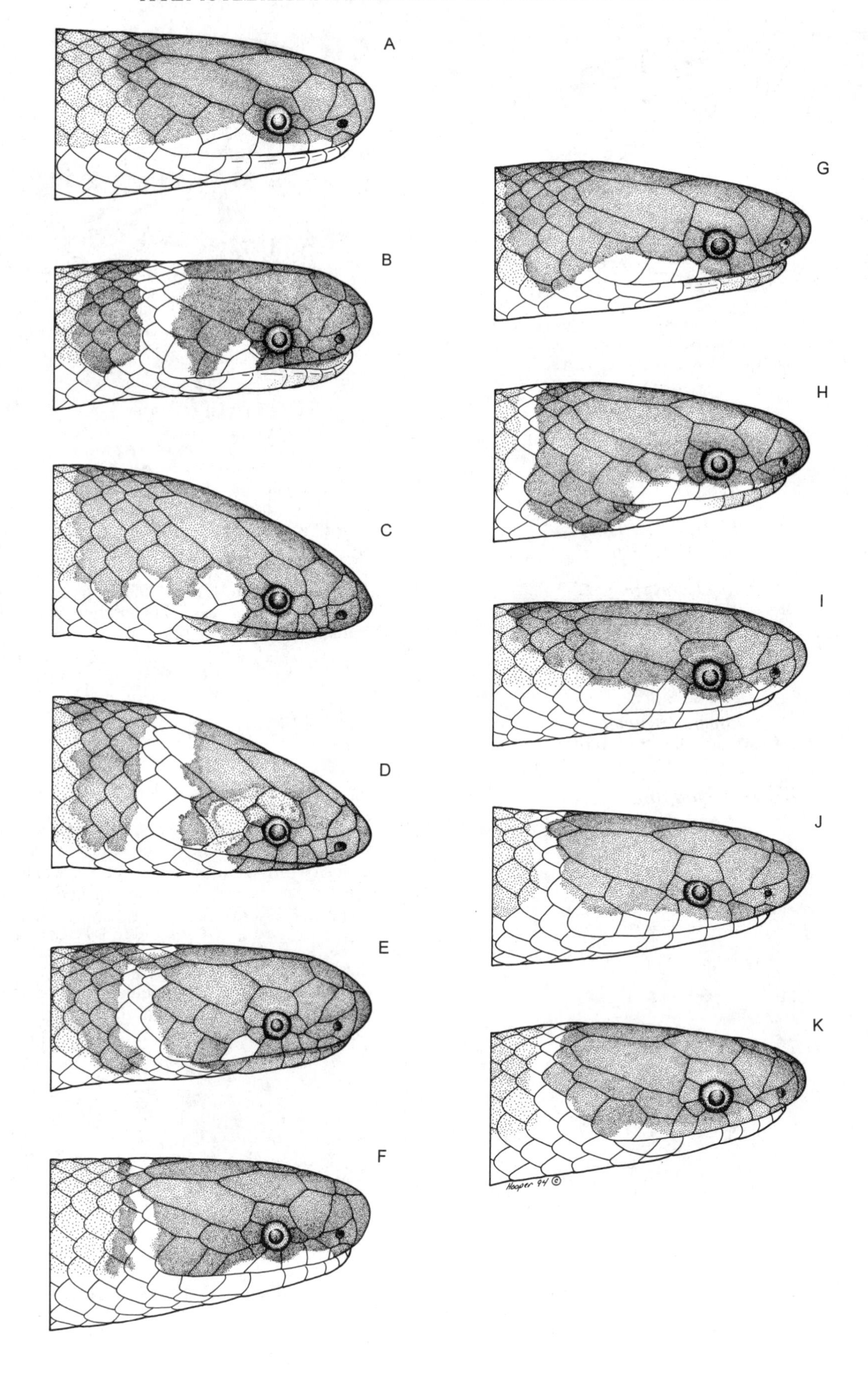
A
B
C
D
E
F
G
H
I
J
K

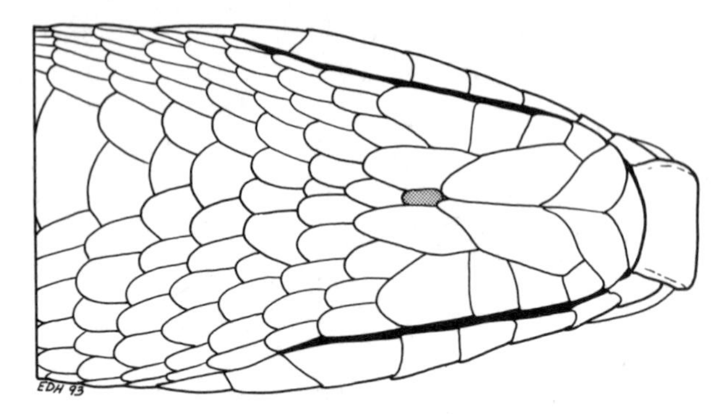

Figure 252. Ventral view of the head of *Salvadora grahamiae*, showing the presence of a single small scale (shaded) between the posterior chin shields. Drawn from a preserved specimen (KU 62889).

Tropidoclonion

Tropidoclonion lineatum

CC 323, pl 24 **S** 163, pl 41

Rhinocheilus

CAAR 175 (1975)

Rhinocheilus lecontei

CC 380, pl 31 **S** 150, pl 37 **CAAR** 175 (1975)

Stilosoma

CAAR 183 (1976)

Stilosoma extenuatum

CC 381, pl 31 **CAAR** 183 (1976)

Cemophora

CAAR 374 (1985)

Cemophora coccinea

CC 380, pl 30, 31, p 372 **CAAR** 374 (1985)

Some authorities have suggested that this taxon consists of two species, *C. coccinea* and *C. lineri*.

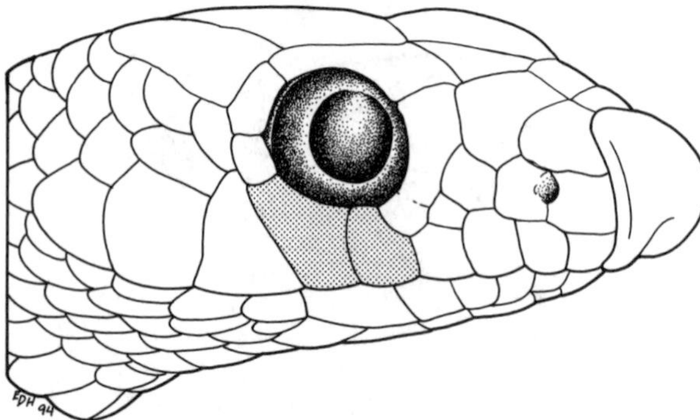

Figure 253. Lateral view of the head of *Salvadora deserticola*, showing two upper labial scales (shaded) in contact with the eye. Drawn from a photograph in Howland (1993).

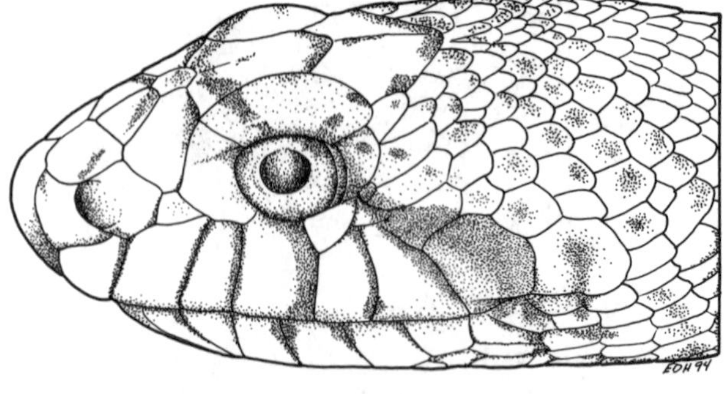

Figure 254. Lateral view of the head of *Pituophis catenifer*, showing a dark line from the eye to the angle of the jaw. Drawn from a photograph by Suzanne L. Collins.

Arizona

CAAR 179 (1976)

Key to Species of ***Arizona***

1a. Tail relatively long (more than 15% total length in males, more than 14% total length in females); range from Great Plains west to eastern Arizona and southern Utah .... ***A. elegans***
CC 362, pl 27 **S** 142 **CAAR** 179 (1976)

1b. Tail relatively short (less than 14% total length in males, less than 13% total length in females); range from Pacific Coast east to central Arizona .... ***A. occidentalis***
CC 362 **S** 142, pl 39 **CAAR** 179 (1976) (as *Arizona elegans occidentalis*)

Drymarchon

CAAR 267 (1981)

Key to Species of ***Drymarchon***

1a. Fifth and 7th supralabial scales meet above the wedge-shaped 6th supralabial scale, which is not in contact with the inferior postocular scale or temporal scale; from extreme southeastern U.S., including peninsular Florida .... ***D. couperi***
CC 349, pl 27, p 349 **CAAR** 267 (1981) (both as *D. corais couperi*)

1b. Sixth supralabial scale is wedge-shaped, but not cut off from the inferior postocular scale or temporal scale by adjacent supralabial scales; from southern Texas south into México .... ***D. corais***
CC 349, pl 27, p 349 **CAAR** 267 (1981)

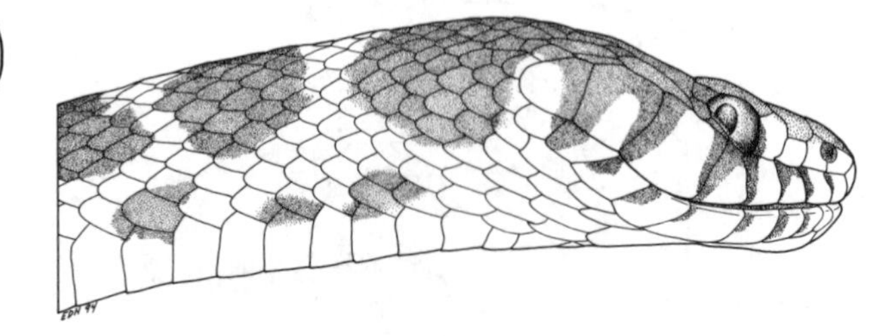

Figure 255. Lateral view of the head and neck of *Thamnophis cyrtopsis*, showing presence of dark blotches on neck. Drawn from a preserved specimen (KU 61180).

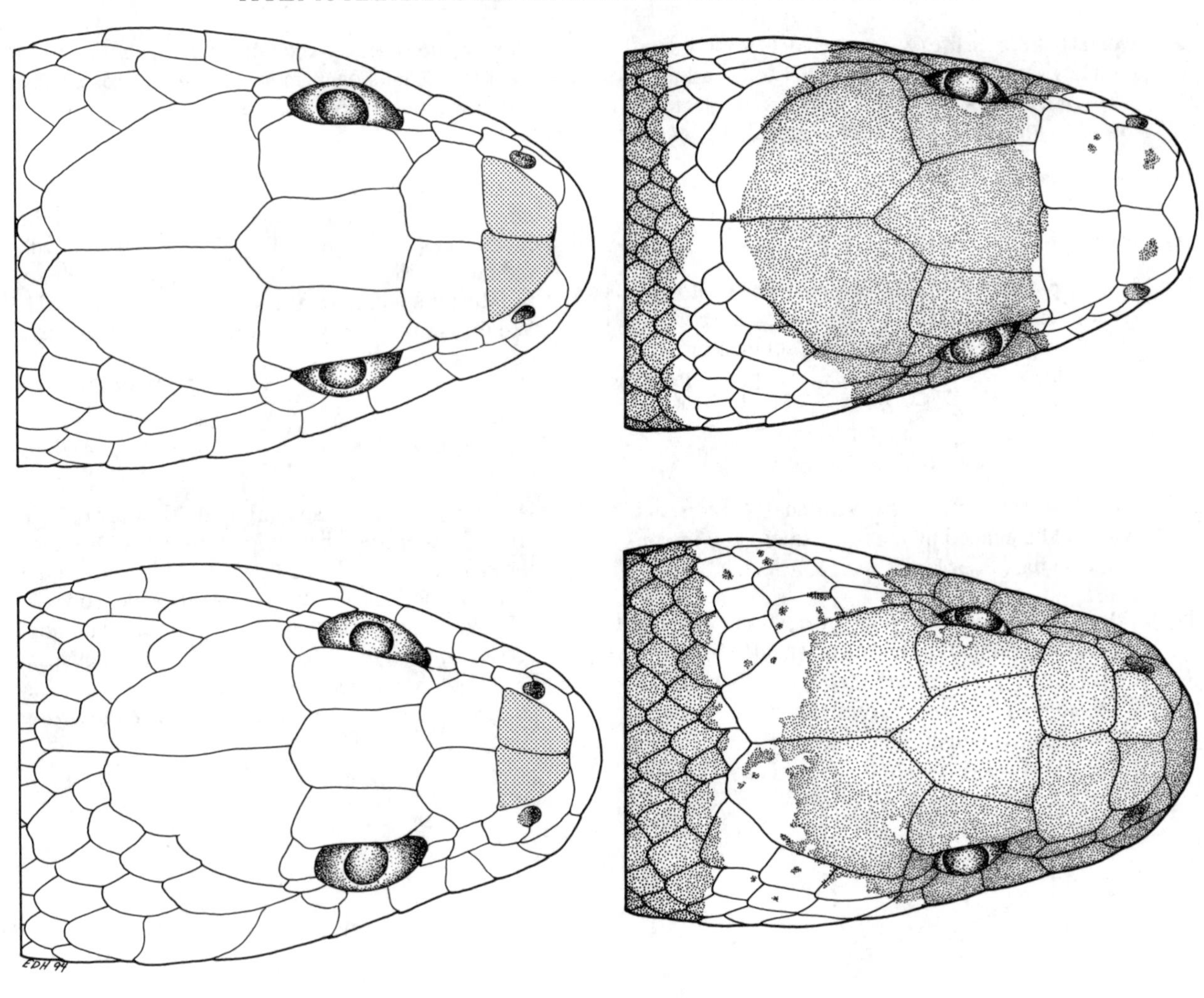

Figure 256. Dorsal view of the heads of two species of the genus *Thamnophis*. Upper: *T. elegans*, showing the internasals broader than long and not tapered anteriorly. Drawn from a preserved specimen (KU 7958). Lower: *T. hammondii*, showing the internasals longer than broad and tapered anteriorly. Drawn from a preserved specimen (KU 5465).

Lampropeltis

CAAR 150 (1973)

Key to Species of *Lampropeltis*

1a. Black-bordered red or orange (in life) dorsal blotches narrower than light-bordered interspaces, or with fewer than fifteen blotches; restricted to southern Texas and New Mexico and adjacent areas of México .... ***L. alterna***
CC 377, pl 32 **CAAR** 55 (1967) (as *L. mexicana*)

1b. Dorsal pattern exceedingly variable, with stripes, rings, crossbands, or blotches, but if with blotches, these are wider than interspaces and/or number fifteen or more; range not restricted to southern Texas and New Mexico and adjacent areas of México 2

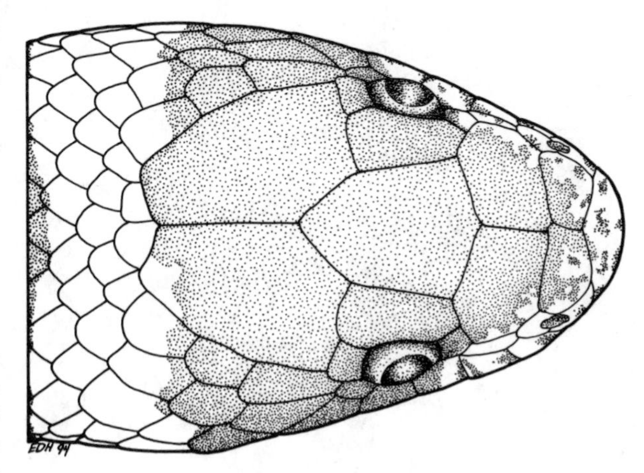

Figure 257. Dorsal view of the heads of three species of the genus *Lampropeltis*. Upper: *L. pyromelana*, showing the uniformly light-colored snout. Drawn from a preserved specimen (KU 182303). Middle: *L. zonata*, showing a uniformly dark-colored snout. Drawn from a preserved specimen (KU 61050). Lower: *L. triangulum*, showing a uniformly dark-colored snout with the anterior edge lighter in color. Drawn from a preserved specimen (KU 204843).

2a. Dorsal pattern (in life) without red or black-bordered blotches; stripes (if present) distinct and confined to western U. S. populations; if unicolored, black and confined to eastern U. S. populations....... ***L. getula***
CC 369, pl 29 **S** 148, pl 37

2b. Dorsal pattern (in life) with red or black-bordered blotches; stripes (if present) indistinct; if unicolored, not black 3

3a. Dorsal pattern of black-bordered brown blotches rarely reaching down to the 5th scale row, or unicolored with indistinct stripes; southeastern and central U. S. .... ***L. calligaster***
CC 377, pl 29, 31 **CAAR** 229 (1979)

Some authorities have suggested that this taxon consists of two species, *L. calligaster* and *L. occipitolineata.*

3b. Dorsal pattern of transverse crossbands or rings, or if with saddle-shaped blotches, these broadly in contact with the 5th or lower scale row 4

4a. Dorsal pattern (body and tail) with forty or more light crossbands; top of head black, snout normally a uniform white (sometimes flecked with black) (Fig. 257); dorsal scale rows at midbody number 23–25; ventral scales number 213–238; Nevada, Utah, Arizona, and New Mexico only .... ***L. pyromelana***
S 149, pl 37 **CAAR** 342 (1983)

4b. Dorsal pattern (body and tail) with fewer than forty light crossbands or, if more than forty crossbands, top of head and snout generally black (Fig. 257)... 5

5a. Dorsal pattern (body and tail) with 31 or more light crossbands that do not widen conspicuously on the lower sides; snout generally black (sometimes with red flecks) (Fig. 257); dorsal scale rows at midbody usually number 23; ventral scales number 194–227; Washington, Oregon, and California only .... ***L. zonata***
S 146, pl 37 **CAAR** 174 (1975)

Some authorities have suggested that this taxon consists of four species, *L. zonata, L. multifasciata, L. parvirubra,* and *L. pulchra.*

5b. Dorsal pattern (body and tail) with 31 or fewer light crossbands or rings that widen conspicuously on the lower sides, or with saddle-shaped blotches rather than crossbands or rings; dorsal scale rows at midbody usually number 21; ventral scales number 152–215; eastern U. S. west to Utah (absent from Washington, Oregon, and California) .... ***L. triangulum***
CC 375, pl 30, p 372 **S** 147, pl 37 **CAAR** 594 (1994)

Some authorities have suggested that this taxon consists of three species, *L. triangulum, L. elapsoides,* and *L. taylori.*

GLOSSARY

ABDOMINALS The fourth set of plates on the plastron of a turtle, bordered anteriorly by the pectorals and posteriorly by the femorals (Fig. 99).

ADPRESSED LIMBS In salamanders (or lizards) the forelimb is laid backward along the body and the hindlimb forward, both at full length; the distance separating the two is usually measured in numbers of costal folds or grooves (or scale rows) (Fig. 1).

ADULT A sexually mature individual. Also used to describe the form of metamorphosed amphibians.

ANALS The most posterior set of plates in the plastron of a turtle, usually paired, bordered anteriorly by the femorals (Fig. 99).

ANAL PLATE (scute or scale) (= cloacal plate) The terminal ventral in snakes, usually larger than other ventrals and free along its posterior margin (Fig. 225).

ANTERIOR (= cranial) Toward the head; may be used in combination with other directional terms (e.g., anteromedial).

AXILLA The armpit.

AXILLARIES Scutes on the anterior margin of the bridge in turtles (Fig. 114).

BARBEL A fleshy extension of skin, usually on the head.

BASAL CONSTRICTION A narrowing of the tail near its origin (Fig. 16).

BICORNUATE TONGUE A configuration of the tongue in ranid frogs, characterized by two posterior projections (Fig. 42).

BLOTCH An irregularly shaped mark of varying size, often with an ill-defined outline.

BOSS A distinct swollen bump on the heads of some frogs and toads (Fig. 52).

BRANCHIAL Refers to gills in larval salamanders.

BRIDGE In turtle shells, the connection between the carapace and plastron.

CANTHALS Scales contributing to the canthal ridge.

CANTHAL RIDGE A sharply defined and angular canthus rostralis.

CANTHUS ROSTRALIS A ridge extending from the snout to the anterior corner of the eye in amphibians or to the eyebrow in reptiles (Fig. 17).

CAPITATE Having a headlike structure, used in describing a snake hemipenis when the enlarged, distal portion is clearly separated from the basal region by a groove or when the head and spinous areas are separated by a ring of tissue.

CARAPACE The dorsal portion of a turtle shell (Fig. 99).

CASQUE A head with bony projections in the form of ridges, crests, and/or tubercles (Figs. 13, 58, 59, 60, 173).

CHEVRON A V-shaped mark, often with the point of the V on the vertebral line.

CHIN SHIELDS Paired, elongate scales on the lower jaw of snakes between the lower labials (Fig. 203).

CHOANA An internal opening of the nasal passage in the roof of the oral cavity (Fig. 29).

CLAW A cornified, usually pointed, appendage at the tip of a digit. Cornification of the tip of the digit itself, such as in *Xenopus laevis,* is not a true claw (Fig. 40).

CLOACA A chamber into which the digestive, reproductive, and urinary tracts empty; and which in turn empties to the outside through a cloacal aperture.

CLOACAL APERTURE The opening of a cloaca (Figs. 98, 144, 225).

CLOACAL PLATE (= anal plate) The terminal ventral scale in snakes, usually larger than other ventrals and free along its posterior margin (Fig. 225).

COSTALS (= pleurals) Scutes in the carapace of turtles lying over the ribs between the vertebrals and marginals (Figs. 99, 105, 107).

COSTAL FOLD The area between costal grooves, usually bulging and distinct.

COSTAL GROOVE A vertical indentation on the sides of salamanders indicating the position of a rib (Fig. 1).

CRANIAL CREST A raised, bony ridge on the tops and sides of the heads of bufonid toads (Figs. 58, 59, 60).

CRUSHING SURFACE A broad plate on the inside of the jaws in some turtles (Fig. 116).

DEWLAP A laterally compressed vertical fold of skin under the throats of some lizards.

DISTAL Away from the origin of an appendage (opposite of **PROXIMAL**).

DORSAL Toward the back or upper part of an animal.

DORSOLATERAL FOLD A raised glandular ridge extending back from the eyes in some frogs (Fig. 39).

DORSUM The back or upper part of an animal.

EPIDERMAL SCALE A cornified skin appendage; scales cover entire surface of the body in many reptiles.

FASCICULUS A concentrated aggregation of vomerine teeth in some amphibians.

FEMORALS The fifth set of plates on the plastron of a turtle, bordered anteriorly by the abdominals and posteriorly by the anals (Fig. 99).

FEMORAL PORE A distinct depression in scales, usually enlarged, on the ventral surfaces of thighs in some lizards (Fig. 137).

FRONTOPARIETALS Scales between the frontal and parietal plates on the heads of some lizards.

FURROW A deep, sharply-defined groove.

GRANULAR SCALE Type of scale that is small, not overlapping, and usually convex.

GULARS 1. The most anterior set of plates on the plastron of a turtle, bordered posteriorly by the humerals (Fig. 99). 2. Scales on the gular fold of some lizards (Fig. 137).

GULAR FOLD A transverse fold of skin across the throat of some lizards (Figs. 137, 149, 168).

HABITUS The general body shape of an animal.

HEAD CAP The area, usually darkly pigmented, on the heads of some Crowned and Blackhead Snakes, which may extend onto the neck (Fig. 251).

HEMIPENIS One of a paired set of intromittent (= copulatory) organs found in lizards and snakes.

HUMERALS The second set of plates on the plastron of a turtle, bordered anteriorly by the gulars and posteriorly by the pectorals (Fig. 99).

INFRALABIALS Lower labials of lizards and snakes.

INFRAMARGINALS Plates between the marginals of the carapace and the plastron in turtles (Fig. 106).

INGUINALS Scutes on the posterior margin of the bridge in turtles (Fig. 114).

INTERCALARY CARTILAGE An element inserted between the last two phalanges in the digits of some frogs and toads, in effect adding an additional joint (Fig. 49).

INTERNASALS Scales lying between the nasals on the heads of some lizards and snakes (Figs. 137, 203).

INTERORBITAL BOSS A distinct swollen bump between the eyes of some frogs and toads (Fig. 52).

INTERPARIETAL A median dorsal head scale in lizards, lying between the parietals (Fig. 137).

KEEL A raised ridge. 1. A median ridge in the scales of some lizards and snakes (Figs. 139, 204). 2. A longitudinal ridge or ridges on the carapace of some turtles. 3. The raised dorsal edge of the tail in some salamanders.

LABIALS Scales that border the lips of lizards and snakes (Figs. 137, 203).

LATERAL Toward the side (away from the longitudinal midline) of an animal.

LICHEN Derived from the symbiotic association of a fungus and a cyanobacterium; they often grow on rocks or other barren surfaces; here in reference to a lightly mottled pattern.

LINGUAL CUSPS The interior (toward the tongue) surfaces of teeth.

LOREALS Scales on the side of the head in some lizards and snakes, situated between the nasal(s) and preocular(s). If only one scale lies between the nasal and eye, it is a loreal scale if longer than high (Figs. 137, 203).

LOREAL PIT A deep depression in the loreal scale of viperid snakes (Fig. 206).

LORILABIALS A longitudinal row of scales lying between the upper labials and loreal or suboculars of some lizards and snakes (Fig. 137).

MARGINAL RIDGE Raised area along the free edge of the carapace in some softshells.

MARGINALS Plates lying along the periphery of the carapace in turtles (Fig. 99).

MAXILLARY TEETH Teeth on the maxillary bones in the upper jaws of some amphibians and reptiles.

MEDIAN Toward the longitudinal midline of an animal.

MELANISM An increase in the amount of dark pigment in some animals, often obscuring the pattern (especially in older individuals).

MENTAL The single scale at the anterior border of the lower jaw in lizards and snakes, bordered laterally by the first lower labials (Figs. 137, 203).

METAMORPHOSIS The transformation in amphibians from larval to subadult form.

METATARSAL TUBERCLES Distinct protrusions or spades, often sharp-edged, on the basal portion of the feet (between the heel and the bases of the digits); usually used in burrowing by some amphibians (Figs. 48, 50).

NASAL (= nasal plate) The scale(s) on the sides of the head in lizards and snakes, in which the openings of the external nares may be found. The nasal scale is entire if the naris is entirely within one scale, and is divided if the naris is in a suture between two scales (Figs. 137, 203).

NASAL SEPTUM The vertical partition separating the nasal passages (Fig. 108).

NASAL TRIDENT The three-pronged, forward-facing mark on the snouts of some turtles in the genus *Graptemys* (Fig. 122).

NASAL VALVE A flap or ridge in the nasal passages that allows movement of air or water in only one direction.

NASOLABIAL GROOVE A depression or shallow line extending from the external nares to the lips in plethodontid salamanders (Fig. 5).

NOTCHED POSTERIOR MARGINALS Marginal scutes with a distinct indentation at rear of the upper shell in some turtles.

NUCHAL (1) (= cervical) The median plate at the anterior edge of the carapace in turtles (Fig. 99). (2) (= occipital) Head scale(s) of some lizards and snakes immediately posterior to the parietals and interparietal (if present) (Figs. 137, 185).

OCCIPITAL See nuchal (Figs. 137, 185, above).

OCCIPITAL SPINES Pointed projections extending posteriorly and sometimes laterally from the occipital region of the head in some lizards (Fig. 185).

OCELLUS An eyelike spot; usually characterized by one or more concentric rings around a central spot.

PAEDOMORPHOSIS The retention of larval characteristics by some adult salamanders.

PALATAL TEETH Teeth on the roof of the mouth in some frogs and toads (Fig. 74).

PALMAR TUBERCLES Small, rounded protuberances on the hands and feet of some amphibians (Fig. 9).

PARAMEDIAN Lying parallel to and on either side of a longitudinal middorsal (or midventral) line.

PARASPHENOID TEETH Teeth on the parasphenoid bones in the roofs of mouths in some amphibians (Fig. 23).

PARATOID GLANDS Large, swollen glandular areas posterior to the eyes of some amphibians (Figs. 7, 34, 41, 53, 55, 57, 58, 60).

PARIETALS Large head scales of lizards and snakes, situated immediately behind the frontals and separated by an interparietal in many lizards (Figs. 137, 203).

PECTORALS The third set of plates in the plastron of turtles, bordered anteriorly by the humerals and posteriorly by the abdominals (Fig. 99).

PECTORAL GLANDS Small glandular masses on the chests of some spadefoots (Fig. 51).

PERRENIBRANCHIATE The retention of gills by some salamanders throughout their lifespan.

PHALANX Any bone or corresponding portion of a finger or toe.

PLASTRON The ventral portion of a turtle shell (Fig. 99).

POSTERIOR (= caudal) Toward the tail, away from the head; may be used in combination with other directional terms (e.g., posteromedial).

POSTFEMORAL POCKETS Infoldings of skin behind the hindlimbs of some lizards (Fig. 189).

POSTLABIALS Scales behind and in line with the labials in lizards (Fig. 161).

POSTMENTALS Scales lying immediately behind the mental along the midventral line of the chin in some lizards and snakes (Fig. 156).

POSTNASALS Scales lying behind the nasal and anterior to the loreal in some lizards and snakes (Figs. 158, 212, 215).

POSTOCULARS Scales bordering the posterior margin of the eye in lizards and snakes (Figs. 137, 203).

POSTROSTRALS Scales immediately behind the rostral in some lizards and snakes (Fig. 137).

PREANALS Scales immediately anterior to the cloacal aperture in lizards (Fig. 144).

PREANAL PORES Extensions of the femoral pore series onto the body anterior to the cloacal aperture in some lizards (in some species only preanal pores are present) (Fig. 163).

PREAURICULARS Scales in the front of the ears of some lizards and snakes.

PREFRONTALS Head scales lying immediately anterior to the frontals in some lizards and snakes (Figs. 137, 203).

PREMAXILLARY TEETH Teeth associated with the premaxilla along the anterior margin of the upper jaw.

PREOCULARS Scales bordering the anterior margin of the eyes in some lizards and snakes (Figs. 137, 203).

RENIFORM Kidney-shaped.

RETICULATION A color pattern with linear markings resembling the meshes of a net.

ROSTRALS Scale(s) at the tip of the snouts of lizards and snakes, separating the two rows of upper labials (Figs. 137, 203).

SERRATE Resembling the teeth of a saw.

SERRATED POSTERIOR MARGIN Indentations that occur between the marginal scutes at the rear of the upper shell in some turtles.

SNOUT-VENT LENGTH (SVL) A measurement of head-body length from the tip of the snout to the cloacal aperture (in salamanders usually measured to the posterior edge of the vent) (Figs. 39, 137).

SPATULATE Shovellike, distinctly flattened.

SPOT A rounded mark of varying size, typically with a distinct outline.

SUBCAUDALS Scales on the ventral surface of the tail in reptiles (Fig. 157).

SUBOCULARS Scales lying directly below the eye in some reptiles (Figs. 137, 228).

SULCUS A groove, sometimes called a suture, that appears on the plastron of turtles.

SUPERCILIARIES Small and usually numerous scales bordering the orbits in some lizards (Fig. 137).

SUPRALABIALS Upper labials in lizards and snakes.

SUPRANASALS Scales lying immediately above the nasals, but lateral to the internasal, in some lizards and snakes (Figs. 152, 183).

SUPRAOCCIPITAL Above the occipital area near the upper rear of the skull (in reference to a median unpaired bone which forms the roof of the opening through which the spinal cord emerges from the skull; in turtles, this bone forms a high median crest that extends posteriorly and serves for muscle attachment).

SUPRAOCULARS Scales above the eye that form shields in snakes; the dorsal surface of the orbit in lizards (Figs. 137, 203).

SUPRAORBITAL SEMICIRCLE The arc formed around the dorsal aspect of the orbit in some lizards (Fig. 198).

SUPRATYMPANIC RIDGE Thick glandular ridge of skin passing above the tympanum in some frogs.

SVL Snout–vent length.

SYMPATRIC Living in the same geographic area as another ecological or taxonomic entity (e.g., population, species, genus) (opposite of **ALLOPATRIC**).

SYNONYM An alternative name for a taxon; occurs when different names are applied to what turns out to be the same species, genus, or family. Only one name (usually the first applied to a taxon) can be correct at any time.

TAXON A taxonomic group of organisms (species, genus, and family are examples) that is evolutionarily distinct (i.e., has its own evolutionary fate) and is distinguished by name and ranked in a definite category.

TEMPORALS Scales behind the postoculars, below the parietal, and above the upper labials (Figs. 137, 203).

TERMINAL DISCS (= toe pads) The expanded tips of digits in some amphibians (Figs. 26, 65, 67, 70, 88).

TL Abbreviation for total length.

TUBERCLE A distinct bump or knob of skin or bone.

TUBERCULATE Characterized by numerous, small, discrete bumps, usually in reference to a region of skin.

TYMPANUM (= ear drum) The membrane covering the external opening of the ear canal (Fig. 39).

VENT The cloacal aperture or anal opening in amphibians and reptiles (Figs. 1, 39, 98, 137, 225, 239).

VENT LOBES Lateral portions of the prominent glandular areas around the vent in some salamanders.

VENTER The underside (belly) of an animal.

VENTRAL Toward the underside (belly) of an animal.

VERTEBRALS Middorsal rows of scales in lizards and snakes or of plates in the carapaces of turtles (Figs. 99, 197).

VOMERINE TEETH Teeth lying on the vomer in the palate of some amphibians (Figs. 23, 63).

WHORL A symmetrical row of enlarged scales circling a caudal segment in some lizards (Fig. 176).

LITERATURE CITED

Ashton, R. E., Jr. and P. S. Ashton. 1985. Handbook of Reptiles and Amphibians of Florida. Part Two. Lizards, Turtles and Crocodilians.Windward Publ., Miami, Florida. 191 pp.

Babcock, H. L. 1919. The Turtles of New England. Mem. Boston Soc. Nat. Hist. 8(3): 325–431 + Pls. 17–32.

Ballinger, R. E. and J. D. Lynch. 1983. How to Know the Amphibians and Reptiles. Wm. C. Brown Co., Dubuque, Iowa. viii + 229 pp.

Baxter, G. T. and M. D. Stone. 1985. Amphibians and Reptiles of Wyoming. 2nd ed. Publ. Wyoming Game and Fish Dept., Cheyenne. v + 137 pp.

Behler, J. L. 1988. Familiar Reptiles and Amphibians. A. A. Knopf, New York. 192 pp.

Behler, J. L. and F. W. King. 1979. Audubon Society Field Guide to North American Reptiles and Amphibians. A. A. Knopf, New York. 719 pp.

Bishop, S. C. 1943. Handbook of Salamanders. Comstock Publ. Co., Ithaca, New York. xiv + 555 pp.

Blair, A. P. 1968. Amphibians. Pp. 167-212. *In* W. F. Blair, A. P. Blair, P. Brodkorb, F. R. Cagle, and G. A. Moore. Vertebrates of the United States. 2nd ed. McGraw-Hill Book Co., New York. ix + 616 pp.

Cagle, F. R. 1968. Reptiles. Pp. 213-268. *In* W. F. Blair, A. P. Blair, P. Brodkorb, F. R. Cagle, and G. A. Moore. Vertebrates of the United States. 2nd ed. McGraw-Hill Book Co., New York. ix + 616 pp.

Capula, M. 1989. Guide to Reptiles and Amphibians of the World. Simon and Schuster, New York. 256 pp.

Carmichael, P. and W. Williams. 1991. Florida's Fabulous Reptiles and Amphibians. World Publ., Tampa, Florida. 120 pp.

Carr, A. F., Jr. 1952. Handbook of Turtles. Comstock Publ. Co., Ithaca, New York. xv + 542 pp.

Collins, J. T. 1997. Standard Common and Current Scientific Names for North American Amphibians and Reptiles. 4th ed. Soc. Study Amphib. Rept. Herpetol. Circ. 19: i-iv + 1-40 pp.

Collins, J. T. 1993. Amphibians and Reptiles in Kansas. 3rd ed. Univ. Kansas Nat. Hist. Mus. Pub. Ed. Ser. 13: xx + 397 pp.

Conant, R. and J. T. Collins. 1998. Peterson Field Guide to Reptiles and Amphibians of Eastern and Central North America. 3rd edition (expanded). Houghton Mifflin Co., Boston. xx + 616 pp.

Dixon, J. R. 1987. Amphibians and Reptiles of Texas. Texas A&M Univ. Press, College Station, Texas. xii + 434 pp.

Ernst, C. H. and R. W. Barbour. 1972. Turtles of the United States. Univ. Press of Kentucky, Lexington. x + 347 pp.

Garrett, J. M. and D. G. Barker. 1987. A Field Guide to Reptiles and Amphibians of Texas. ix + 225 pp.

Howland, J. M. 1993. Herps of Arizona. Arizona Wildlife Views 36(3): 36 pp.

Johnson, T. R. 1977. The Amphibians of Missouri. Univ. Kansas Nat. Hist. Mus. Pub. Ed. Ser. 6: ix + 134 pp.

Johnson, T. R. 1987. The Amphibians and Reptiles of Missouri. Publ. Missouri Dept. Conserv., Jefferson City. x + 368 pp.

Leviton, A. E. 1971. Reptiles and Amphibians of North America. Doubleday and Co., New York. 252 pp.

Martof, B. S., W. M. Palmer, J. R. Bailey, and J. R. Harrison III. Amphibians and Reptiles of the Carolinas and Virginia. Univ. North Carolina Press, Chapel Hill. 264 pp.

Mount, R. H. 1975. The Reptiles and Amphibians of Alabama. Publ. Auburn Univ. Agric. Exp. Sta., Auburn, Alabama. vii + 347 pp.

Niering, W. A. 1985. Wetlands. A. A. Knopf, New York. 638 pp.

Ortenburger, A. I. 1928. The Whipsnakes and Racers: Genera *Masticophis* and *Coluber*. Mem. Univ. Michigan Mus. 1: xviii + 247 pp.

Passmore, N. I. and V. C. Carruthers. 1979. South African Frogs. Witwatersrand Univ. Press, Johannesburg, South Africa. xviii + 270 pp.

Pfingsten, R. A. and F. L. Downs. 1989. Salamanders of Ohio. Bull. Ohio Biol. Surv. 7(2): xx + 345 pp.

Pope, C. H. 1956. The Reptile World. A. A. Knopf, Inc., New York. xxv + 339 pp.

Pritchard, P. C. H. 1979. Encyclopedia of Turtles. TFH Publ., Neptune City, New Jersey. 895 pp.

Shaw, C. E. and S. Campbell. 1974. Snakes of the American West. A. A. Knopf, New York. xii + 332 pp.

Smith, H. M. 1956. Handbook of Amphibians and Reptiles of Kansas. 2nd ed. Univ. Kansas Nat. Hist. Mus. Misc. Publ. 9: 356 pp.

Smith, H. M. 1978. Amphibians of North America. Western Publ. Co., New York. 160 pp.

Smith, H. M. and E. D. Brodie, Jr. 1982. Reptiles of North America. Western Publ. Co., New York. 240 pp.

Stebbins, R. C. 1951. Amphibians of Western North America. Univ. California Press, Berkeley. ix + 539 pp.

Stebbins, R. C. 1985. Peterson Field Guide to Western Reptiles and Amphibians. 2nd ed. Houghton Mifflin Co., Boston.xiv + 336 pp.

Tennant, A. 1984. The Snakes of Texas. Texas Monthly Press, Austin, Texas. 561 pp.

Vogt, R. C. 1981. Natural History of Amphibians and Reptiles of Wisconsin. Publ. Milwaukee Pub. Mus., Milwaukee, Wisconsin. 205 pp.

Zappalorti, R. T. 1976. The Amateur Zoologist's Guide to Turtles and Crocodilians. Stackpole Books, Harrisburg, Pennsylvania. 208 pp.

THE AUTHORS

Robert Powell received a Bachelors of Arts Degree in Zoology in 1970, an M.A. in Biology in 1971, and a Doctoral Degree in Zoology in 1984. Since 1972, he has been on the faculty of Avila College in Kansas City, Missouri. An ardent believer in the tenet that good biologists must experience biology by "getting their hands dirty," he has been leading field trips since 1973, taking students to various sites throughout the Midwest as well as to the southwestern deserts, southern swamps, and eastern forests. From 1978 through 1985, the varied habitats of México became the focus of many trips. Since 1986, Dr. Powell and his students have been working in the West Indies.

First and foremost a teacher, Robert Powell brings his experiences into the classroom and laboratory. Working exclusively with undergraduates at Avila and in summer research programs, many of his nearly two hundred publications have been coauthored by his students, a large number of whom have gone on to pursue postgraduate work at institutions from coast to coast.

Robert Powell has been a Research Associate at the University of Kansas Natural History Museum since 1988. Efforts that began in the late 1970s to write a useful set of keys for his own herpetology classes led eventually to the association with Joseph T. Collins and Errol D. Hooper, Jr. The results of that collaboration are displayed in this publication. Robert Powell can be contacted at the Department of Natural Sciences, *Avila College,* Kansas City, Missouri 64145 (email: powellr@mail.avila.edu).

Joseph T. Collins is one of the nation's most prolific authors about wildlife in general, and amphibians and reptiles in particular. His titles include the *Peterson Field Guide to Reptiles and Amphibians of Eastern and Central North America* (co-authored with Roger Conant), *Amphibians and Reptiles in Kansas* (three editions), *Fishes in Kansas* (two editions with Frank B. Cross), *Turtles in Kansas* (with Janalee Caldwell), *Natural Kansas, Kansas Wildlife* (with Suzanne L. Collins, Bob Gress, and Gerald J. Wiens), *Kansas Wetlands: A Wildlife Treasury* (with Suzanne L. Collins and Bob Gress), and *A Guide to Great Snakes of Kansas* (featuring the photography of Suzanne L. Collins). He also co-edited (with Richard A. Seigel and Susan S. Novak) *Snakes: Ecology and Evolutionary Biology* and co-edited (with Richard A. Seigel) *Snakes: Ecology and Behavior.* He is the author of eleven other books on amphibians and reptiles as well as over two hundred articles on lower vertebrates, and is the Director of The Center for North American Amphibians and Reptiles, a non-profit foundation based in Lawrence, Kansas. In April of 1996 at a ceremony in the state capitol, Collins was invested by the governor as The Wildlife Author Laureate of Kansas, the first Author Laureate ever recognized by the state. Collins is Adjunct Herpetologist with the Kansas Biological Survey and Herpetologist Emeritus at the University of Kansas Natural History Museum, both in Lawrence. Joseph T. Collins can be contacted at *The Center for North American Amphibians and Reptiles,* 1502 Medinah Circle, Lawrence, Kansas 66047, and the *Kansas Biological Survey,* 2041 Constant Avenue, Lawrence, Kansas 66047-2906 (email: jcollins@ukans.edu).

Errol D. Hooper, Jr. is a naturalist, outdoorsman, and scientific illustrator, who earned a Bachelor of Science Degree in Organismal Biology in 1986 and a Master of Historical Administration and Museum Studies Degree in 1989, both from the University of Kansas. He is a Research Associate in the Exhibits Department and the Division of Mammalogy at the University of Kansas Natural History Museum. His most recent work has been with the Missouri Department of Conservation Riparian Ecosystem Assessment and Management Project, surveying and monitoring amphibians and reptiles at two conservation areas in northern Missouri. Errol Hooper began his illustrating career at The University of Kansas, and has published over 330 scientific images. He currently resides in northern Missouri with his wife and two children where, among other things, he coaches high school wrestling. Errol D. Hooper, Jr. can be contacted at Route 2, Box 158, Greentop, Missouri 63546 (email: lhooper@truman.edu).